Lecture Notes in Computer Science 14749

Founding Editors

Gerhard Goos
Juris Hartmanis

Editorial Board Members

Elisa Bertino, *Purdue University, West Lafayette, IN, USA*
Wen Gao, *Peking University, Beijing, China*
Bernhard Steffen⬤, *TU Dortmund University, Dortmund, Germany*
Moti Yung⬤, *Columbia University, New York, NY, USA*

The series Lecture Notes in Computer Science (LNCS), including its subseries Lecture Notes in Artificial Intelligence (LNAI) and Lecture Notes in Bioinformatics (LNBI), has established itself as a medium for the publication of new developments in computer science and information technology research, teaching, and education.

LNCS enjoys close cooperation with the computer science R & D community, the series counts many renowned academics among its volume editors and paper authors, and collaborates with prestigious societies. Its mission is to serve this international community by providing an invaluable service, mainly focused on the publication of conference and workshop proceedings and postproceedings. LNCS commenced publication in 1973.

Kevin Buzzard · Alicia Dickenstein ·
Bettina Eick · Anton Leykin · Yue Ren
Editors

Mathematical Software – ICMS 2024

8th International Conference
Durham, UK, July 22–25, 2024
Proceedings

Editors
Kevin Buzzard
Imperial College London
London, UK

Bettina Eick
TU Braunschweig
Braunschweig, Germany

Yue Ren
Durham University
Durham, UK

Alicia Dickenstein
University of Buenos Aires
Buenos Aires, Argentina

Anton Leykin
Georgia Institute of Technology
Atlanta, GA, USA

ISSN 0302-9743 ISSN 1611-3349 (electronic)
Lecture Notes in Computer Science
ISBN 978-3-031-64528-0 ISBN 978-3-031-64529-7 (eBook)
https://doi.org/10.1007/978-3-031-64529-7

© The Editor(s) (if applicable) and The Author(s), under exclusive license to Springer Nature Switzerland AG 2024
Chapter "Integrating Mathematical Data and Resources: Advancements in zbMATH Open for Enhanced Mathematical Research Accessibility and Reproducibility" is licensed under the terms of the Creative Commons Attribution 4.0 International License (http://creativecommons.org/licenses/by/4.0/). For further details see license information in the chapter.

This work is subject to copyright. All rights are solely and exclusively licensed by the Publisher, whether the whole or part of the material is concerned, specifically the rights of translation, reprinting, reuse of illustrations, recitation, broadcasting, reproduction on microfilms or in any other physical way, and transmission or information storage and retrieval, electronic adaptation, computer software, or by similar or dissimilar methodology now known or hereafter developed.
The use of general descriptive names, registered names, trademarks, service marks, etc. in this publication does not imply, even in the absence of a specific statement, that such names are exempt from the relevant protective laws and regulations and therefore free for general use.
The publisher, the authors and the editors are safe to assume that the advice and information in this book are believed to be true and accurate at the date of publication. Neither the publisher nor the authors or the editors give a warranty, expressed or implied, with respect to the material contained herein or for any errors or omissions that may have been made. The publisher remains neutral with regard to jurisdictional claims in published maps and institutional affiliations.

This Springer imprint is published by the registered company Springer Nature Switzerland AG
The registered company address is: Gewerbestrasse 11, 6330 Cham, Switzerland

If disposing of this product, please recycle the paper.

Preface

It is our pleasure to present the proceedings of the 8th International Congress on Mathematical Software (ICMS), which was held during July 22–25, 2024, at the Department of Mathematical Sciences, Durham University (UK).

The development of mathematical software is an integral part of research in mathematics and its applications. The design and analysis of new algorithms goes hand-in-hand with the development of new theoretical results. Practical implementations are a central tool for applications in mathematics and beyond. The International Congress on Mathematical Software takes place every 2 years and is a venue were developers of mathematical software from different research areas meet, exchange ideas and initiate new projects.

The program of the 2024 meeting consisted of 13 topical sessions with 121 talks and three plenary lectures. Each of the sessions provided an overview of the challenges and the progress made in the subfield of the session with a focus on the development of software and its applications. The topical sessions formed the core of the program. Session contributors and plenary speakers were given the option to submit their work for publication in these proceedings, and 37 papers have been selected through a peer reviewing process that relied on 167 reviews of the different versions of the manuscripts.

The three plenary lectures of the Congress were given by

Matthias Köppe (UC Davis, USA)
Heather Macbeth (Fordham University, USA)
Mohab Safey El Din (Sorbonne University, France)

The 13 topical sessions had the topics: Number Theory and Related Areas, Novel Formalisations of Mathematics in Lean, Software for the Applications of Group Theory to Combinatorics, Classical Algebraic Geometry & Modern Computer Algebra, Advancing Computer Algebra with Massively Parallel Methods, Computer Algebra Applications in the Life Sciences, Machine Learning Within Computer Algebra Systems, Numerical Software for Special Functions, Algorithms for Relative Equilibria in the N-Body Problem, Mathematical Research Data, Symbolic-Numeric Methods in Algebraic Geometry, Polyhedral Geometry and Combinatorics, and the General Contributed Session.

We thank all the speakers of the 2024 ICMS as well as session organizers, program committee members, local organizers and the members of the advisory board for helping to make this conference a success. Finally, we thank the sponsors for their financial support of the event.

May 2024

Kevin Buzzard
Alicia Dickenstein
Bettina Eick
Anton Leykin
Yue Ren

Organization

General Chair

Alicia Dickenstein   Universidad de Buenos Aires, Argentina

Program Committee Chairs

Kevin Buzzard	Imperial College London, UK
Bettina Eick	TU Braunschweig, Germany
Anton Leykin	Georgia Institute of Technology, USA

Program Committee

John Abbott	RPTU Kaiserslautern, Germany
David Ang	University College London, UK
Anton Betten	Kuwait University, Kuwait
Martin Bies	RPTU Kaiserslautern, Germany
Janko Boehm	RPTU Kaiserslautern, Germany
Hong Duong	University of Birmingham, UK
Matthew England	Coventry University, UK
Anne Frühbis-Krüger	University of Oldenburg, Germany
Amparo Gil	Universidad de Cantabria, Spain
Marshall Hampton	UMN Duluth, USA
Jeroen Hanselman	RPTU Kaiserslautern, Germany
Anders Jensen	Aarhus, Denmark
Alexander M. Kasprzyk	University of Nottingham, UK
Lars Kastner	TU Berlin, Germany
Kisun Lee	Clemson University, USA
Fangming Li	Imperial College London, UK
Amelia Livingston	University College London, UK
AmirHosein Sadeghimanesh	Coventry University, UK
Javier Segura	Universidad de Cantabria, Spain
Leonard Soicher	Queen Mary University of London, UK
Thomas Yahl	University of Wisconsin-Madison, USA
Matthias Zach	RPTU Kaiserslautern, Germany
Jujian Zhang	Imperial College London, UK

Advisory Board

Michael Joswig (Chair)	TU Berlin & MPI-MiS, Germany
Anna Maria Bigatti	University of Genoa, Italy
Jacques Carette	McMaster University, Canada
James Davenport	University of Bath, UK
Jonathan Hauenstein	Notre Dame University, USA
Manuel Kauers	Johannes Kepler University Linz, Austria
George Labahn	University of Waterloo, Canada
Josef Urban	Czech Technical University in Prague, Czech Republic
Timo de Wolff	TU Braunschweig, Germany

Local Organizers

Yue Ren (Chair)
Oliver Daisey
Jeff Giansiracusa
Iolo Jones
David Lanners
Julio Quijas Aceves
Victoria Schleis
Daniele Turchetti

Sponsors

Durham University
UKRI Future Leaders Fellowship Programme

Topical Sessions

Number Theory and Related Areas

Session Chair

John Abbott	RPTU Kaiserslautern, Germany

Novel Formalisations of Mathematics in Lean

Session Chairs

David Ang	University College London, UK
Fangming Li	Imperial College London, UK
Amelia Livingston	University College London, UK
Jujian Zhang	Imperial College London, UK

Software for the Applications of Group Theory to Combinatorics

Session Chairs

Anton Betten	Kuwait University, Kuwait
Leonard Soicher	Queen Mary University of London, UK

Classical Algebraic Geometry and Modern Computer Algebra: Innovative Software Design and Its Applications

Session Chairs

Martin Bies	RPTU Kaiserslautern, Germany
Lars Kastner	TU Berlin, Germany
Matthias Zach	RPTU Kaiserslautern, Germany

Advancing Computer Algebra with Massively Parallel Methods

Session Chairs

Janko Boehm	RPTU Kaiserslautern, Germany
Anne Frühbis-Krüger	University of Oldenburg, Germany

Computer Algebra Applications in the Life Sciences

Session Chairs

AmirHosein Sadeghimanesh Coventry University, UK
Hong Duong University of Birmingham, UK

Machine Learning Within Computer Algebra Systems

Session Chairs

Matthew England Coventry University, UK
Alexander M. Kasprzyk University of Nottingham, UK

Numerical Software for Special Functions

Session Chairs

Amparo Gil U. Cantabria, Spain
Javier Segura U. Cantabria, Spain

Algorithms for Relative Equilibria in the N-Body Problem

Session Chairs

Marshall Hampton UMN Duluth, USA
Anders Jensen Aarhus University, Denmark

Mathematical Research Data

Session Chair

Jeroen Hanselman RPTU Kaiserslautern, Germany

Symbolic-Numeric Methods in Algebraic Geometry

Session Chairs

Kisun Lee   Clemson University, USA
Thomas Yahl   University of Wisconsin - Madison, USA

Polyhedral Geometry and Combinatorics

Session Chair

Victoria Schleis   Durham University, UK

General Session

Session Chair

Yue Ren   Durham University, UK

Posters/Software Demonstrations

Jakub Malinowski (Wrocław University of Science and Technology, Poland): *Computing growth functions of periodic tessellation.*

Mikelis Emils. Mikelsons (RPTU Kaiserslautern, Germany), Martin Bies (RPTU Kaiserslautern, Germany), Andrew P. Turner (University of Pennsylvania, USA) and the OSCAR Collaboration: FTHEORYTOOLS, *A Computer Tool for Singular Elliptic Fibrations.*

Plenary Lectures

The Reformation of Sage

Matthias Köppe

Department of Mathematics, University of California, Davis, One Shields Ave, Davis, CA, 95616 USA
mkoeppe@math.ucdavis.edu

SageMath, originally created by William Stein in 2005 as the "System for Algebra and Geometry Experimentation", is an actively developed, open-source, Python-based mathematical software system released under the GNU General Public License. The two initial streams of development, experimental research in arithmetic geometry and free software for lower-division mathematics classes, were widened over the years to cover many areas of mathematics, thanks to contributions by hundreds of individuals.

The Sage library comprises about 3000 first-party Python and Cython modules; but as a result of a long-standing development principle, SageMath also makes use of hundreds of third-party, separately maintained packages written either in Python/Cython or in C, C++, Common Lisp, Fortran, and various custom or domain-specific languages. This makes Sage a major integrating force in the mathematical software world, downstream of many tributaries.

The past five years have brought a major reformation to the Sage project. SageMath completed its migration to Python 3 (2020); abandoned the pioneering Sage Notebook (2006–2019) in favor of the Jupyter Notebook and JupyterLab; embraced the modern Python packaging methodology; redirected its development workflow to GitHub (2023); and developed a Continuous Integration system that can travel upstream to the projects that Sage depends on.

In my talk, I will focus on the next step, which has the potential to be a watershed moment for the project. It addresses the fact that the rapid early growth of SageMath and the desired close integration of the numerous upstream systems had been achieved by putting a rigid system of dependencies in place. This canalization has severely restricted the development of an ecosystem downstream of the Sage library, and it has cemented a barrier isolating SageMath from the vibrant environment of scientific Python.

In an ecosystem restoration effort ongoing since 2020, the Sage library is being reformed as a modular system of several types of pip-installable packages, which can be built, developed, used, and tested separately. The integration of features and coverage of various subfields and communities is achieved by the interplay of several of the modularized packages at runtime. Packages named after a dependency, for example sagemath-gap or sagemath-singular, localize the build-time dependencies, but also give attribution and greater visibility to the upstream project and are intended to invite and inspire renewed collaboration. The new reusability of parts of the Sage library in the

Python ecosystem may finally bring the fortunate choice of Python – now the most widely used programming language – as the main implementation and surface language for SageMath to its full potential.

Algorithm and Abstraction in Formal Mathematics

Heather Macbeth

Fordham University, New York, NY, 10023 USA
hmacbeth1@fordham.edu

I will analyse differences in style between traditional prose mathematics writing and computer-formalised mathematics writing, presenting five case studies. My discussion will focus on two aspects where good style seems to differ between the two: in their incorporation of computation and of abstraction. I will argue that this reflects a different mathematical aesthetic for formalised mathematics.

Polynomial System Solving with the Msolve Library

Mohab Safey El Din

Sorbonne Université, LIP6, UMR CNRS 7606, PolSys Team,
Institut Universitaire de France (IUF)
Mohab.Safey@lip6.fr

Solving multivariate polynomial systems arises in a wide range of areas in information theory (e.g. cryptography, error-correcting codes) and engineering sciences (e.g. robotics, biology). The msolve library is a C library which provides implementations of a number of algorithms based on Gröbner bases computations for its algebraic component and bisection algorithms for real root isolation. We will present its design and current capabilities and discuss further developments. Joint work with J. Berthomieu, C. Eder and V. Neiger.

Contents

Plenary Lectures

The Reformation of Sage 3
   Matthias Köppe

Algorithm and Abstraction in Formal Mathematics 12
   Heather Macbeth

Number Theory and Related Areas

Computing the Determinant of a Dense Matrix over $\mathbb{Z}$ 29
   John Abbott and Claus Fieker

FastECPP over MPI 36
   Andreas Enge

Attacking a Levelled Fully Homomorphic Encryption System
with Topological Data Analysis 46
   Aaruni Kaushik

Novel Formalisations of Mathematics in Lean

Formalising Families of ℓ-adic Galois Representations in Lean 4 57
   Ivan Farabella

Formalization of the Existence of Frobenius Elements 63
   Jou Glasheen

Formalising Analysis in Lean: Compactness and Dimensionality 72
   Dawid Lipinski

Formalisation of the Category of Hopf Algebras in Lean4 78
   Jujian Zhang, Yunzhou Xie, Yichen Feng, and Yanqiao Zhou

Software for the Applications of Group Theory to Combinatorics

Computing the Group of an Algebraic Variety over a Finite Field 89
   Abdullah Alazemi and Anton Betten

Computer Classification of Linear Codes Based on Lattice Point
Enumeration 97
    Sascha Kurz

Software for Proper Vertex-Colouring Exploiting Graph Symmetry 106
    Leonard H. Soicher

Classical Algebraic Geometry and Modern Computer Algebra: Innovative Software Design and Its Applications

Localization in Gromov—Witten Theory of Toric Varieties in a Computer
Algebra System 115
    Giosuè Muratore

Advancing Computer Algebra with Massively Parallel Methods

Massively Parallel Methods for Free Resolutions 127
    Santosh Gnawali

Towards Parallel Methods in Birational Geometry 135
    Benjamin Mirgain

Towards Parallel Algorithms for Gromov-Witten Invariants of Elliptic
Curves 145
    Ali Traore

Computer Algebra Applications in the Life Sciences

A SageMath Package for Elementary and Sign Vectors with Applications
to Chemical Reaction Networks 155
    Marcus S. Aichmayr, Stefan Müller, and Georg Regensburger

Machine Learning Within Computer Algebra Systems

Symbolic Integration Algorithm Selection with Machine Learning:
LSTMs Vs Tree LSTMs 167
    Rashid Barket, Matthew England, and Jürgen Gerhard

Exploring Alternative Machine Learning Models for Variable Ordering
in Cylindrical Algebraic Decomposition 176
    Rohit John and James Davenport

Constrained Neural Networks for Interpretable Heuristic Creation
to Optimise Computer Algebra Systems 186
    Dorian Florescu and Matthew England

Machine Learning for Number Theory: Unsupervised Learning
with L-Functions 196
    Thomas Oliver

Numerical Software for Special Functions

Approximation of an Inverse of the Incomplete Beta Function 207
    Michael B. Giles and Casper Beentjes

DLMF Standard Reference Tables on Demand 215
    Bonita V. Saunders, Sean Brooks, Ron Buckmire,
    Rachel E. Vincent-Finley, Franky Backeljauw, Stefan Becuwe,
    Bruce Miller, Marjorie McClain, and Annie Cuyt

Mathematical Research Data

Integrating Mathematical Data and Resources: Advancements
in zbMATH Open for Enhanced Mathematical Research Accessibility
and Reproducibility 225
    Maxence Azzouz-Thuderoz, Madhurima Deb, Matteo Petrera,
    Moritz Schubotz, and Olaf Teschke

A FAIR File Format for Mathematical Software 234
    Antony Della Vecchia, Michael Joswig, and Benjamin Lorenz

Predefined Software Environment Runtimes as a Measure
for Reproducibility 245
    Aaruni Kaushik

Towards a FAIR Documentation of Workflows and Models in Applied
Mathematics 254
    Marco Reidelbach, Björn Schembera, and Marcus Weber

Symbolic-Numeric Methods in Algebraic Geometry

Monodromy Coordinates 265
    Taylor Brysiewicz

Effective Alpha Theory Certification Using Interval Arithmetic: Alpha
Theory over Regions 275
   Kisun Lee

Gröbner Degenerations of Determinantal Ideals with an Application
to Toric Degenerations of Grassmannians 285
   Fatemeh Mohammadi

Polyhedral Geometry and Combinatorics

Eigenvalue Methods for Sparse Tropical Polynomial Systems 299
   Marianne Akian, Antoine Béreau, and Stéphane Gaubert

Dynamic Decomposition of Tropical Prevarieties for Celestial Mechanics 313
   Anders Nedergaard Jensen

Regular Flips in `mptopcom` 322
   Lars Kastner

A Framework for Generalized Tropical Homotopy Continuation 331
   Oliver Daisey and Yue Ren

General Session

Integrating GeoGebra with React and WebAssembly: A Web-Based
Approach for Mathematical Software Development 343
   Mitsushi Fujimoto

DetGB: A Software Package for Computing Gröbner Bases
of Determinantal Ideals 354
   Chenqi Mou, Qiuye Song, and Yutong Zhou

Extrapolating Solution Paths of Polynomial Homotopies Towards
Singularities with PHCpack and Phcpy 365
   Jan Verschelde and Kylash Viswanathan

Author Index 375

Plenary Lectures

The Reformation of Sage

Matthias Köppe[(✉)]

University of California, Davis, Department of Mathematics, One Shields Avenue, Davis CA 95616, USA
mkoeppe@math.ucdavis.edu

Abstract. The monolithic open-source mathematical software system Sage, developed since 2005, is being transformed into a modular system of Python libraries with a renewed focus on collaboration and vertical integration. This paper is the author's personal account of this modularization project, led by him since 2020.

Keywords: Python (computer programming) · Open-source software · Computer algebra systems · Mathematical software

1 Introduction: What Is Sage?

SageMath [9], originally created by W. Stein in 2005 as the *System for Algebra and Geometry Experimentation* [5], is an actively developed, open source, Python-based mathematical software system released under the GNU General Public License. Its initial dual focus, on enabling experimental research in arithmetic geometry and on providing facilities for lower-division mathematics classes, was widened over the years to cover many other areas of mathematics.

The Sage project had been one of the pioneers of open-source computational notebooks, inspired by the proprietary computational notebook solutions that had been available since the 1990s in Mathematica and Maple. According to first-hand accounts by W. Stein (Sage) [4] and F. Perez (IPython) [3], development of the Sage notebook started in 2006, and it was usable by Sage Days 2, held later that year. Development of the IPython notebook based on zeromq commenced in 2010, and it was included in the ipython 0.12 release (2011). Sage 6.4 (2014) made the IPython (later, Jupyter) notebook available as an alternative to sagenb. Although the Jupyter notebook soon surpassed the Sage notebook, some missing features hindered a full adoption; but when Sage switched to Python 3 with the release of version 9.0 (2020), sagenb was finally abandoned. As of 2024, Sage uses notebook 7, based on JupyterLab technology, and *user interface development is no longer an activity of the Sage project*.

The (belated) replacement of sagenb by the industry-standard Jupyter notebook adheres to one of the principles of the Sage project. Following its motto *we are building the car, not reinventing the wheel* has made SageMath a major integrating force in the mathematical software landscape.

Partially supported by grant DMS-2012764 of the National Science Foundation.

Sage makes use of hundreds of third-party, separately maintained packages [8] written either in Python or in other languages (C, C++, Common Lisp) and facilitates installing them in two ways:

1. Sage provides installation scripts that automatically build the packages from their source tarballs; this aspect of Sage is referred to as *Sage-the-distribution*.
2. Sage is able to detect and use suitable versions of installed packages and includes an information system that can tell users which system packages to install on Arch Linux, conda-forge, Debian/Ubuntu, Fedora, Homebrew, OpenSuSE, Void Linux, etc.

Scripts and metadata for the third-party packages are located in the GitHub repository sagemath/sage in `SAGE_ROOT/build/pkgs`. The other content of the repository is the source code of ca. 3000 first-party Python and Cython modules, which together form the *Sage library*.

2 The Modularization Project

2.1 Problem Description and Goals

1. The Python and Cython modules of the Sage library have an extensive web of interdependencies ("everything depends on everything"). These dependencies range from build-time to module-level to method-level to test-only dependencies.
2. The modules of the Sage library also have numerous dependencies on third-party non-Python packages: To just build the Cython modules, installations of at least the following packages are required: `boost`, `brial`, `ecl`, `eclib`, `ecm`, `fflas_ffpack`, `flint`, `libgd`, `gap`, `giac`, `givaro`, `glpk`, `gmp`, `gsl`, `iml`, `lcalc`, `libbraiding`, `libhomfly`, `libpng`, `linbox`, `m4ri`, `m4rie`, `mpc`, `mpfi`, `mpfr`, `ntl`, `pari`, `rw`, `singular`, `symmetrica`, as well as a number of Python packages. Then, to be able to start and use Sage, more Python packages need to be installed.
3. As a consequence of the above, it was impossible to build or run parts of the Sage library. Moreover, it was impossible to install the whole Sage library using just Python infrastructure (pip).
4. This is the key mechanism that isolates the Sage developer and user communities from the (vastly larger and more active) Python community. Developers of new code face the difficult question: *Are the facilities that a portion of the Sage library offers really important enough for my project—that it makes sense to accept being locked in to the Sage system and isolated from the rest of the Python world?* Outside of a narrow context of research mathematics, the answer will often be *no*.

I argue that this is a fundamental, severe bug in the setup of the Sage community. My modularization project, announced in May 2020 in the sage-devel post [2], has set out to fix this bug by creating viable distributions of portions of the Sage library. Viable here means: Separately buildable with minimal dependencies, separately runnable, and separately testable—and thus making it possible to reach many more users and attract many more developers.

2.2 Origins

The problems laid out in the previous section had been discussed in the Sage community earlier. For brevity, I will restrict myself to quoting a few voices from the lively 2016 sage-devel thread [6] to sketch these earlier discussions.

W. Stein noted *the friction involved in making code that depends on the Sage distribution available to the world. [...] this friction is a very, very significant problem to the growth of Sage.*

As a possible solution, V. Braun noted that *some core feature set [...] that must be importable (and usable) by itself [...] could be shipped as a separate package even if it's not a separate repository* and suggested the use of the *namespace package support* in Python 3.

However, V. Braun also cautioned about possible negative effects on the documentation: *Let's say we break out all elliptic curve stuff into a separate package. Just like with* cysignals, *this will remove all elliptic curve documentation from the Sage reference manual.*

W. Stein also raised concerns regarding the lack of reusability, on the example of the 3-dimensional plotting facilities of Sage: *[...] by developing it as a separate library, we can clarify how it depends on the rest of sage (and vice versa). We might also be able to make it available outside Sage, and it could suddenly be of huge value to the Python world. [...] It is a [...] waste that this 3d functionality is only available in Sage?*

W. Stein argued that not all aspects of the problems were of technical nature, but rather that *problems were partly caused by us not supporting and encouraging the creation of code outside of standard Sage. [...] we as a community (not me, but certainly many others) are shockingly discouraging and negative toward any code that isn't officially in the core Sage library.*

K. Crisman suggested developing *more infrastructure for supporting the advertising of packages of this kind*, and K. Lee suggested including *guidelines for development of these external Sage packages in the Sage developer manual.*

2.3 The Tools

I will now turn to describing the design of the modularization project. The interested reader will find many details and references in the GitHub issue [1].

The monorepository. Sage development will continue to use the same GitHub repository sagemath/sage for all development. Also the structure of the source tree (`SAGE_ROOT/src/sage`) will remain monolithic, for the convenience of Sage developers—at least until a separate community of developers forms that wishes to maintain a distribution packages in its own repository. The modularization project only adds source tree stubs and metadata of Python distribution packages in a directory `SAGE_ROOT/pkgs/`. To facilitate development, the top-level `bootstrap` script is extended to take care of synchronizing metadata such as the version constraints for dependencies, based on centralized information kept with the metadata of the Sage distribution. After running `./bootstrap`, each of the directories in SAGE_ROOT/pkgs/ contains the source of a complete distribution package, which can be built and installed using standard Python tools.

Documentation and doctests as integration and attribution tools. The documentation will not be modularized. The modularization project recognizes the value that the comprehensive Sage reference manual provides to users. In particular, the project recognizes the important roles of doctests that cross modularization boundaries—both for mathematical exposition and for integration testing purposes. To make modularized distributions testable separately, we annotate doctests in the source code with tags like # needs sage.rings.number_field or # needs sage.libs.flint. Annotations like the latter give a secondary benefit, namely attribution for the libraries that Sage uses for particular types of computations. This alleviates the conflict between abstraction and attribution. The tool ./sage --fixdoctests assists developers with managing the placement of the doctest tags in the source code.

Modularized testing in venvs and integration testing. We test the modularized distributions using automatically created virtual environments. The technical tool for this is the standard Python tool tox. To run the tests in the Sage repository, developers can use additional convenience commands such as make SAGE_CHECK=yes pypi-wheels. Also the entire Sage library will continue to be tested as a whole, both on Sage developers' machines (./sage -t --all) and on GitHub Actions. When preparing a modularized distribution, at the beginning, there will be lots of doctest failures when testing it separately. Using ./sage -t --baseline-stats-path known-test-failures.json, we can monitor modularization regressions.

Modularization import linter and fixer. For example, the package sage.misc is split over 5 pip-installable distributions, so importing from sage.misc.all is not a good practice in the Sage libary. The linter ./sage --tox -e relint helps by flagging such imports; and the script ./sage --fix-imports rewrites them as more specific import statements. This enables a convenient development workflow in which developers type from sage.all import ... and then have it automatically replaced by the correct import statements.

Information system for features and distributions. The Sage Reference Manual contains an automatically generated section for each of the packages that are part of Sage-the-distribution. Each modularized distribution package (defined in SAGE_ROOT/pkgs/) likewise has a section in the reference manual.

Automatic deployment to PyPI. The modularized distributions are automatically built (as source distributions and wheels) and deployed to the Python Package Index on every release. For building the wheels of distributions that have non-Python dependencies, we reuse the scripts of the Sage distribution.

2.4 The Rules

Here I informally summarize the technological constraints of modern Python packaging that a design of modularized distributions must respect.

1. *Two granularities:* The smallest unit is a Python/Cython module. Several modules are packaged together in a distribution package ("distribution", "pip-installable package"). There are no technical constraints on what modules can go into which distribution. We can partition a package like `sage.misc` and ship the modules from it as parts of 5 different distributions.
2. *Separation of Phases:* Any installation of a distribution in source format ("sdist") first builds a wheel package (which contains the compiled versions of any Cython modules); distributions can run arbitrary code during the build. Then the wheel is installed, which means that simply a zip file is unpacked into the `site-packages` directory. There is no such thing as a script that would run at the time of installing a wheel.
3. *Dependencies:* Distribution packages declare build dependencies and runtime dependencies (on other Python distribution packages). These two can be entirely unrelated to each other. Additionally, it is possible to declare optional runtime dependencies; for example, those necessary for running tests, or for providing certain advertised extra features. However, the set of modules that form a distribution package should be static, as there is simply no place in the name (or metadata) of a wheel package that could indicate different configurations of the distribution.
4. As a result of the above, it is possible to create distributions that contain some modules that cannot be imported because some of their dependencies are missing. That's OK; they can become importable simply by the presence of other distributions, in particular those declared as extra dependencies. All of this is discovered at runtime, more specifically, at the time of importing a module. There is no ahead-of-time linking step of any sort.

Some desirable criteria for designing the set of modularized distributions are the following.

5. *Not too fine-grained:* As one of the extremes in the design space, each of the Sage library's ca. 3000 modules could, of course, be shipped in a separate distribution. This would guarantee to allow users to install exactly the modules needed for a specific task, but would leave the user alone with the task of putting together (and maintaining) a collection that works.
6. *Not too many non-Python dependencies per distribution:* This is so that when users of a modularized distribution package make their first step to contributing to it (perhaps fixing a bug), they can do so locally with minimal preparation. Requiring a full installation of the Sage distribution for development, whether local or in the cloud, is too much friction.
7. *Compilation time matters:* Each distribution should include much fewer Cython modules than the 570 that make up the Sage library. Cython modules need to be compiled before they can be used. Compilation times matter

to users and developers. On the other hand, the number of Python modules shipped by a distribution does not matter all that much.
8. *Discoverability:* The modularized distributions should make it easy for users to understand which of them will be needed and should make it convenient to install extra features. In particular, at the Sage or IPython prompt, we need to make it convenient for users to discover functionality that has not been installed yet; both ahead of time and after an exception due to an uninstalled feature is raised.

2.5 The Blocks

Instead of defining a distribution package for every possible community or subfield of mathematics that Sage supports, the modularization project introduces three types of distribution packages:

Packages named after a basic mathematical structure. The packages may also cover a wide range of generalizations/applications of the structure after which they are named. Users who work in a specialized research area will, of course, recognize what structures they need. The down-to-earth naming also creates discoverability by a broader audience.

- sagemath-combinat: provides "everything combinatorial", except for graphs.
- sagemath-graphs: also provides posets, combinatorial designs, abstract simplicial complexes, quivers, etc.
- sagemath-groups
- sagemath-modules: also has matrices, tensors, homology, coding theory, etc.
- sagemath-polyhedra: also provides fans, hyperplane arrangements, polyhedral complexes, linear and mixed-integer optimization, lattice point sets, and toric varieties
- sagemath-schemes
- sagemath-symbolics: symbolic expressions and symbolic calculus (provided by maxima running in ecl)

Packages named after a third-party non-Python dependency. This makes technical sense because the dependencies will be localized to this distribution package, but it also helps give *attribution and visibility* to these libraries and projects that Sage depends on. Examples include:

- sagemath-flint
- sagemath-gap
- sagemath-giac
- sagemath-linbox
- sagemath-ntl
- sagemath-pari
- sagemath-singular

The project uses several complementary techniques to take care of Cython modules that depend on multiple libraries. In many cases, such multiple dependencies are gratuitous and can be resolved simply by splitting a Cython module into several parts. In other cases, such as for basic ring element implementations that are highly encumbered with multi-library dependencies, it can be resolved by moving certain conversion methods out to separate new Cython modules, which will be imported rather than cimported. The project has already eliminated many build-time dependencies on pari from the basic ring element implementations such as the Integers in this way. An extreme example that is still to be done at the time of writing is the class NumberFieldElement_quadratic.

When dependencies have been reduced to, say, two simultaneous libraries per module, then we can ship the module as part of a distribution that has only one required runtime dependency but two build-time dependencies that are used for the affected modules. To be able to import the affected module, another distribution has to be present. For example, the distribution sagemath-modules uses a build-time dependency on numpy for building matrix implementations such as Matrix_real_double_dense (which cimports numpy). But numpy is not a declared runtime dependency of sagemath-modules; it is only an extra that can be activated by using pip install "sagemath-modules[RDF]".

Packages named after a technical functionality.

- sagemath-objects: Sage extends Python's object system by dynamic mix-in classes that are driven by categories and axioms. It is loosely modeled on concepts of category theory and inspired by Scratchpad/Axiom/FriCAS, Magma, and MuPAD. This distribution package makes Sage objects, the element/parent framework, basic categories and functors, the coercion system and the related metaclasses available.
- sagemath-categories: This distribution package contains the full set of categories defined by Sage, as well as basic mathematical objects such as integers and rational numbers, a basic implementation of polynomials, and affine spaces. None of this brings in additional dependencies.
- sagemath-plot: Plotting facilities, depending on matplotlib.
- sagemath-repl: The top-level interactive environment with the preparser that defines the surface language of SageMath. This distribution also includes the doctesting facilities, as the doctests are written in the surface language.
- sagemath-standard: Everything as provided by a standard installation of the Sage distribution. This is an empty meta-package.

Fig. 1 shows the build-time dependencies and runtime dependencies for selected modularized distribution packages. Each of the distribution packages can define a list of "extras" (nicknames for sets of optional dependencies that provide additional advertised features). When using pip to install a distribution or when declaring a dependency in their own package, users can use square bracket notation to "adjoin" these extras. As an example, users who need fast linear algebra over $GF(p^n)$ and over cyclotomic fields can use pip install "sagemath-modules[GFpn,CyclotomicField]".

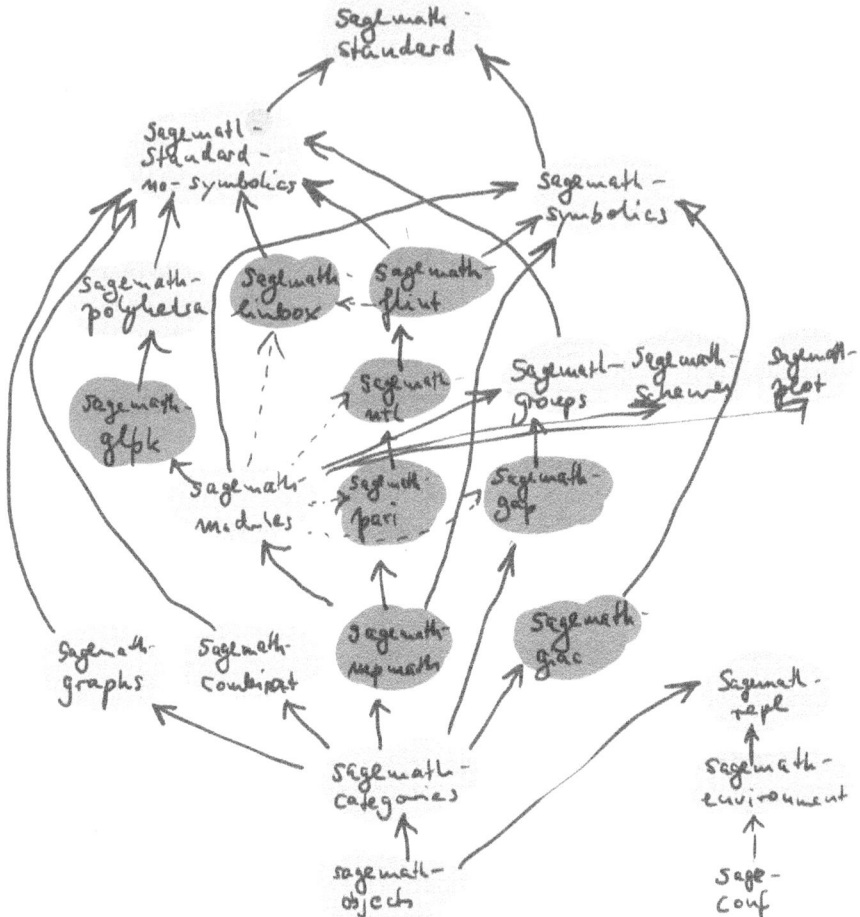

Fig. 1. Modularized distribution packages (selection). Solid arrows are build dependencies, dashed arrows are runtime dependencies.

Acknowledgments. SageMath is the product of two decades of work by hundreds of contributors [7] and works on top of hundreds of separately maintained open-source software packages [8].

I wish to thank in particular the members of the Sage community that have shaped the path to my current role in the project: *Anne Schilling* and *Nicolas M. Thiéry* for their inspiring enthusiasm for bringing algebraic combinatorics to Sage (2008–2015); *Dmitrii V. Pasechnik* for welcoming and reviewing many of my earliest contributions to SageMath (2015–2017); and my collaborators on polyhedral geometry and optimization in Sage, in particular *Vincent Delecroix, Moritz Firsching, Simon King, Jonathan Kliem, Jean-Philippe Labbé, Thierry Monteil, Travis Scrimshaw,* and *Yuan Zhou* (2017–2019).

Since 2020, when I stepped up to take a larger role in Sage development, I have interacted with many members of the Sage community. Regarding their impact on the modularization project, I wish to highlight in particular *Isuru Fernando* for making me aware of implicit namespace packages (2020); *Dmitrii V. Pasechnik* and *Tobias Diez* for a period of constructive interactions on build system and development tools (2020–2022); *François Bissey, Jonathan Kliem, Kwankyu Lee, John H. Palmieri,* and *Travis Scrimshaw* for many discussions, collaborations, and reviews (2020–); and *William Stein* for his encouragement and support for the modularization project.

References

1. Köppe, M.: Meta-ticket: modularize sagelib into separate distributions (pip-installable packages) sagemath-... for Sage 10.x (2024), GitHub issue. https://github.com/sagemath/sage/issues/29705
2. Köppe, M., et al.: Proposal for Sage 9.3: modularization of sagelib; in particular, splitting out a sage_objects package, sage-devel discussion thread (2020). https://groups.google.com/g/sage-devel/c/M9QTWtln6zU
3. Perez, F.: Blog post (2012). https://web.archive.org/web/20190817173145/http://blog.fperez.org/2012/01/ipython-notebook-historical.html
4. Stein, W.: History of notebooks, sage-devel discussion thread (2015). https://groups.google.com/g/sage-devel/c/uc9HIMREh9Y/m/Ihpq4vomBgAJ
5. Stein, W., Joyner, D.: SAGE: System for algebra and geometry experimentation. ACM SIGSAM Bull. **39**(2), 61–64 (2005). http://www.sagemath.org/files/sage_stein2005.pdf
6. Stein, W., et al.: How we develop Sage, sage-devel discussion thread (2016). https://groups.google.com/g/sage-devel/c/29ndCD8z94k
7. The SageMath Developers: Developers around the world (2024). https://www.sagemath.org/development-map.html
8. Packages and features. In: The SageMath Developers: Sage 10.3 Reference Manual (2024). https://doc.sagemath.org/html/en/reference/spkg/index.html
9. The SageMath Developers: SageMath, the Sage Mathematics Software System, version 10.4.beta0 (2024). https://doi.org/10.5281/zenodo.593563. https://github.com/sagemath/sage

Algorithm and Abstraction in Formal Mathematics

Heather Macbeth

Fordham University, New York, NY 10023, USA
hmacbeth1@fordham.edu

Abstract. I analyse differences in style between traditional prose mathematics writing and computer-formalised mathematics writing, presenting five case studies. I note two aspects where good style seems to differ between the two: in their incorporation of computation and of abstraction. I argue that this reflects a different mathematical aesthetic for formalised mathematics.

1 Introduction

In the last twenty years, formalisation—building up proofs as line-by-line logical deductions from the axioms of mathematics, with the help of specialised computer systems[1]—has seen increasing interest from mathematicians. The rapidly increasing coverage of the mathematical literature in these systems is very much a social process: their mathematical libraries are built collaboratively by hundreds of people, and code contributed by one person will be reviewed in detail by another, and often thoroughly re-worked a year later by a third.

In this kind of human and social process, culture develops spontaneously. The back-and-forth of discussion in this process includes frequent comment on a formalised proof's beauty, elegance, cleverness, and other abstract properties generally associated with mathematical aesthetics. The communities of mathematicians doing this work consider computer-formalised proofs to be, not simply utilitarian certificates for the correctness of logical claims, but a fully-fledged medium for mathematical exposition.

In this article I describe (necessarily very subjectively) some aspects of this aesthetic of computer-formalised mathematics writing. Much of this aesthetic is inherited from traditional prose mathematics writing, on which there is a vast literature [2,4,15,16,22,25,30–32]. I therefore focus on cases in which good style in formalised mathematics seems to differ from good style in traditional prose mathematics. I present five case studies,[2] grouped by theme: how to integrate computation (Sect. 2) and how much use to make of abstraction (Sect. 3).

---

[1] Examples include Agda, Coq, Lean, HOL Light, Isabelle, Metamath and Mizar.
[2] Disproportionately drawn from Lean's [26,27] *Mathlib* [8], of which I am a maintainer.

2 Computation

A faithful computer-formalised translation of a traditional prose proof will commonly use computation "in the small:" a proof step which seems obvious to humans often represents a whole chain of strict logical reasoning, and in most modern systems automation is used to help construct such chains.

Interestingly, such a translation will sometimes also use computation "in the large:" several notable formalisations' [11,14,21] targets are theorems whose published proofs rely on the result reported by a computer program.

So what about using computation "in the middle?" In this section I explore proofs where there is no absolute need to outsource a calculation to computer—and where, in traditional writing, simple inertia would prevent one from doing this—but which are arguably improved by increased reliance on computation.

2.1 Classification of Wallpaper Groups

My first example arises in classifying the 17 wallpaper groups. This classification is heavily dependent on case analysis, one branch of which is to consider wallpaper groups which contain translations and rotations but no reflections. These can be classified according to the orbit types of centres of symmetry. For example, one of these wallpaper groups, which in our classification we will associate to the tuple $(2, 3, 6)$, has three centres of symmetry, at which the stabilisers are generated by rotations of $\frac{2\pi}{2}$, $\frac{2\pi}{3}$, and $\frac{2\pi}{6}$.

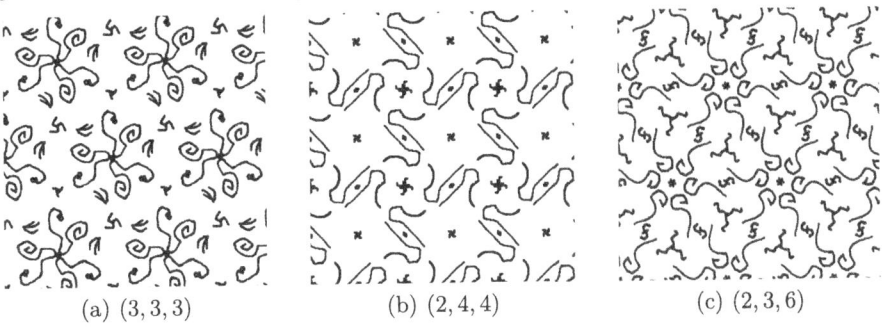

(a) $(3,3,3)$ (b) $(2,4,4)$ (c) $(2,3,6)$

Fig. 1. The wallpaper groups containing rotations but no reflections. (Images via [10])

The following arithmetic lemma classifies the possible tuples which can arise. The wallpaper groups associated to these tuples are depicted in Fig. 1.

Lemma 1. *Let* $2 \leq p \leq q \leq r$ *be natural numbers, with*

$$\frac{1}{p} + \frac{1}{q} + \frac{1}{r} = 1. \qquad (\star)$$

Then (p, q, r) *is one of* $(3, 3, 3)$, $(2, 4, 4)$, $(2, 3, 6)$.

I first present a traditional prose proof lifted from a textbook [9].

Proof. We get $p = q = r = 3$ if all of $\frac{1}{p}$, $\frac{1}{q}$ and $\frac{1}{r}$ have their mean value of $\frac{1}{3}$. Otherwise p must be 2.

If r and q have *their* mean value of $\frac{1}{4}$, we get $p = 2$, $q = r = 4$.

If not, q must be 3, and r is forced to be 6, by $(\star)$.

Secondly, I describe a proof of this lemma that I wrote in Lean with Anne Baanen. I am trying to translate the Lean code fairly literally.

Proof. The inequalities $0 < \frac{1}{r} \leq \frac{1}{q} \leq \frac{1}{p} \leq \frac{1}{2}$ and the equality $(\star)$ yield that

$$\frac{1}{3} \leq \frac{1}{p} \leq \frac{1}{2} \tag{1}$$

$$\frac{1}{2}\left[1 - \frac{1}{p}\right] \leq \frac{1}{q} < \frac{1}{2} \tag{2}$$

$$\frac{1}{r} = 1 - \frac{1}{p} - \frac{1}{q}. \tag{3}$$

There are finitely many natural numbers p satisfying (1); case-split on these. For each of these there are finitely many natural numbers q satisfying (2); case-split on these. For each of these, r can be determined from (3).

There is an algorithm implicit in these proofs. The second (formalised) proof looks almost like a recipe for cooking the first (textbook) proof: it describes the steps to be carried out, rather than actually performing those steps visibly for the reader (i.e. documenting the available choices at each case split).

This is very typical: as mentioned, proof-writing in systems such as Lean frequently invokes "tactics," small computer programs to construct parts of proofs. But once we start to describe proofs via the recipes which would construct them, there is no need to stick to the original recipe. This was noted by Hales et al. [14]:

> In the original, computer calculations were a last resort after as much was done by hand as feasible. In the [formalisation], the use of computer has been fully embraced. As a result, many laborious lemmas of the original proof can be automated or eliminated altogether.

I will argue that an aesthetically pleasing formal proof is one which has a short and simple recipe. As the next two examples will show, this is not the same thing as a proof which is itself short and simple.

2.2 The Kochen–Specker Paradox

I next consider a theorem from quantum mechanics.

Theorem 1 (Kochen–Specker [24]). *There does not exist a boolean (say red or green) colouring of the vectors in $\mathbb{R}^3$, such that all triples $u, v, w \in \mathbb{R}^3$ of nonzero mutually-orthogonal vectors are coloured green, red, red in some order.*

I will discuss a streamlined proof due to Peres [28]. The approach is to deduce a contradiction from the colouring of the following 33 nonzero vectors[3] in $\mathbb{R}^3$:

---

[3] Down from 117 vectors in the original Kochen–Specker proof. Following Peres' notation, $\bar{1}$ is shorthand for -1, 2 is shorthand for $\sqrt{2}$, and $\bar{2}$ is shorthand for $-\sqrt{2}$.

$$|\bar{1}\bar{1}2|\bar{1}02|\bar{1}12|\bar{1}2\bar{1}|\bar{1}20|\bar{1}21|0\bar{1}2|002|012|02\bar{2}|02\bar{1}| \\ |020|021|022|1\bar{1}2|102|112|12\bar{1}|120|121|2\bar{2}0|2\bar{1}\bar{1}| \\ |2\bar{1}0|2\bar{1}1|20\bar{2}|20\bar{1}|200|201|202|21\bar{1}|210|211|220| \quad (4)$$

The basic driver of the proof is that, once enough of the 33 vectors have been coloured, the colours of the rest can be determined greedily: a vector orthogonal to a green vector must be red, and a vector orthogonal to two orthogonal red vectors must be green.

Here is an outline of Peres' proof [28] of the impossibility.

Proof. We can determine the colours of some vectors without loss of generality.

By the symmetry	and by the known facts	we can assume
choice of z-axis		001 green; 100, 010 red
choice of x vs $-x$	010 red	101 green; $\bar{1}01$ red
choice of y vs $-y$	100 red	011 green; $0\bar{1}1$ red
choice of x vs y	001 green, thus 110 red	$1\bar{1}2$ green; $\bar{1}12$ red

Now a suitable greedy sequence of deductions [depicted in Fig. 2, written out explicitly in the original] forces a contradiction.

For comparison, here is an outline of a proof formalised by John Harrison in HOL Light in 2005.[4] It is really a brute force search.

Proof. Perform the following binary search: split on a vector whose colour is not yet known; then in each case (red or green) greedily make all possible deductions. Stop if a contradiction is found. Recurse if not.

The result of this process is that every branch terminates in a contradiction.

The most notable difference from Peres' prose proof is to abandon his symmetry argument (which reduces to only one configuration on which the greedy algorithm need be run) and instead just run the greedy argument at each stage of a binary search.[5] In effect, both arguments amount to the implementation of a search algorithm. The search algorithm used in the formalised proof is very simple, whereas the symmetry considerations incorporated in the original proof can be considered as baroque optimisations to the search algorithm to get its

---

[4] https://github.com/jrh13/hol-light/blob/e736197/Tutorial/Custom_tactics.ml.

[5] A second difference is that when running the greedy algorithm from a partial colouring produces a contradiction, the formalised version does not write out the *certificate*: an explicit path of deductions leading to the concluding contradiction. But this is less controversial. In the case of the configuration in Fig. 2, when the path of deductions is written out, it does not appear to contain any particular insight. See Harrison [17, section 3.4] for a similar example.

"runtime" within the scale of human readability. Harrison [19, section 3] defends this choice in his formalisation on the grounds of convenience: the simpler algorithm is easier to implement. But I argue that it is also defensible on the grounds of aesthetics: the simpler algorithm is easier for the reader to grasp.[6]

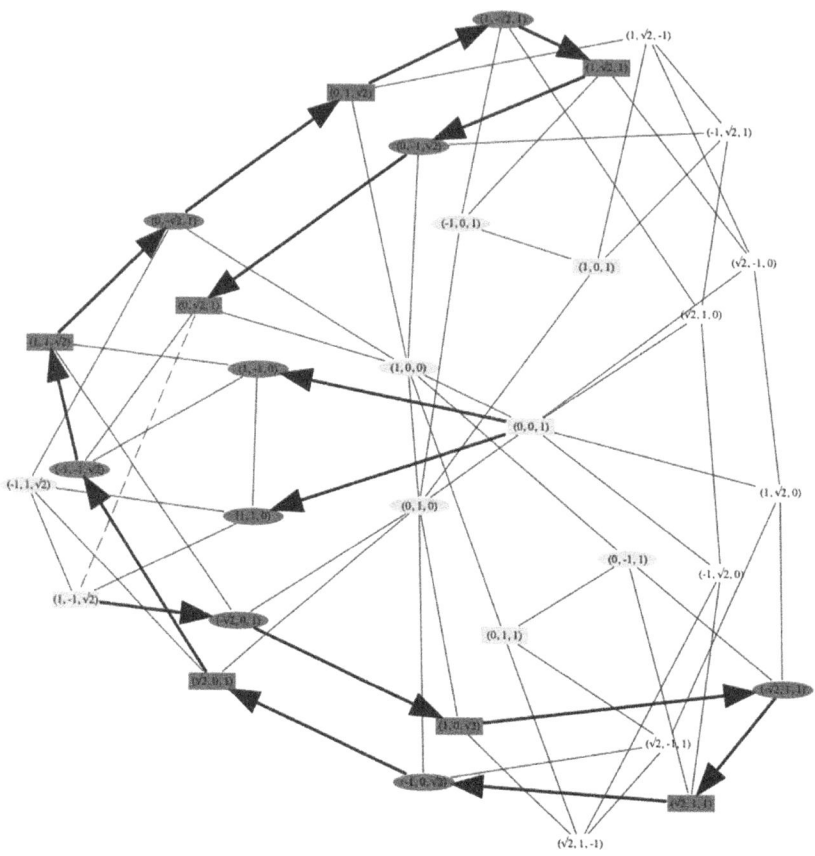

Fig. 2. Peres' argument [28] for the impossibility of a suitable boolean colouring of the 33 vectors (4). Edges connect orthogonal vectors. Oval (respectively, rectangular) labels denote vectors found to be red (resp. green). Light-shaded (resp. dark-shaded) labels denote vectors whose colour is chosen by an initial symmetry argument (resp. is forced as a consequence of adjacent vectors' colours). Arrows indicate the order in which these deductions occur. The dotted edge indicates two orthogonal vectors which are both green, contradiction.

---

[6] Gonthier et al. [12, section 4.3] report a similar example, in which an appeal to a logical decision procedure produces an "intellectually more satisfying" proof than the original argument involving a detailed combinatorial case analysis.

2.3 Multiplication of Chebyshev Polynomials

My last example on the theme of computation highlights a different kind of computation. Let $T_n(x)$ denote the n-th Chebyshev polynomial of the first kind. Recall these polynomials satisfy a recurrence relation

$$T_{n+2}(x) = 2xT_{n+1}(x) - T_n(x).$$

Lemma 2 (Multiplication formula for Chebyshev polynomials). *For all natural numbers m and k, $2T_m T_{m+k} = T_{2m+k} + T_k$.*

A purely algebraic[7] proof of this lemma is necessarily an induction. The inductive step of the scheme that works has us prove a statement for $m+2$,

$$\forall k : \mathbb{N}, 2T_{m+2}T_{(m+2)+k} = T_{2(m+2)+k} + T_k,$$

given the corresponding statements for m and $m+1$.

The following is how we have been trained to write rigorous proofs of equalities in mathematics articles: as a transitive chain of reasoning.

Proof. Indeed,

$$\begin{aligned}
2T_{m+2}T_{m+k+2} &= 2[2xT_{m+1} - T_m]T_{m+k+2} \\
&= 2x[2T_{m+1}T_{(m+1)+(k+1)}] - 2T_m T_{m+(k+2)} \\
&= 2x[T_{2(m+1)+(k+1)} + T_{k+1}] - [T_{2m+(k+2)} + T_{k+2}] \\
&= [2xT_{2m+k+3} - T_{2m+k+2}] + [2xT_{k+1} - T_{k+2}] \\
&= T_{2m+k+4} + T_k.
\end{aligned}$$

Halmos [15, section 16], writing long before interactive proof assistants were widespread, calls out the "proof that consists of a chain of expressions separated by equal signs" as an example of lazy writing,

> unhelpful [symbolism] that merely reports the result of the act and leaves the reader to guess how they were obtained,

and advocates for replacing such proofs by a "recipe for action" (a metaphor I already borrowed in Sect. 2.1). Here is an alternate proof of the Chebyshev lemma which precisely consists of such a recipe. This approach follows a formalisation of mine, contributed to Mathlib.[8]

Proof. Indeed, two applications of the inductive hypothesis give

$$2T_{m+1}T_{(m+1)+(k+1)} = T_{2(m+1)+(k+1)} + T_{k+1} \qquad (\star_1)$$

$$2T_m T_{m+(k+2)} = T_{2m+(k+2)} + T_{k+2} \qquad (\star_2)$$

---

[7] There is an alternative approach using trigonometric identities.
[8] Mathlib [8], RingTheory/Polynomial/Chebyshev, line 209.

and three applications of the recurrence relation give

$$T_{m+2} = 2xT_{m+1} - T_m \qquad (\star_1)$$

$$T_{(2m+k+2)+2} = 2xT_{(2m+k+2)+1} - T_{2m+k+2} \qquad (\star_2)$$

$$T_{k+2} = 2xT_{k+1} - T_k \qquad (\star_3)$$

A Gröbner basis computation[9] shows that LHS − RHS of the desired result,

$$2T_{m+2}T_{m+k+2} = T_{2m+k+4} + T_k,$$

is in the ideal generated by LHS − RHS of $(\star_1)$, $(\star_2)$, $(\star_1)$, $(\star_2)$, $(\star_3)$.

In a traditional prose proof, there is a high barrier to outsourcing this kind of computation to a specialised computer algebra system. The code performing the calculation must set up (under some names) the 11 variables[10]

$$\begin{array}{lll} T_{m+2}, \ T_{k+2}, \ T_{m+k+2}, & T_{2m+k+4}, \ x, \\ T_{m+1}, \ T_{k+1}, & T_{2m+k+3}, \\ T_m, \ T_k, & T_{2m+k+2}, \end{array}$$

the five polynomials in these 11 variables which generate the ideal, and a sixth polynomial whose membership in the ideal is to be checked. This process is tedious and error-prone; it will demand close attention from both author and reader. By contrast, when formalising, there is no such barrier: the problem statement is already available in a suitable electronic format.

The point is not just that in formalisation the second proof becomes feasible; I argue it is also more elegant. It is easier to grasp at high level: it is clear upfront what facts are being used, and the reader can check by eye that the goal appears to be within the scope of the Gröbner basis algorithm as run on these facts. Its black-boxing of the routine algorithm also makes the ideas more transparent—in this case, the choice of specialisations of the two inductive hypotheses.[11]

---

[9] This ability to send a computation to the Gröbner basis algorithm is a standard offering in formalisation languages [6,18,29]. In Lean this is performed via an external call to Sage; it was implemented by Dhruv Bhatia and Rob Lewis in 2022.

[10] We normalise indices before the computation.

[11] Indeed, let $k+b$ be the chosen instantiation of the $(m+1)$-inductive hypothesis and $k+a$ that of the m-inductive hypothesis:

$$2T_{m+1}T_{(m+1)+(k+b)} = T_{2(m+1)+(k+b)} + T_{k+b},$$
$$2T_m T_{m+(k+c)} = T_{2m+(k+c)} + T_{k+c}.$$

In order for there to be a nontrivial polynomial relation among these, the goal

$$2T_{m+2}T_{(m+2)+k} = T_{2(m+2)+k} + T_k,$$

and some uses of the recurrence, we need to arrange that the T-indices which appear,

$$m + \{2, 1, 0\}, \quad k + \{0, b, c\}, \quad m + k + \{2, b+1, c\}, \quad 2m + k + \{4, b+2, c\},$$

are all either (a) sets of three consecutive numbers (in which case the recurrence relation provides an identity connecting them) or (b) all the same. This forces $c = 2$, and that forces $b = 1$, leading to the instantiations $(\star_1)$, $(\star_2)$ chosen.

In summary, I argue that in formalisation the threshold for switching to full automation should lower, with many "mid-sized" computations automated away.

3 Abstraction

I now turn to the second realm in which I argue that there is a stylistic difference between prose and formal mathematics: the question of abstraction.

The principle that every mathematical argument should be generalised to exactly its proper context dates at least to Bourbaki [5, section 2]:

> Where the superficial observer sees only two, or several, quite distinct theories, lending one another "unexpected support" ... [we advocate] to look for the deep-lying reasons for such a discovery, to find the common ideas of these theories, buried under the accumulation of details properly belonging to each of them ... and to put them in their proper light.

This idea was profoundly influential. But though widely agreed on in principle, it is not followed universally in practice. For example, Halmos [15] advises writers,

> The observation that a proof proves something a little more general than it was invented for can frequently be left to the reader.

The main reason is psychological: abstractions seem to be a cognitive barrier for readers. A secondary, related reason is practical: you can't expect your reader to be confident in the application of an abstraction that she has never seen before, and so it's courteous to her to specialise it.

In formalised mathematics the trade-offs are different. The practical obstruction to abstraction nearly disappears,[12] though the psychological one remains. Moreover, as the examples in this section will explore, the usual arguments in favour of abstraction apply somewhat more strongly than in prose mathematics writing. All told, formal mathematics favours decidedly more use of abstraction.

3.1 Lax–Milgram Theorem

I first consider the Lax–Milgram theorem, a functional analysis result which turns up in the standard approach to linear elliptic partial differential equations.

Let H be a real Hilbert space, $B : H \times H \to \mathbb{R}$ a bilinear form.

Theorem 2 (Lax–Milgram). *Suppose there exist constants $\alpha, \beta > 0$ so that*

- *(boundedness) for all $u, v \in H$, $|B[u,v]| \leq \alpha \|u\| \|v\|$*
- *(coercivity) for all $u \in H$, $B[u,u] \geq \beta \|u\|^2$.*

Then for each $f \in H^*$, there exists a unique $u \in H$ so that for all $v \in H$, $B[u,v] = f(v)$.

---

[12] Your reader has immediate access to a full exposition of an unfamiliar abstraction; moreover, thanks to verification, she can trust you when you state that all the preconditions hold for that abstraction to be applicable in the context at hand.

The proof of this theorem begins by constructing a bounded linear map $A : H \to H$ such that for all $u, v \in H$, we have $B[u, v] = \langle A(u), v \rangle$. By the coercivity of B, we have for all u

$$\beta \|u\|^2 \leq B[u, u] = \langle A(u), u \rangle \leq \|A(u)\| \|u\|,$$

so (by the above if $u \neq 0$ and trivially if $u = 0$)

$$\beta \|u\| \leq \|A(u)\|. \tag{5}$$

It suffices to show that the operator A is bijective. I will concentrate on one step of the bijectivity argument: the step where we exploit (5) to establish that A is injective and has closed range. As usual I present two proofs.

Proof ([20], slightly compressed). If $u_1, u_2 \in H$, then $\|A(u_1 - u_2)\| \geq \beta \|u_1 - u_2\|$, from which it's clear that A is injective.

To see that the range of A is closed in H, let $\{u_j\}_{j=1}^\infty \subset H$ satisfy $Au_j \to w$ for some $w \in H$. We need to show that there exists $u \in H$ so that $Au = w$.

For this, we notice that

$$\|u_i - u_j\| \leq \beta^{-1} \|Au_i - Au_j\|.$$

The sequence $\{Au_j\}_{j=1}^\infty$ converges, so it must be Cauchy, so we see that $\{u_j\}_{j=1}^\infty$ must be Cauchy, and so must converge to some $u \in H$. Since A is bounded,

$$\|Au - w\| = \lim_{j \to \infty} \|Au - Au_j\| \leq \alpha \lim_{j \to \infty} \|u - u_j\| = 0.$$

That is, $Au = w$.

A close read of this proof snippet suggests that it doesn't seem to use the Hilbert space structure very much. And indeed, it is possible to extract the work as a lemma in general metric spaces. The appropriate abstraction is the following property of a function $f : X \to Y$ between metric spaces: that there exists a constant $\beta > 0$ such that for all x_1 and x_2, $\beta d_X(x_1, x_2) \leq d_Y(f(x_1), f(x_2))$.

As it turns out, the same argument appears in the proof of the Contraction Mapping Theorem, in a different special case (the case $Y = X$). When Yury Kudryashov formalised the Contraction Mapping Theorem for Mathlib in 2020, he recognised the appropriate context for the argument,[13] and wrote a self-contained theory development in Mathlib for such functions,[14] for which he introduced the name *antilipschitz maps*. In fact, I would not be surprised to learn that this fairly short (1–2 pages of text) and easy theory has been rediscovered and redeveloped many times, under many names.

With that abstraction and theory development available, the snippet of the Lax–Milgram theorem we are discussing reduces simply to the following:

---

[13] https://github.com/leanprover-community/mathlib/pull/1859#discussion_r36549 0281.
[14] Mathlib [8], Topology/MetricSpace/Antilipschitz.

Proof. A is uniformly continuous and by (5) it is antilipschitz, so it is injective and has closed range.

Daniel Roca González contributed this efficient proof of the Lax–Milgram theorem to Mathlib in 2022.[15] (The theorem had earlier been formalised in Coq [3], following a somewhat different proof.)

In this example we see illustrated Bourbaki's original argument in favour of abstraction: deduplication. Formal mathematics is done at scale: it is written from the axioms up, so nontrivial proofs form parts of a vast corpus; writing formal mathematics is much more like writing an encyclopaedia than like writing an article. At this scale, the "two, or several theories" united by an abstraction are very likely all to turn up, and simple efficiency favours using the abstraction.[16]

3.2 Smooth Vector Bundles

My last example (a bit more technical than the others in this article) is taken from the theory of smooth vector bundles in Lean, which is joint work of mine with Floris van Doorn in 2022–23.

In this example, the particular definition of smooth vector bundle we chose for our theory matters.

Definition 1. *A smooth vector bundle with fibre F over a smooth manifold B consists of*

- *a collection of topological vector spaces indexed by B;*
- *a topology on the total space, i.e. on their disjoint union;*
- *a collection of trivialisations, each identifying the fibre-union over some open set $U \subseteq B$ homeomorphically with $U \times F$, commuting via projections with the identity on U, and fibrewise an isomorphism of topological vector spaces;*
- *with the property that for two trivialisations in the collection the induced map $U \cap V \to \mathrm{End}(F)$ is smooth.*

I will discuss two approaches to the proof of the following statement.

Proposition 1. *The total space of a smooth vector bundle is a smooth manifold.*

Note that the fact that this is a theorem to be proved, rather than part of the definition, is a consequence of our choice of definition.

Here is how you might prove this theorem in prose. Since I didn't find a presentation of the theory of vector bundles in the literature which started with precisely our definition, this proof is not taken directly from real life.

---

[15] Mathlib [8], Analysis/InnerProductSpace/LaxMilgram.
[16] To believe that habitual abstraction really will avoid the repetition of proofs at large scale, you must be something of a Platonist: you must believe (as I do!) that the "natural context" of an argument is sufficiently unambiguous that others who need it will formulate it in the same way, and thus be led to stumble across your version.

Proof. Let H be the model space for the smooth manifold B. Given a trivialisation $\psi = (\psi_b, \psi_f) : \pi^{-1}(U) \to U \times F$ and a chart $\varphi : V \xrightarrow{\sim} \varphi(V) \subseteq H$ for B, define a candidate chart

$$\Phi_{\psi,\varphi} : \pi^{-1}(U \cap V) \to \varphi(U \cap V) \times F,$$

$$\Phi_{\psi,\varphi}(p) := (\varphi(\psi_b(p)), \psi_f(p)).$$

We need to check that for any two trivialisations ψ_1, ψ_2 and any two charts φ_1, φ_2 the transition function $\Phi_{\psi_2,\varphi_2} \circ \Phi_{\psi_1,\varphi_1}^{-1}$ is smooth. This works out since $\psi_2 \circ \psi_1^{-1}$, φ_1 and φ_2 are all smooth.

Our Lean formalisation uses Kobayashi–Nomizu's abstraction of a *structure groupoid* [23] for a way in which a space is modelled on another space, which is used there as the approach to the definition of smooth manifolds. Sébastien Gouëzel developed this theory in Mathlib in 2019.[17]

The advantage of our chosen definition of smooth vector bundle is that, following a suggestion of Gouëzel,[18] it too can be expressed using this structure groupoid abstraction. Let H be the model space for the smooth manifold B. Let E be a smooth vector bundle over B with fibre F. We consider the sequence

$$E \xrightarrow{\text{modelled on}} B \times F \xrightarrow{\text{modelled on}} H \times F :$$

1. E is modelled on $B \times F$ with the charts being the trivialisations, and our vector bundle definition amounts to the condition that the transition functions between these charts lie in the *smooth fibrewise-linear groupoid*;
2. $B \times F$ is in turn is modelled on $H \times F$ with the charts being the usual product manifold charts, and with the transition functions between these charts lying in the usual smooth manifold structure groupoid.

In this language, here is an outline of our formalisation[19] of the theorem.

Proof. "Modellings" can be composed, so the modellings of E on $B \times F$ and of $B \times F$ on $H \times F$ yield a modelling of E on $H \times F$. Structure groupoid properties can also be composed, so the transition functions between these induced charts lie in the smooth manifold structure groupoid for $B \times F$.

This composition theorem for structure groupoids was formulated by us for the project;[20] to our knowledge it does not appear in the literature.

This hierarchy of undigested abstractions is certainly a more obscure approach to this material than would be acceptable in a traditional prose presentation. But it has a certain elegance, and it brings organisational assistance: some work can be done cleanly at high level, and the more painful direct manipulation of partially-defined smooth functions appears only when checking the various preconditions for the abstractions to apply.

---

[17] Mathlib [8], Geometry/Manifold/ChartedSpace.
[18] Mathlib [8], Geometry/Manifold/ChartedSpace, line 139.
[19] Mathlib [8], Geometry/Manifold/VectorBundle/Basic, line 486.
[20] Mathlib [8], Geometry/Manifold/LocalInvariantProperties, line 698.

This is very much a slogan of formalisation: that it incentivises abstraction to cope with the demands of writing proofs in full detail [7]. Gonthier [11] similarly notes that his formalisation of the Four-Colour Theorem produced several abstractions, "new and rather elegant nuggets of mathematics," as a byproduct.

4 Conclusion

In this article I discuss only the question of, given a fixed statement, what constitutes a good proof (formal or informal) of that statement. An orthogonal question is how to best express the development of a whole mathematical theory.[21] This is a big question and it has produced an interesting literature [1, 13].

I have argued that good computer-formalised writing differs from good prose writing in two aspects: its incorporation of algorithms and of abstractions. These two aspects have an interesting commonality: in prose writing, both represent breaks in tone, or even in the very experience of reading—moments at which the reader is sent to a reference in order to read up on an unfamiliar abstraction, or to her computer to study and run a piece of code. But in formalised writing these are not breaks: prerequisites, computation and main argument form an integrated whole.

Montaño [25] argues that we experience a proof as beautiful according to the narrative experience of reading it, the "quality of its storytelling." In formalised mathematical writing, more kinds of thinking can be incorporated without causing breaks in the narrative flow. Our storytelling will be all the richer in consequence.

Acknowledgements. I am grateful to Isabelle Petersen for assistance in typesetting the notes for this article. I also thank the audience of the first version of this material (a talk at the 2023 workshop "Machine Assisted Proof" at the Institute for Pure and Applied Mathematics), whose stimulating comments helped to sharpen the argument, and Tom Hales, John Harrison and Kim Morrison for useful comments on a draft.

Disclosure of Interests. The author has no competing interests to declare that are relevant to the content of this article.

References

1. Affeldt, R., Cohen, C., Kerjean, M., Mahboubi, A., Rouhling, D., Sakaguchi, K.: Competing inheritance paths in dependent type theory: a case study in functional analysis. In: Peltier, N., Sofronie-Stokkermans, V. (eds.) IJCAR 2020. LNCS (LNAI), vol. 12167, pp. 3–20. Springer, Cham (2020). https://doi.org/10.1007/978-3-030-51054-1_1

---

[21] A crude analogy is to consider the statements of a mathematical theory as a digraph, with edges denoting easy implications (some implications are easy in both directions and their edges are bidirectional). To design a mathematical theory development, you must select a spanning tree for this digraph.

2. Arnold, V.I.: On teaching mathematics. Russ. Math. Surv. **53**(1), 229–236 (1998). https://doi.org/10.1070/RM1998v053n01ABEH000005
3. Boldo, S., Clément, F., Faissole, F., Martin, V., Mayero, M.: A Coq formal proof of the Lax-Milgram theorem. In: Proceedings of the 6th ACM SIGPLAN Conference on Certified Programs and Proofs, CPP 2017, pp. 79–89. Association for Computing Machinery, New York, NY, USA (2017). https://doi.org/10.1145/3018610.3018625
4. Bonsall, F.F.: A down-to-earth view of mathematics. Am. Math. Mon. **89**(1), 8–15 (1982). https://doi.org/10.2307/2320989
5. Bourbaki, N.: The architecture of mathematics. Am. Math. Mon. **57**(4), 221–232 (1950). https://doi.org/10.2307/2305937
6. Chaieb, A., Wenzel, M.: Context aware calculation and deduction. In: Kauers, M., Kerber, M., Miner, R., Windsteiger, W. (eds.) Calculemus/MKM -2007. LNCS (LNAI), vol. 4573, pp. 27–39. Springer, Heidelberg (2007). https://doi.org/10.1007/978-3-540-73086-6_3
7. Commelin, J., Topaz, A.: Abstraction boundaries and spec driven development in pure mathematics. Bull. Am. Math. Soc. **61**(2), 241–255 (2024). https://doi.org/10.1090/bull/1831
8. mathlib community, T.: The Lean mathematical library. In: Proceedings of the 9th ACM SIGPLAN International Conference on Certified Programs and Proofs, CPP 2020, pp. 367–381. Association for Computing Machinery, New York, NY, USA (2020). https://doi.org/10.1145/3372885.3373824
9. Conway, J.H., Burgiel, H., Goodman-Strauss, C.: The Symmetries of Things, A K Peters, Wellesley, MA (2008). https://doi.org/10.1201/b21368
10. Eck, D.: Wallpaper symmetry sketchpad. https://math.hws.edu/eck/js/symmetry/wallpaper.html. Accessed 25 March 2024
11. Gonthier, G.: Formal proof - the four color theorem. Notices Am. Math. Soc. **55**(11), 1382–1393 (2008). www.ams.org/notices/200811/tx081101382p.pdf
12. Gonthier, G., et al.: A machine-checked proof of the odd order theorem. In: Blazy, S., Paulin-Mohring, C., Pichardie, D. (eds.) ITP 2013. LNCS, vol. 7998, pp. 163–179. Springer, Heidelberg (2013). https://doi.org/10.1007/978-3-642-39634-2_14
13. Gouëzel, S.: A formalization of the change of variables formula for integrals in mathlib. In: Buzzard, K., Kutsia, T. (eds.) Intelligent Computer Mathematics: 15th International Conference, CICM 2022, Tbilisi, Georgia, September 19–23, 2022, Proceedings, pp. 3–18. Springer, Cham (2022). https://doi.org/10.1007/978-3-031-16681-5_1
14. Hales, T., et al.: A formal proof of the Kepler conjecture. Forum Math. Pi **5**, e2 (2017). https://doi.org/10.1017/fmp.2017.1
15. Halmos, P.R.: How to write mathematics. Enseign. Math. **2**(16), 123–152 (1970)
16. Hardy, G.H.: A mathematician's apology. Cambridge University Press (1940)
17. Harrison, J.: Formalized mathematics. Technical Report 36, Turku Centre for Computer Science (TUCS), Lemminkäisenkatu 14 A, FIN-20520 Turku, Finland (1996). http://www.cl.cam.ac.uk/~jrh13/papers/form-math3.html
18. Harrison, J.: Automating elementary number-theoretic proofs using Gröbner bases. In: Pfenning, F. (ed.) CADE 2007. LNCS (LNAI), vol. 4603, pp. 51–66. Springer, Heidelberg (2007). https://doi.org/10.1007/978-3-540-73595-3_5
19. Harrison, J.: Formalizing an analytic proof of the prime number theorem (dedicated to Mike Gordon on the occasion of his 60th birthday). J. Autom. Reason. **43**, 243–261 (2009). https://doi.org/10.1007/s10817-009-9145-6
20. Howard, P.: Lecture notes for M612: Partial Differential Equations (2020). https://people.tamu.edu/~phoward/M612.html

21. Immler, F.: A verified ODE solver and the Lorenz attractor. J. Autom. Reason. **61**(1), 73–111 (2018). https://doi.org/10.1007/s10817-017-9448-y
22. Inglis, M., Aberdein, A.: Beauty is not simplicity: an analysis of mathematicians' proof appraisals. Philosophia Math. **23**(1), 87–109 (07 2014). https://doi.org/10.1093/philmat/nku014
23. Kobayashi, S., Nomizu, K.: Foundations of Differential Geometry, Volume I, Issue 15, Volumes 1-2 of Interscience Tracts in Pure and Applied Mathematics, Interscience Publishers, New York, NY (1963)
24. Kochen, S., Specker, E.P.: The problem of hidden variables in quantum mechanics. J. Math. Mech. **17**(1), 59–87 (1967)
25. Montaño, U.: Ugly mathematics: why do mathematicians dislike computer-assisted proofs? Math. Intell. **34**(4), 21–28 (2012). https://doi.org/10.1007/s00283-012-9325-9
26. Moura, L., Ullrich, S.: The Lean 4 theorem prover and programming language. In: Platzer, A., Sutcliffe, G. (eds.) CADE 2021. LNCS (LNAI), vol. 12699, pp. 625–635. Springer, Cham (2021). https://doi.org/10.1007/978-3-030-79876-5_37
27. de Moura, L., Kong, S., Avigad, J., van Doorn, F., von Raumer, J.: The Lean theorem prover (system description). In: Felty, A.P., Middeldorp, A. (eds.) CADE 2015. LNCS (LNAI), vol. 9195, pp. 378–388. Springer, Cham (2015). https://doi.org/10.1007/978-3-319-21401-6_26
28. Peres, A.: Two simple proofs of the Kochen-Specker theorem. J. Phys. A Math. Gen. **24**(4), 1175–1178 (1991). https://doi.org/10.1088/0305-4470/24/4/003
29. Pottier, L.: Connecting Gröbner bases programs with Coq to do proofs in algebra, geometry and arithmetics. In: Rudnicki, P., Sutcliffe, G., Konev, B., Schmidt, R.A., Schulz, S. (eds.) Proceedings of the LPAR 2008 Workshops, Knowledge Exchange: Automated Provers and Proof Assistants, and the 7th International Workshop on the Implementation of Logics, Doha, Qatar, November 22, 2008. CEUR Workshop Proceedings, vol. 418, CEUR-WS.org (2008). https://ceur-ws.org/Vol-418/paper5.pdf
30. Rota, G.C.: The phenomenology of mathematical beauty. Synthese **111**(2), 171–182 (1997). https://doi.org/10.1023/A:1004930722234
31. Tao, T.: What is good mathematics? Bull. Am. Math. Soc. New Ser. **44**(4), 623–634 (2007). https://doi.org/10.1090/S0273-0979-07-01168-8
32. Wells, D.: Are these the most beautiful? Math. Intell. **12**(3), 37–41 (1990). https://doi.org/10.1007/BF03024015

Number Theory and Related Areas

Computing the Determinant of a Dense Matrix over $\mathbb{Z}$

John Abbott[✉] and Claus Fieker

Rheinland-Pfälzische Technische Universität Kaiserslautern, Kaiserslautern, Germany
{John.Abbott,Claus.Fieker}@rptu.de

Abstract. We present a new, practical algorithm for computing the determinant of a non-singular dense, uniform matrix over $\mathbb{Z}$; the aim is to achieve better practical efficiency, which is always at least as good as currently known methods. The algorithm uses randomness internally, but the result is guaranteed correct. The main new idea is to use a modular HNF in cases where the Pauderis–Storjohann HCOL method performs poorly. The algorithm is implemented in OSCAR 1.0.

Keywords: Determinant · integer matrix · unimodularity · 15-04 · 15A15 · 15B36 · 11C20

1 Introduction

We present a new, practical algorithm for computing the non-zero determinant of a matrix over $\mathbb{Z}$. The algorithm is fully general but is intended for dense, uniform matrices (see Sect. 1.1) We exclude matrices with zero determinant since verifying that a matrix has zero determinant involves different techniques; and with high probability we can quickly detect whether the determinant is zero (*e.g.* checking modulo two random 60-bit primes).

The best algorithm for computing the determinant depends on a number of factors: *e.g.* dimension of the matrix, size of the entries (in relation to the dimension), size of the determinant (usually not known in advance). Currently, the heuristic algorithm in [1] is the best choice for large dimension, dense, uniform matrices which have "very few", *i.e.* $O(1)$, non-trivial Smith invariant factors: the "expected" complexity is then $O(n^3 \log(n)(\log n + \log ||A||_{\max})^2)$. The extra condition on the Smith invariant factors very likely holds for matrices with entries chosen from a uniform distribution over an interval of width $\Omega(n)$ — see [2]. The conclusion of [1] gives an indication of the difficulty in assessing the theoretical complexity in the case that the number of non-trivial Smith invariant factors is not $O(1)$; they do nevertheless indicate good performance in practice when there are few non-trivial Smith invariant factors.

The main idea in [1] is the HCOL step which "divides" the working matrix by another matrix in Hermite normal form, eventually arriving at a unimodular matrix, which is then verified to be unimodular. There is at least one HCOL step for each non-trivial Smith invariant factor. We address the case where there are many non-trivial Smith invariant factors, and the largest of these is not too large. Instead of several rounds of division by HCOL matrices, we do a single division by a modular HNF: the choice of strategy depends on the size of the denominator of the solution a random linear system (as in [3]).

1.1 Uniform Dense Matrices

For many sorts of special matrix there are specific, efficient algorithms for computing the determinant. However, it can be hard to recognize whether a given matrix belongs to one of these special classes: e.g. deciding if a matrix is permuted triangular is NP-complete [4]. We shall concern ourselves with "uniform dense" matrices of integers whose *entries are mostly of the same size*, and make no attempt to recognize any special structure.

2 Notation, Terminology, Preliminaries

Here we introduce the notation and terminology we shall use.

Definition 1. *We define the **entrywise maximum** of $A \in \text{Mat}_{r \times c}(\mathbb{C})$ as*

$$||A||_{\max} = \max\{A_{ij} \mid 1 \leq i \leq r \text{ and } 1 \leq j \leq c\}$$

This is useful for specifying complexity. We also have the following relations $||A||_{\max} \leq ||A||_1 \leq c||A||_{\max}$ and $||A||_{\max} \leq ||A||_\infty \leq r||A||_{\max}$.

Definition 2. *We say that $A \in \text{Mat}_{n \times n}(\mathbb{Z})$ is **unimodular** iff $\det(A) = \pm 1$.*

Remark 1. In [5] there is an efficient algorithm to verify that a matrix is unimodular: it can also produce a "certificate" in the form of a "sparse product" representing the inverse of the matrix. Its complexity is $O(n^\omega \log n \, M(\log n + \log ||A||_{\max}))$ where ω is the exponent of matrix multiplication.

Definition 3. *Let $A \in \text{Mat}_{n \times n}(\mathbb{C})$. Then **Hadamard's row bound** for the determinant is*

$$H_{row}(A) = \prod_{i=1}^n r_i$$

*where $r_i \in \mathbb{R}_{\geq 0}$ satisfying $r_i^2 = \sum_{j=1}^n |A_{ij}|^2$ is the "euclidean length" of the i-th row of A. Clearly $H_{row}(A) \leq n^{n/2} ||A||_{\max}^n$. One may analogously define **Hadamard's column bound**, $H_{col}(A)$. We may combine these to obtain*

$$H(A) = \min(H_{row}(A), H_{col}(A)) \geq |\det(A)|$$

Remark 2. If $A \in \text{Mat}_{n \times n}(\mathbb{Z})$ and all entries are of similar magnitude then $H_{row}(A)$ and $H_{col}(A)$ are typically "of similar size", so there may be little benefit in computing both of them. Complexity is $O^\sim(n^2 \log ||A||_{\max})$.

Example 1. The Hadamard row bound, $H_{\text{row}}(A)$, equals $|\det A|$ iff the rows of A are mutually orthogonal; similarly for $H_{\text{col}}(A)$. For instance, if A is diagonal or a Hadamard matrix then both bounds are equal to the determinant.

In general, neither Hadamard bound is exact: for instance, there exist unimodular matrices with arbitrarily large entries, such as $\begin{pmatrix} F_{k-1} & F_k \\ F_k & F_{k+1} \end{pmatrix}$ where F_k is the k-th element of the Fibonacci sequence. Also, if $n > 1$ and all entries $A_{ij} = 1$ then $\det(A) = 0$ but $H(A) = n^{n/2}$.

Definition 4. *Let $A \in \text{Mat}_{n \times m}(\mathbb{Z})$. The **row Hermite Normal Form** (abbreviated: **row-HNF**) of A is $H = U_L A$ where $U_L \in \text{Mat}_{n \times n}(\mathbb{Z})$ is invertible, and $H \in \text{Mat}_{n \times m}(\mathbb{Z})$ is in "upper triangular" row echelon form with positive pivots and such that for each pivot column j and each row index $i < j$ we have $0 \leq H_{ij} < H_{ii}$ — here we adopt the usual left-to-right convention for the columns. Observe that the row-HNF is an echelon $\mathbb{Z}$-basis for the $\mathbb{Z}$-module generated by the rows of A.*

*The **col-HNF** may be defined analogously, but is not needed in this article. For brevity we shall write just **HNF** to refer to the "row" version.*

Definition 5. *Let $d \in \mathbb{Z}_{\neq 0}$. Then the d-**modular row-HNF** comprises the first m rows of the row-HNF of $\binom{dI_m}{A}$. This is a full-rank, upper triangular matrix in $\text{Mat}_{m \times m}(\mathbb{Z})$ whose diagonal entries divide d. Clearly the d-modular HNF is an echelon basis for the $\mathbb{Z}$-module sum of $d\mathbb{Z}^m$ with the module generated by the rows of A.*

*The d-**modular col-HNF** may be defined analogously, but is not needed in this article. For brevity we shall write just **modular HNF** to refer to the "row" version (with the usual left-to-right convention for the columns).*

Remark 3. The modular HNF can be computed essentially by following the standard HNF algorithm and reducing values modulo d — except we must not reduce the rows of dI_m. Thus computation of the modular HNF has bit complexity $O(n^3(\log d)^{1+\epsilon} + n^2 \log \|A\|_{\max})$ assuming "soft linear" basic arithmetic (incl. gcd computation).

Remark 4. Let $A \in \text{Mat}_{n \times n}(\mathbb{Z})$ be non-singular, let p be a prime dividing $\det(A)$, and let s_n be the greatest Smith invariant factor of A; so $s_n \neq 0$ since A is non-singular. Then for any exponent $e \in \mathbb{N}_{>0}$ every diagonal element of the p^e-modular HNF of A divides $p^{\min(e,k)}$ where $k = \nu_p(s_n)$. Also the product of the diagonal elements divides $\det(A)$.

Example 2. Let $d = 8$ and $M = \begin{pmatrix} 3 & -5 & 7 \\ 1 & 1 & -7 \\ 1 & 9 & 5 \end{pmatrix}$. Then the d-modular HNF is $H = \begin{pmatrix} 1 & 1 & 1 \\ 0 & 8 & 0 \\ 0 & 0 & 4 \end{pmatrix}$. From the diagonal of H we conclude that 32 divides $\det(A)$; indeed $\det(A) = -320$. Observe that $AH^{-1} = \begin{pmatrix} 3 & -1 & 1 \\ 1 & 0 & -2 \\ 1 & 1 & 1 \end{pmatrix}$ has integer entries — naturally, since the $\mathbb{Z}$-module generated by the rows of A is a submodule of that generated by the rows of H. Clearly $\det(A)/\det(H) = \det(AH^{-1}) \in \mathbb{Z}$.

Definition 6. *Let $A \in \mathrm{Mat}_{n \times m}(\mathbb{Z})$. The **Smith Normal Form** of A is $S = U_L A U_R$ where $U_L \in \mathrm{Mat}_{n \times n}(\mathbb{Z})$ and $U_R \in \mathrm{Mat}_{m \times m}(\mathbb{Z})$ are invertible, and $S \in \mathrm{Mat}_{n \times m}(\mathbb{Z})$ is diagonal such that each $S_{i+1,i+1}$ is a multiple of S_{ii}. By convention, the diagonal entries S_{ii} are non-negative, and are known variously as elementary divisors, invariant factors or Smith invariants; we call any $S_{ii} \neq 1$ a **non-trivial Smith invariant factor**. We use the standard abbreviation **SNF**, and shall write simply s_j for the j-th Smith invariant factor.*

Remark 5. With the non-negativity convention the SNF is unique; in contrast, the matrices U_L and U_R are not unique. Be aware that some authors reverse the divisibility criterion of the diagonal elements; in which case the tuple of non-zero Smith invariants is reversed.

2.1 Unimodular Matrices

For any square $n \times n$ matrix M we write $\mathrm{adj}(M)$ for its adjoint (aka. adjugate); it is well-known that $M \, \mathrm{adj}(M) = \det(M) I_n$. Thus for every unimodular matrix U we have $U^{-1} = \pm \mathrm{adj}(U)$. The entries of $\mathrm{adj}(U)$ are determinants of certain minors; this suggests that there could exist unimodular matrices whose inverses contain large entries. Indeed, some explicit families were given in [6]: we recall one such family: let $N \in \mathbb{Z}$ be "large", and let

$$U = \begin{pmatrix} 1 & \epsilon & \epsilon & \cdots & \epsilon & N \\ 0 & 1 & 0 & \cdots & 0 & 0 \\ 0 & N & 1 & \cdots & 0 & 0 \\ 0 & \epsilon & N & \cdots & 0 & 0 \\ 0 & \epsilon & \epsilon & \cdots & 0 & 0 \\ \vdots & \vdots & \vdots & \ddots & \vdots & \vdots \\ 0 & \epsilon & \epsilon & \cdots & N & 1 \end{pmatrix}$$

where taking $\epsilon < N/\sqrt{n}$ ensures that the inverse contains at least one entry whose size is near the limit predicted by Hadamard's bound; we may also "perturb slightly" the individual N and ϵ values in the matrix. Such matrices are almost worst cases for the Pauderis–Storjohann algorithm for unimodularity verification recalled in Sect. 2.2.

2.2 Unimodularity Verification

An important step in obtaining a guaranteed result from the HCOL algorithm is verification that a matrix is unimodular: an efficient algorithm with good asymptotic complexity to achieve this was presented in [5]. The key to its speed is *double-plus-one lifting* — a variant of quadratic Hensel lifting which cleverly limits entry growth while computing the inverse p-adically.

The algorithm has worst-case performance when the input matrix is not unimodular; so we want to avoid applying it unless we are "quite certain" that the matrix is indeed unimodular. Also a unimodular matrix whose inverse has large entries (*e.g.* the family from Sect. 2.1) leads to nearly worst-case performance.

3 Determinant Algorithms for Integer Matrices

There are currently three practical and efficient algorithms for computing the guaranteed determinant of an integer matrix: multi-modular chinese remaindering (CRT), solving a random linear system followed by some CRT steps [3], and the HCOL method [1]. The first two algorithms use a bound for the determinant as guarantee, whereas the HCOL method uses *unimodular verification* (see Sect. 2.2). The HCOL method has better asymptotic complexity (and practical performance) provided that there are only very few non-trivial Smith invariant factors. Our new algorithm in Sect. 4 addresses the case where there are more than a few non-trivial Smith invariant factors by dividing by a modular HNF.

3.1 Using the Modular HNF

The cost of computing a modular HNF depends (softly linearly) on the size of the modulus. If the matrix A has several non-trivial Smith invariant factors then s_n is likely a not-too-large factor of $\det(A)$, and we can obtain a "large factor" of s_n by solving a random linear system as in [3] — the factor, d, appears as the common denominator of the solution. If d is "large", we just do an HCOL iteration. But if d is small, we compute the d-modular HNF, H_d, and replace $A \leftarrow AH_d^{-1} \in \mathrm{Mat}(\mathbb{Z})$ noting the factor $\det(H_d)$, which is just the product of the diagonal entries — this effectively condenses several HCOL iterations into a single step. Moreover, after updating, $\det(A)$ is very likely to be small; to make a precise probabilistic claim we would have to know the distribution of matrices A with which we compute. Given A and d we can estimate quickly and reasonably accurately how long it will take to compute H_d.

4 Algorithm for Computing Determinant

We present an efficient method for computing the determinant of non-singular $A \in \mathrm{Mat}_{n \times n}(\mathbb{Z})$ with $n > 2$; we ignore any "special structure" A may have (and assume it was detected and handled separately during preprocessing).

- (0) Input matrix $A \in \mathrm{Mat}_{n \times n}(\mathbb{Z})$ ⟵ *may be modified during algorithm*
- (1) Let $e \leftarrow \lceil \log_2 \|A\|_{\max} \rceil$
- (2) Let $h = 1 + \lceil \log_2 H \rceil$ where $H \geq |\det(A)|$, *e.g.* Hadamard's bound.
- (3) Using CRT, compute $d \leftarrow \det(A) \bmod m$ where m is a product of wordsize-bit primes; stop when either $\log_2 m > e$ or $\log_2(|d|) + 60 < \log_2 m$
- (4) $D \leftarrow 1$ ⟵ *always a factor of* $\det(A)$
- (5) **Main loop**
- (5.1) if $|d| = 1$ and A is verified as unimodular then return D
- (5.2) if $60 + \log_2 |d| < \log_2 m$ and $\log_2 |d| <$ HNF threshold
- (5.2.1) Compute $|d|$-modular HNF, H_d; set $D_H = \det(H_d)$ and $A_{new} \leftarrow AH_d^{-1}$
- (5.2.2) If $D_H \neq 1$ update $A \leftarrow A_{new}$ and $D \leftarrow D_H D$; goto **Main loop**

(5.3) Solve linear system $Ax = b$ where $b \in \mathrm{Mat}_{n \times 1}(\mathbb{Z})$ is chosen randomly
(5.4) Let D_x be the common denominator of the solution x
(5.5) if $h - \log_2(D_x d)$ less than CRT threshold
(5.5.1) Continue chinese remaindering from step (3) until $\log_2(m) > h/D_x$.
(5.5.2) Return $D_x d$ where d is the symmetric remainder of $\det(A) \mod m$.
(5.6) if $\log_2(D_x) <$ HNF threshold
(5.6.1) Compute D_x-modular HNF, H_{D_x}; set $D_H = \det(H_{D_x})$
(5.6.2) Update $A \leftarrow A H_{D_x}^{-1}$ and $D \leftarrow D_H D$; goto **Main loop**
(5.7) From x compute the "HCOL" matrix H_x
(5.8) Update $D \leftarrow D_x D$ and $A \leftarrow A H_x^{-1}$; goto **Main loop**

5 Practical Tests

We give some timings to illustrate that our method can be substantially better than other known practical algorithms. The examples were constructed so that the new HNF-reduction will always take place (otherwise our method is merely a refined version of that presented in [1]). To simplify the presentation we vary just a single parameter: the matrix dimension. Each $n \times n$ matrix is constructed to have $n/2$ non-trivial Smith invariant factors (being a random 11-bit prime number), and entries of size roughly 1000 bits (Table 1).

Table 1. Timings for determinant computation

n	New	HCOL	ABM	NTL	CoCoA
30	0.03	0.13	0.06	0.04	0.05
60	0.16	2.85	0.80	0.24	0.41
120	0.88	5.7	3.9	2.3	4.0
240	3.6	44	42	19	41
480	14.7	346	390	220	500
960	67	3570	4600	2640	6500

Our new algorithm is already the fastest with relatively modest 30×30 matrices, and the advantage becomes more marked as the matrix dimension increases. The "HCOL" method from [1] performs poorly here because our test cases were chosen to exhibit this. NTL [7] and CoCoA [8] both use chinese remaindering, but NTL clearly has a more refined implementation. ABM refers to the native det function of OSCAR which just delegates the computation to FLINT [9] which uses the method from [3]; for these test cases, the algorithm essentially reduces to chinese remaindering.

6 Conclusion

We have presented a new, practical algorithm for computing determinant of a matrix with integer entries which exhibits good performance already for modestly sized matrices. The new algorithm has been implemented as part of the system OSCAR [10], and will be part of the next major release.

Acknowledgements. Both authors are supported by the Deutsche Forschungsgemeinschaft, specifically via Project-ID 286237555 – TRR 195

References

1. Pauderis, C., Storjohann, A.: Computing the invariant structure of integer matrices: fast algorithms into practice. In: Proceedings International Symposium on Symbolic and Algebraic Computation, ACM Press, pp. 307–314 (2013)
2. Eberly, W., Giesbrecht, M., Villard, G.: On computing the determinant and Smith form of an integer matrix. In: Proceedings 41st Annual IEEE Symposium on Foundations of Computer Science, 2000, pp. 441–458 (2000)
3. Abbott, J., Bronstein, M., Mulders, T.: Fast deterministic computation of determinants of dense matrices. In: Proceedings International Symposium on Symbolic and Algebraic Computation, ACM Press, 1999, pp. 197–204 (1999)
4. Fertin, G., Rusu, I., Vialette, S.: Obtaining a triangular matrix by independent row-column permutations. In: Proceedings 26th International Symposium on Algorithms and Computation, 2015, pp. 165–175 (2015). https://hal.science/hal-01189621
5. Pauderis, C., Storjohann, A.: Deterministic unimodularity certification. In: Proceedings International Symposium on Symbolic and Algebraic Computation, ACM Press, 2012, pp. 281–288 (2012)
6. Nishi, T., Rump, S., Oishi, S.: On the generation of very ill-conditioned integer matrices. Nonlinear Theor. Appl. IEICE **2**(2), 226–245 (2011)
7. Shoup, V.: NTL: a Library for doing Number Theory. Website https://www.libntl.org, version 11.5.1 (2024)
8. Abbott, J., Bigatti, A.: CoCoALib: a C++ library for doing Computations in Commutative Algebra. Available at https://cocoa.dima.unige.it/, versions: CoCoALib-0.99820, CoCoA-5.4.1u (2024)
9. The FLINT team. Flint: fast library for number theory, Website https://flintlib.org, part of OSCAR 1.0.0 (2024)
10. The OSCAR team, Oscar – open source computer algebra research system, Website https://www.oscar-system.org, version 1.0.0 (2024)

FastECPP over MPI

Andreas Enge[✉]

INRIA, Université de Bordeaux, CNRS, CANARI, 33400 Talence, France
andreas.enge@inria.fr
https://enge.math.u-bordeaux.fr/

Abstract. The FastECPP algorithm is currently the fastest approach to prove the primality of general numbers, and has the additional benefit of creating certificates that can be checked independently and with a lower complexity. This article shows how by parallelising over a linear number of cores, its quartic time complexity becomes a cubic wall-clock time complexity; and it presents the algorithmic choices of the FastECPP implementation in the author's CM software https://www.multiprecision.org/cm/ which has been written with massive parallelisation over MPI in mind, and which has been used to establish a new primality record for the "repunit" $(10^{86453} - 1)/9$.

1 FastECPP and Its Complexity

Since its inception, ECPP has become the fastest practical algorithm for proving the primality of arbitrary numbers N; additionally, it creates a certificate that can be verified independently and in less time. It is based on the key observation that if one can find an elliptic curve over $\mathbb{Z}/N\mathbb{Z}$ and a point on the curve of sufficiently large order N', then the primality of N' implies the primality of N. One step of the certificate creation process consists of searching for a suitable value of N', smaller than N by at least one bit, the elliptic curve and the point; then it continues recursively with N'. If $L = \lceil \log_2(N) \rceil$, this requires $O(L)$ steps. Verification goes through these $O(L)$ steps again, but checking their correctness is much faster than finding them.

To find a suitable curve, Goldwasser and Kilian in their original algorithm [14] use point counting on random curves, which results in a (heuristic, under the assumption that all occurring numbers behave randomly with respect to the sizes of their prime factors) time complexity of $\tilde{O}(L^6)$ using asymptotically fast arithmetic, where the $\tilde{O}$ notation hides logarithmic factors. Atkin and Morain observe in [2] that one can find suitable elliptic curves using the complex multiplication method, which lowers the heuristic complexity to $\tilde{O}(L^5)$; this is now known as the ECPP algorithm.

The FastECPP version, attributed to Shallit in [18, §5.10] and worked out in [13,19], improves a bottleneck, lowering the heuristic complexity to $\tilde{O}(L^4)$.

In this section, we present the FastECPP algorithm and its heuristic analysis following [19]; for a comprehensive overview of primality testing and proving algorithms, see [20]. Basic ECPP can essentially be recovered from FastECPP by dropping substep (1) of the first phase below and by computing the square roots of the discriminants in substep (2) one by one. Since this is slower in all cases and does not even substantially simplify the implementation, the basic variant is only of historical interest.

First Phase. One step of the certificate creation process determines a probable prime number N' that is smaller than N and then recursively creates a certificate for N'. Each of the steps consists of the following substeps.

(1) Write down the set $\mathcal{Q} = \{q_1^*, q_2^*, \ldots\}$ of $\tilde{\Theta}(L)$ smallest "signed primes" with $\left(\frac{q^*}{N}\right) = 1$, where $q^* = q$ if q is a prime that is 1 mod 4, $q^* = -q$ if q is a prime that is 3 mod 4, or $q^* \in \{-4, \pm 8\}$. Compute their square roots modulo N, in time $\tilde{O}(L^2)$ each or $\tilde{O}(L^3)$ altogether.

(2) Fundamental quadratic discriminants are exactly the products of signed primes (without multiplicities); create a set of $\tilde{\Theta}(L^2)$ negative discriminants D with $|D| \leqslant D_{\max} \in \tilde{O}(L^2)$ and their square roots modulo N as products of the precomputed roots (products of two elements of $\mathcal{Q}$ are enough). Try to solve the Pell equation $4N = t^2 - v^2 D$ for all D using Cornachia's algorithm, in time $\tilde{O}(L)$ per problem or $\tilde{O}(L^3)$ altogether. The success probability is of the order of $1/\sqrt{|D|} \in \tilde{\Omega}(1/L)$, so $\tilde{\Theta}(L)$ values t are expected to be obtained; then if N is prime, there are elliptic curves modulo N with complex multiplication by D and $m = N + 1 \pm t$ points.

(3) Trial factor the m as $m = cN'$, where the cofactor $c \geqslant 2$ is B-smooth for some bound B (for instance, $B = 2$), and all primes dividing N' are larger than B; each factorisation takes $\tilde{O}(L^2)$ for $B \in \tilde{O}(L^3)$ for a total of $\tilde{O}(L^3)$ (more details are given in §2).

(4) Test the N' for primality, in time $\tilde{O}(L^2)$ each for a total of $\tilde{O}(L^3)$; the expected number of remaining primes N' is in $\Theta(1)$.

So each step, of which there are $O(L)$, has a complexity of $\tilde{O}(L^3)$.

Second Phase. For each of the $O(L)$ steps of the first phase, carry out the following substeps: Construct the class polynomial of degree $h \in \tilde{O}\left(\sqrt{|D|}\right)$, where h is the class number of D, in time $\tilde{O}(|D|) \subseteq \tilde{O}(L^2)$ [6]. Find a root modulo N in time $\tilde{O}(L^3)$ and write down the two corresponding CM elliptic curves; take random points on the curves in time $\tilde{O}(L^2)$ for the square roots yielding the Y-coordinates and multiply them by the cofactors c to obtain a point of prime order N'. So this phase also has $O(L)$ independent steps of complexity $\tilde{O}(L^3)$ each.

Notice that all substeps above admit a trivial parallelism of $\Theta(L)$; so using $\tilde{\Omega}(L)$ cores, the wall-clock time complexity of FastECPP becomes $\tilde{O}(L^3)$. Some of the substeps could be further parallelised to $\Theta(L^2)$ cores, but it is probably not possible to improve the complexity of a modular square root or a primality test beneath $\tilde{\Omega}(L^2)$ even in parallel.

The final certificate is of bit size $O(L^2)$ and can be verified in sequential time $\tilde{O}(L^3)$, or in wall-clock time $\tilde{O}(L^2)$ on $\tilde{\Omega}(L)$ cores.

2 Implementation Choices

The only previously available free implementation of the FastECPP algorithm was written by Jared Asuncion within PARI/GP [1] as the `primecert` function, with numbers of around 1000 digits in mind. The CM software [5], on the other hand, encapsulated over two decades of my research on algorithms for CM of elliptic curves; so I decided to add first a sequential FastECPP implementation and eventually an MPI version tuned for massive parallelism and with record computations in mind. Both are available in CM since version 0.4.0 as the binaries `ecpp` and `ecpp-mpi`, respectively. This section documents the choices made in the implementation and the parameter selection depending on the input size L and the available computing power. Indeed the MPI version dedicates the main process to the coordination of the computation and uses the w remaining ones to do the actual work. It adapts to the number of available cores by choosing parameters depending on w; so also the final certificate depends on w, see §5.

First Phase. Substep (0). Inspired by PARI/GP, the class numbers of all discriminants up to $D_{\max}$ are precomputed in time $\tilde{O}(D_{\max}^{1.5})$ by looping over all quadratic forms of discriminant in the desired range. Also, a number of prime products used for trial division are precomputed. Both are parallelised over all w workers, and the results can be stored on disk for reuse over several runs.

The following substeps (1) to (4) may need to be repeated over several rounds in the unlucky case that no candidate N' remains at the end.

Substep (1). Let $D_{\max} = \min\left(2^{35}, \max(2^{20}, L^2/2)\right)$ (or, more precisely, the smallest power of 2 not below this), $h_{\max} = 100000$ and $p_{\max} = \max(29, \lfloor L/2^{10} \rfloor)$. The values of $h_{\max}$ and $p_{\max}$ are intended to speed up the second phase: Only discriminants with class number at most $h_{\max}$ are considered, which is an upper bound on the degree of class polynomials constructable in reasonable time; and the maximal prime dividing the class number is bounded by $p_{\max}$, so that after expressing the class field as a tower of prime degree extensions using [8], it is enough to determine roots of polynomials of degree at most $p_{\max}$.

The main process computes the set $\mathcal{Q}$ of kw smallest signed primes and the discriminants divisible by up to 7 primes from $\mathcal{Q}$ under the additional restrictions above. The integer k is chosen minimally such that the expected number of remaining N' after substep (4), obtained using the formula for s in [13, §4], the formula in the last line of [19, §2] for the success probability of solving the Pell equation and the precomputed class numbers, is at least 3 in the first round (to allow for some choice), or 1 in later rounds (to avoid computing too many square roots). Then each worker computes k square roots modulo N.

Substep (2). The implementation evenly distributes *all* discriminants among the workers to solve the Pell equation. If the expected number of N' is much larger than 1, this may lead to unnecessary work being executed, but on the other hand, it may also lead to more choice and thus a smaller N' and ultimately fewer steps. In the absence of a clear optimality criterion, the design decision was to favour smaller certificates over faster running times; see the comparison with PARI/GP in §5.

Substep (3). Let P be the product of all primes up to B; then $|\log P/B - 1| \in o(1)$ by [21, Th. 4]. The numbers m_i of size $\Theta(L)$ obtained in the previous substep need to be written as $m_i = c_i N_i'$ with c_i having only prime factors dividing P, and N_i' coprime to P. The implementation follows [13, §4] and proceeds in batches to first compute all the $P \bmod m_i$. In a first step, it constructs bottom-up a binary tree with the m_i on the leaves and each inner node being the product of its two descendants; then it replaces the root $M = m_1 m_2 \cdots$ by $P \bmod M$, and top-down each node by its parent modulo the node, ending up with $P \bmod m_i$ on the leaves. We may split the m_i into batches, handled independently, such that $M \leqslant P$. Then each batch contains $O(\log_2 P/L)$ numbers m_i, and the tree computation takes time $\tilde{O}(\log P)$. If we have $\tilde{\Omega}(\log_2 P/L)$ numbers m_i, then the amortised cost per m_i is in $\tilde{O}(L)$. But we may also use larger values of B, such that $\log P \gg \log M$; the sequential version ecpp uses $B \in \Theta(L^3)$ for $\tilde{\Theta}(L)$ values m_i with an amortised cost of $\tilde{O}(L^2)$ per value. Now we compute the square-free part of c_i as $c_i' = \gcd(m_i, P \bmod m_i)$ in time $\tilde{O}(L)$, the square part of c_i as $c_i'' = \gcd(m_i/c_i', c_i')$ and so on. This gcd part usually has a total complexity of $\tilde{O}(L)$, but in unlikely cases (for instance if $m_i = 2^L$) may require $\tilde{O}(L^2)$. (A complexity of $\tilde{O}(L)$ could be achieved by using the algorithm of [3], which is expected to be slightly slower on average.)

Notice that the tree-based approach requires $\tilde{\Theta}(\log P)$ of memory, and that Moore's law acts similarly on the number of cores and on the memory; indeed one has observed over the last decades that the memory per core has increased only very moderately. Otherwise said, the total memory available over $\Theta(L)$ cores is of the order of $\tilde{O}(L)$, which implies that $B \in \tilde{O}(L)$: So asymptotically in the parallel setting, the effect of batch trial division vanishes.

The MPI implementation is parallelised in two dimensions, with respect to the m_i and to the primes in P. It splits the m_i into $b \approx 16$ batches, each of which is sent to an MPI communicator with $w' = \lfloor w/b \rfloor$ workers. Then worker i handles the product P_i of primes between $(i-1) \cdot 2^{29}$ and $i \cdot 2^{29}$, so that the effective smoothness bound becomes $B = w' \cdot 2^{29}$. By the already cited [21, Th. 4], each of the P_i has about the same bit size of $\log_2(P_i) \approx 2^{29}/\log(2) \approx 775 \cdot 10^6$, so that each level of the tree requires about 97 MiB of memory. In the 86453 digit record, there are up to $2^{29}/(86453 \log(10)) \approx 2700$ values m_i on the leaves, and thus $\lceil \log_2(2700) \rceil + 1 = 13$ levels, so the total memory requirements are about 1.3 GiB per core.

Substep (4). The implementation performs Miller–Rabin tests in batches of w numbers, starting with the smallest ones. So of all the non-smooth parts computed in substep (3), the smallest prime one will be retained as N'.

3 Details of the 86453 Digit Record

The second phase of FastECPP is built upon a number of research results that were already implemented in the CM software [5]. It computes class polynomials by complex approximations as described in [6]. Optimal class invariants are chosen derived from Weber functions [22], simple [9] or double eta quotients [10], including cases where it is enough to compute lower-degree subfields of the class field [11]. The evaluation of modular functions, which is the most important part of the class polynomial computation, is optimised following [6,7]. To ease the step of factoring class polynomials modulo primes, the class fields are then represented as towers of cyclic Galois extensions of prime degree following [8]. The software relies on a number of libraries from the GNU project, notably GMP [15], MPFR [16] and MPC [12], and on PARI/GP [1] for computations with class groups and Flint [17] for root finding modulo a prime.

Computations have been carried out on the PlaFRIM cluster in Bordeaux, https://www.plafrim.fr/, to prove primality of the "repunit" $(10^{86453}-1)/9$. The certificate is available in PARI/GP format at

https://www.multiprecision.org/downloads/ecpp/cert-r86453.bz2.

An independent verification of the certificate has been carried out by the factordb server, see

http://factordb.com/index.php?id=1100000000046752372

The first phase computed 2980 steps in about 103 d of wall-clock time and 383 years of CPU time, in several runs with 759 to 2639 cores depending on machine availability. Of this CPU time, about 13% were devoted to the computation of square roots in substep (1), 47% to solving Pell equations in substep (2), 10% to batch trial factoring in substep (3) and 30% to primality tests in substep (4). The largest occurring discriminant in absolute value was -34223767071, the largest prime q^* was 240869 of the discriminant -3329532187, the largest prime of a class number was 277, appearing for 16 different discriminants. The effective trial factor bound $B = w' \cdot 2^{29}$ varied depending on the number w of cores assigned to a run, with $43 \leqslant w' \leqslant 172$.

The second phase was run on a machine with 96 cores and 1 TB of memory and took about 25 years of CPU time (wall-clock time was lost), that is, only about 6% of the total CPU time of both phases. Inside the phase, 2.5% of the time was devoted to computing tower representations for class fields, 2.8% to verifying orders of points on elliptic curves, and close to 95% to finding roots of class polynomials. Depending on the largest factor of the class number, which determines the degrees of the polynomials, running times for this dominant step vary considerably. The longest step took close to 42 d for factoring a class polynomial for the discriminant -2083578323 of class number $11920 = 2^4 \cdot 5 \cdot 149$ for an intermediate prime of 82089 digits; this would also have been the wall-clock time for this phase had it been run sufficiently in parallel.

Verification of the certificate took about 48 h using PARI/GP on the same machine with 96 cores.

4 Sizes of Smooth Parts

It is shown in [13, §4] that the probability of obtaining a prime quotient after removing the B-smooth part from an L-bit number is asymptotically $e^\gamma \log_2 B/L$, where $\gamma = 0.577\ldots$ is the Euler–Mascheroni constant, and that in this case the expected number of binary digits of the B-smooth part is $\log_2 B \in \tilde{O}(1)$ for B polynomial in L. The following computations give more precise estimates. They rely on approximations from analytic number theory for which very tight error bounds are available, for instance in [21]. As a consequence, the estimates are relevant for the sizes of numbers we consider, say for $B \geqslant 2^{29}$ and L of the order of a few thousands. To simplify the exposition, however, only the main terms are given, using $\sim$ to denote asymptotic equality up to lower order terms.

The probability that a number between 2^{L-1} and 2^L is a B-smooth number times a prime is given by

$$\frac{1}{2^{L-1}} \sum_{f B\text{-smooth}} \sum_{p\text{ prime},\ 2^{L-1}/f < p \leqslant 2^L/f} 1 \sim \frac{1}{2^{L-1}} \sum_{f B\text{-smooth}} \frac{2^{L-1}}{f \log(2^L/f)}$$

$$\sim \frac{1}{\log(2^L)} \sum_{f B\text{-smooth}} \frac{1}{f},$$

using the prime number theorem and $\log f \in o(L)$.

For $0 < \alpha \leqslant 1$, all $f \leqslant B^\alpha$ are B-smooth, and their contribution to the sum is $\sum_{f \leqslant B^\alpha} (1/f) \sim \alpha \log B$ using the main term of the harmonic number. So the conditional probability that $f \leqslant B^\alpha$ for an L-bit number assumed to be a B-smooth number f times a prime (for L tending to infinity and B growing slowly) is α/e^γ, and the probability density function for this event with respect to α is the constant $1/e^\gamma$. In particular, $1/e^\gamma \approx 56\%$ of the primes N' remaining after substep (4) gain less than $\log_2 B$ bits compared to N.

For $1 < \alpha \leqslant 2$, numbers f with $B < f \leqslant B^\alpha$ that are *not* B-smooth are divisible by exactly one prime $P > B$, and what remains is B-smooth. So their contribution to the sum is

$$\sum_{B < f \leqslant B^\alpha} \frac{1}{f} - \sum_{B < P \leqslant B^\alpha,\ P\text{ prime}} \frac{1}{P} \sum_{f \leqslant B^\alpha/P} \frac{1}{f}$$

$$\sim (\alpha - 1) \log B - \sum_{B < P \leqslant B^\alpha,\ P\text{ prime}} \frac{1}{P}(\alpha \log B - \log P)$$

$$\sim (\alpha - 1) \log B - \alpha \log B \big(\log \log(B^\alpha) - \log \log(B)\big) + \big(\log(B^\alpha) - \log(B)\big)$$

$$= (2\alpha - 2 - \alpha \log \alpha) \log B$$

using [21, Thm. 5 and 6]. So the probability density function is $(1 - \log \alpha)/e^\gamma$ for this range of α. In particular, $(2 - 2\log 2)/e^\gamma \approx 34\%$ of the primes N' remaining

42   A. Enge

after substep (4) gain between $\log_2 B$ and $2\log_2 B$ bits compared to N, and only 10% of the numbers gain more.

For higher values of α, exact computations become unwieldy, but the previous computation still yields a *lower bound* (since non-smooth numbers with more than one large prime factor are subtracted multiple times) on the contribution of B-smooth $f \leqslant B^\alpha$ to the sum as

$$\sum_{f \leqslant B^\alpha} \frac{1}{f} - \sum_{B < P \leqslant B^\alpha,\, P\text{ prime}} \frac{1}{P} \sum_{f \leqslant B^\alpha/P} \frac{1}{f} \sim (2\alpha - 1 - \alpha \log \alpha) \log B.$$

The function reaches its maximum $(e-1)\log B$ for $\alpha = e$. Otherwise said, the probability that an N' gains more than $2.72 \log_2 B$ bits over N is less than $1 - (e-1)/e^\gamma \leqslant 3.6\%$.

5   Comparison

As seen in the previous section, the probability that a candidate for N' gains much more than $\log_2 B$ bits is rather small, so maybe spending almost half of the time for solving Pell equations in the record to obtain more prime candidates for N' is not optimal. On the other hand, probabilities of gaining more digits in one step (and thus requiring fewer steps and less time) are amplified by the maximum statistics. For instance, the expected number of prime candidates for N' in the first step of the record was 8.9; so the probability of gaining at least $2 \log_2 B$ bits was about $1 - \left((3 - 2\log 2)/e^\gamma\right)^{8.9} \approx 58\%$, and in fact $77 = 2.2 \cdot \log_2 B$ bits were gained. Precisely, with $w = 1135$ workers, 41440753 discriminants were considered in substep (2), 40788 curve cardinalities were trial factored in substep (3), and 3405 primality tests were carried out in substep (4) before finding a prime N' (which is consistent with the expected number of 8.9 primes in a sample of 40788).

The following table compares the wall-clock times (in minutes) and the lengths of the certificates between PARI/GP [1] (which is parallelised using threads) and CM [5] for the first prime after 10^n for different values of n. The first two column blocks show figures for the parallel versions on a machine with 128 cores. The last column block provides the results for the serial code in CM on the same type of machine. The initial smoothness bound B decreases (down to 2^{20}) for PARI/GP as the N' become smaller, and remains fixed for CM over the course of the algorithm. Notice that the storage requirement and the verification time of a certificate for a number of given size are proportional to the number of steps, so that shorter certificates are preferable.

n	Pari/Gp			Cm ecpp-mpi			Cm ecpp		
	$\log_2 B$	#steps	time	$\log_2 B$	#steps	time	$\log_2 B$	#steps	time
1000	24	131	0.30	32	33	5.2	22	88	0.80
2000	26	220	3.6	32	76	15	25	157	15
4000	30	344	41	32	166	47	28	274	210
5000	30	399	92	32	204	51	29	342	510
10000	30	740	820	32	444	220	32		

Let us first compare the parallel implementations. The Cm certificates are considerably shorter with an advantage that decreases as Pari/Gp chooses larger smoothness bounds B. The number of steps increases roughly linearly as expected (slightly more for Cm, slightly less for Pari/Gp). One notices that the average gain of bits per step, computed as $\log_2(10^n)$ divided by the number of steps, is above $2.3 \log_2 B$ for Cm, illustrating the effect of the maximum statistic.

Somewhat surprisingly, neither of the two implementations shows the quartic asymptotic running time for a fixed number of cores, which may be due to "overparallelisation" and thus less than optimal CPU times for the smaller instances: Cm spends 78% of the wall-clock time on trial factorisation for 5000 digits, but only 43% for 10000 digits.

The running times of the serial implementation of Cm, however, reflect closely the quartic complexity; the computations for the 10000 digit number are expected to take close to a week and were not carried out. When compared to the parallel version, it also becomes apparent how the latter one profits from part of the additional computing power not for decreasing the wall-clock time, but for shortening the certificates.

The FastECPP implementation in Cm has been adopted by the primality proving community, which has used it for all ECPP records since the first release of the code in Cm 0.4.0 in May 2022, that is, for all but two out of the 20 entries at

https://t5k.org/top20/page.php?id=27.

Of the AKS/cyclotomy type competitors, the one with the best complexity known to date is the (probabilistic) algorithm in [4]. It computes an (Nd)-th power in the ring

$$\big((\mathbb{Z}/N\mathbb{Z})[Y]/(f(Y))\big)[X]/(X^e - r(Y))$$

with f of degree $d \in (\log L)^{O(\log \log \log L)} \subseteq L^{o(1)}$ and $e \in L^{2+o(1)}$, in time $L^{4+o(1)}$. While its sequential complexity is comparable to that of FastECPP, it does not seem possible to reach a wall-clock time complexity of $L^{3+o(1)}$ by parallelising it to $\Omega(L)$ cores, so that FastECPP appears to remain the algorithm of choice for proofs of large general primes in a parallel setting.

References

1. [SW Rel.] Allombert, B., Belabas, K.: PARI/GP version 2.15.4 (2024). LIC: GPL-2-or-later. URL: https://pari.math.u-bordeaux.fr/, SWHID: ⟨swh:1:dir:8e76e2daa122f03e6a9206e18a62aa7ab48efb93;origin=https://pari.math.u-bordeaux.fr/git/pari.git;visit=swh:1:snp:cd7a1ce7663980b27dfdfc3e96c97fb073271c02;anchor=swh:1:rel:68863db68dd3d346ce5685ad767360c01dfca26a⟩
2. Atkin, A.O.L., Morain, F.: Elliptic curves and primality proving. Math. Comput. **61**(203), 29–68 (1993)
3. Bernstein, D.J.: How to find smooth parts of integers (2004). Preprint. https://cr.yp.to/factorization/smoothparts-20040510.pdf
4. Bernstein, D.J.: Proving primality in essentially quartic random time. Math. Comput. **76**(257), 389–403 (2007)
5. [SW Rel.] Enge, A.: CM version 0.4.3, Feb. 2024. LIC: GPL-3-or-later. URL: https://www.multiprecision.org/cm/, SWHID: ⟨swh:1:dir:056f ba450fbd9406efd86a 5db93895fa63d212df;origin=https://gitlab.inria.fr/enge/cm;visit=swh:1:snp:d0a38 ff75431aab4c91e1a50a51b26 a573e17784;anchor=swh:1:rev:7a6567cf2d98aa9a37166 b2c4c99f7dfcecfea58⟩
6. Enge, A.: The complexity of class polynomial computation via floating point approximations. Math. Comput. **78**(266), 1089–1107 (2009)
7. Enge, A., Hart, W., Johansson, F.: Short addition sequences for theta functions. J. Integer Sequences **18**(2), 1–34 (2018)
8. Enge, A., Morain, F.: Fast decomposition of polynomials with known Galois group. In: Fossorier, M., Høholdt, T., Poli, A. (eds.) AAECC 2003. LNCS, vol. 2643, pp. 254–264. Springer, Heidelberg (2003). https://doi.org/10.1007/3-540-44828-4_27
9. Enge, A., Morain, F.: Generalised weber functions. Acta Arith. **164**(4), 309–341 (2014)
10. Enge, A., Schertz, R.: Constructing elliptic curves over finite fields using double eta-quotients. Journal de Théorie des Nombres de Bordeaux **16**(3), 555–568 (2004)
11. Enge, A., Schertz, R.: Singular values of multiple eta-quotients for ramified primes. LMS J. Comput. Math. **16**, 407–418 (2013)
12. [SW Rel.] Enge, A., et al.: GNU MPC — A library for multiprecision complex arithmetic with exact rounding version 1.2.1 (2020). LIC: LGPL-3-or-later. URL: https://www.multiprecision.org/mpc/, SWHID: ⟨swh:1:dir:ebd0a7bca44757a5e4939545d52 cc68ef882a306;origin=https://gitlab.inria.fr/mpc/mpc;visit=swh:1:snp:a9728932e a1e8286f2634cac6ba18a340a184977;anchor=swh:1:rev:1d9a8349c839cf1935 3568ecfa32ace2223084d1⟩
13. Franke, J., Kleinjung, T., Morain, F., Wirth, T.: Proving the primality of very large numbers with fastECPP. In: Buell, D. (ed.) ANTS 2004. LNCS, vol. 3076, pp. 194–207. Springer, Heidelberg (2004). https://doi.org/10.1007/978-3-540-24847-7_14
14. Goldwasser, S., Kilian, J.: Almost all primes can be quickly certified. In: Proceedings of the 18th Annual ACM Symposium on Theory of Computing, pp. 316–329 (1986)
15. [SW Rel.] Granlund, T., et al.: GMP — The GNU Multiple Precision Arithmetic Library version 6.2.1 (2020). LIC: LGPL-3-or-later. URL: https://gmplib.org/, SWHID: ⟨ swh:1:dir:31da2a73b2e10e765fb52996d15e6f5f453453a3;origin=https://gm plib.org/repo/gmp-6.2/;visit=swh:1:snp:f40ef7cd40cb4ccec48e3d4291d44ca8def8f5 92;anchor=swh:1:rel:acd9a44abc3f7a2a39ab039d3d4ac81eb57e5943⟩

16. [SW Rel.] Hanrot, G., et al.: GNU MPFR — A library for multiple-precision floating-point computations with exact rounding version 4.1.0 (2020). LIC: LGPL-3-or-later. URL: https://www.mpfr.org/, SWHID: ⟨swh:1:dir:0e32d50b65ab886c5bcd44f63e3394980ad2fcdb;origin=https://gitlab.inria.fr/mpfr/mpfr;visit=swh:1:snp:22ecd8c12efff 07509da63e0f9ce6dddda4e3b8b;anchor=swh:1:rel:e956e33703ea78dd0f263ed6b872e5c9bf83d010⟩
17. [SW Rel.] Hart, W., et al.: FLINT: fast library for number theory version 2.9.0 (2022). LIC: LGPL-2.1-or-later. URL: https://flintlib.org/, SWHID: ⟨swh:1:dir:d7c1dec3fb591a70205462058c842a245c68dd31;origin=https://github.com/flintlib/flint;visit=swh:1:snp:c82faab4eaecd9393bff3fcd224d3cf65699e51b;anchor=swh:1:rev:e143df4b0f19d2f841e36234a12b69f48c4359b9⟩
18. Lenstra, A.K., Lenstra Jr, H.W.: Algorithms in number theory. In: Algorithms and Complexity. Ed. by Jan van Leeuwen. vol. A. Handbook of Theoretical Computer Science. Amsterdam: Elsevier, pp. 674–715 (1990)
19. Morain, F.: Implementing the asymptotically fast version of the elliptic curve primality proving algorithm. Math. Comput. **76**(257), 493–505 (2007)
20. Morain, F.: La primalité en temps polynomial [d'après Adleman, Huang; Agrawal, Kayal, Saxena]. In: *Astérisque* **294**(917), 205–230 (2004)
21. Rosser, J.B., Schoenfeld, L.: Approximate formulas for some functions of prime numbers. Ill. J. Math. **6**, 64–94 (1962)
22. Schertz, R.: "Die singulären Werte der Weberschen Funktionen $\mathfrak{f}, \mathfrak{f}_1, \mathfrak{f}_2, \gamma_2, \gamma_3$". In: Journal für die reine und angewandte Mathematik 286/287, pp. 46–74 (1976)

Attacking a Levelled Fully Homomorphic Encryption System with Topological Data Analysis

Aaruni Kaushik[✉][iD]

University of Kaiserslautern-Landau, Kaiserslautern, Germany
aaruni.kaushik@math.rptu.de

Abstract. In this paper, we study the Fully Homomorphic Encryption system proposed by Craig Gentry, Amit Sahai, and Brent Waters in [3]. We present a restated version of the proposed cryptosystem, and consider the ciphertexts resulting from the use of this system, to find patterns in them. For this task, we train a machine learning pipeline, utilizing Topological Data Analysis. We show that for secure parameters chosen according to the available literature, this machine learning approach can simply guess what the plain text should have been in a majority of cases (accuracy $> 70\%$), in very good time complexity ($< O(n^3)$). This attack was ineffective against the base Learning With Errors (LWE) problem, which hints at a possible flaw somewhere in the reduction from LWE to the cryptosystem.

Keywords: Fully Homomorphic Encryption · Machine Learning · Lattice Based Encryption · Craig Gentry

1 Introduction

Fully homomorphic encryption (FHE) is often considered to be the "holy grail" of encryption systems. It allows one to offload heavy computations to an external computer environment (eg, cloud computing) without sacrificing on security or privacy. But by definition, it introduces more structure into the solution space of the ciphertext. It is possible that this additional structure might be exploitable, and an attack could be mounted against such schemes.

The Shortest Vector Problem (SVP) in Lattices is known to be a NP-hard problem [1]. Learning With Errors (LWE) is (only) conjectured to be a hard problem: its security is based on the assumption that Gap-SVP is NP-hard [5]. Some lattice based cryptosystems base their hardness on this LWE problem, that is, state that solving their system is at least as hard as solving the LWE instance. But now, we are no longer looking at the original hard problem, but something (cryptosystem) derived from something (LWE) derived from something (Gap-SVP) derived from the original hard problem. If proper care is not maintained in this reduction process, the security of the resulting system may be undermined.

2 Background

2.1 Helper Functions

In this section we define some helpful functions we will use in our cryptosystem. For a given l, we define $q = 2^{l-1}$. Then we work in $\mathbb{Z}_q$.

BitDecomp: BitDecomp takes a vector of integers from $\mathbb{Z}_q$, and does an element wise binary decomposition. It produces a "reverse" decomposition, with the least significant bit on the left.

Definition 1 (BitDecomp). *For any vector $\vec{a} = (a_1, a_2, \cdots, a_k) \in \mathbb{Z}_q^k$*

$$BitDecomp(\vec{a}) = (a_{1,0}, a_{1,1}, \cdots, a_{1,l-1}, \cdots, a_{k,0}, \cdots, a_{k,l-1})$$

where $a_{1,0}$ is $LSB(a_1)$, and $a_{k,l-1}$ is the MSB in the l-padded binary representation of a_k, and may be 0.

BitDecomp^{-1}: BitDecomp^{-1} is the inverse function to BitDecomp. It turns a vector of $k \times l$ numbers into k numbers. If the input is k groups of l bits, the output is a vector of k decimal numbers which the bits form.

Definition 2 (BitDecomp^{-1}). *For any vector $\vec{a'} = (a_{1,0}, a_{1,1}, \cdots, a_{1,l-1}, \cdots, a_{k,0}, \cdots, a_{k,l-1}) \in \mathbb{Z}_q^N$, where $N = k \cdot l$,*

$$BitDecomp^{-1}(\vec{a'}) = (\sum 2^j \cdot a_{1,j}, \cdots, \sum 2^j \cdot a_{k,j})$$

Remark 1. We abuse our notation when we define BitDecomp^{-1}, as it is not a true inverse function of BitDecomp. If one starts with $\vec{a} \in \{0,1\}^N$, then these operations invert each other, but BitDecomp$^{-1}(\vec{a})$ is well-defined even when $\vec{a} \in \mathbb{Z}_q^N$.

Flatten: We define Flatten as the binary decomposition of the inverse binary decomposition of a vector. If $\vec{a'} \in \{0,1\}^N$, Flatten is simply an identity function. Otherwise, it serves as our trapdoor. It "flattens" a vector of integer into a vector of bits, while also being one-way.

Definition 3 (Flatten). *For any vector $\vec{a'} \in \mathbb{Z}_q^N$, where $N = k \cdot l$,*

$$Flatten(\vec{a'}) = BitDecomp(BitDecomp^{-1}(\vec{a'}))$$

PowersOf2: PowersOf2 expands a vector $\vec{b}$ of length k to a vector of length $k \times l$, by giving a vector consisting of $\vec{b}$ multiplied element wise with powers of 2 from 0 until $l-1$.

Definition 4 (PowersOf2). *For any vector $\vec{b} = (b_1, b_2, \cdots, b_k) \in \mathbb{Z}_q^k$,*

$$PowersOf2(\vec{b}, l) = (2^0 b_1, 2^1 b_1, \cdots, 2^{l-1} b_1, \cdots, 2^0 b_k, 2^1 b_k, \cdots, 2^{l-1} b_k)$$

Remark 2. Technically, PowersOf2 also depends on l, but we skip writing it as a formal parameter of the function as l is a fixed parameter for a given cryptosystem, and we only use PowersOf2 within the context of a cryptosystem.

Remark 3. We extend the above operations to matrices by applying them row wise to the matrix.

2.2 Cryptosystem

We now state the cryptosystem, not as given in [3], but as reformulated in [4]. This is exactly the same cryptosystem, just penned slightly more mathematically.

Setup (n, λ): Choose q, the lattice dimension parameter, according to the desired security level and set $l = \log_2(q) + 1$.
Choose $\chi = N_{\text{discrete}}(\text{mean} = \frac{q}{2^{3+\lambda}m}, \sigma^2)$, where σ, the standard deviation, is chosen according to the desired security level.
Set $m = (2 \times n \times \log_2(q)) + 1$, and $N = (n+1) \times l$.
Return parameters

$$params = (q, n, m, l, N, \chi)$$

Refer to [4] for details on the choice of parameters according to security level.

SecretKeyGen(*params*): Sample $t \leftarrow \mathbb{Z}_q^n$ uniformly.
Return secret key

$$\vec{sk} = (1, -t_1, \ldots, -t_n) \in \mathbb{Z}_q^{n+1}$$

PublicKeyGen(*params*, sk): Sample a matrix $B \leftarrow \mathbb{Z}_q^{m \times n}$ and an error vector $\vec{e} \leftarrow \chi^m$.
Recall that $\vec{sk} = (1, -t_1, \ldots, -t_n)$. Set $\vec{b} = B \cdot \vec{t} + \vec{e}$ and $A = (\vec{b}, B) \in \mathbb{Z}_q^{m \times (n+1)}$.
Return public key

$$\vec{pk} = A$$

$\mathcal{C}_{pk}(\textbf{\textit{params}}, \mu)$: Sample $R \leftarrow \{0,1\}^{N \times m}$ uniformly.
Return ciphertext of plaintext μ

$$C = \text{Flatten}(\mu \cdot I_N + \text{BitDecomp}(R \cdot A)) \in \{0,1\}^{N \times m}$$

$\mathcal{D}_{\{0,1\}sk}(\textbf{params, C})$: Parity decryption algorithm. $\mathcal{D}_{\{0,1\}sk}$ decrypts only the parity of μ, or, decrypts μ correctly only if $\mu \in \{0,1\}$.
Set $\vec{v} = \text{PowersOf2}(\vec{sk})$. Choose an i such that $2^i \in (\frac{q}{4}, \frac{q}{2}]$. Then set $x = \langle C_i, v \rangle$,

where C_i is the ith row of C when counting from 0.
Return plaintext
$$\mu = \lfloor \frac{x}{v_i} \rceil \in \{0, 1\}$$
where v_i is the i-th element of $\vec{v}$ when counting from 0.

$\mathcal{D}_{sk}(\boldsymbol{params}, \mathbf{C})$: General case decryption algorithm. $\mathcal{D}_{sk}$ correctly recovers $\mu \in \mathbb{Z}_q$.
Set $\vec{v} = \text{PowersOf2 } \vec{sk}$.
Let the first $l-1$ coefficients of $\langle C \cdot \vec{v} \rangle$ be $(c_{l-1}, c_{l-2}, \cdots, c_1)$.
Recover $\mu_1 = \text{LSB}(\mu) = \lfloor \frac{c_1}{2^{l-2}} \rceil$.
Then recover the next significant bit $\mu_2 = \lfloor \frac{c_2 - 2^{l-3}}{2^{l-2}} \rceil$, and so on.
Finally, return the plaintext as
$$\mu = 2^0 \mu_1 + 2^1 \mu_2 + \cdots + 2^{l-2} \mu_{l-1} \in \mathbb{Z}_q$$

Add$(\boldsymbol{C_1}, \boldsymbol{C_2}, +)$: For ciphertexts C_1 and C_2, output
$$C = \text{Flatten}(C_1 + C_2)$$

MultConst$(\boldsymbol{C}, \boldsymbol{\alpha}, \cdot)$: Let C be a ciphertext, and $\alpha \in \mathbb{Z}_q$.
Set $M_\alpha := \text{Flatten}(\alpha \cdot I_n)$. Then, output
$$C = Flatten(M_\alpha \cdot C)$$

Remark 4. MultConst is not a "true" homomorphic evaluation function, as one of its operands is an unencrypted number α. But it behaves in the expected manner, that is, $\mathcal{D}(\text{MultConst}(C, \alpha, \cdot)) = \mathcal{D}(C) \cdot \alpha$.

Remark 5. The given cryptosystem is correct, i.e.,
$$\mathcal{D}_{sk}(params, \mathcal{C}_{pk}(params, \mu)) = \mu$$

3 Cryptanalysis

Homomorphic cryptosystems are by definition malleable. But also, their "homomorphiness" gives the entire ciphertext space additional structure. We speculate that this additional structure is enough to undermine indistinguishability property, and it is feasible to leverage this with a Chosen Plaintext Attack (CPA).

3.1 Basic Idea

In the system under consideration, all our ciphertexts are big square matrices. Imagine, if one could squash all square matrices of parameter n into a single linear spectrum. Then, we imagine all encryptions of a plaintext (given some constant parameters for the system) form a distinct pile on this spectrum. This way, for a parameter n, we have 2^n piles in a spectrum of 2^{N^2} width (like before, $N = (n+1) \times l$), as we have only 2^n plaintexts to encrypt, but our ciphertext space is $\{0,1\}^{N \times N}$, which has 2^{N^2} members. In particular, we theorize that there must be a way to find these piles such that they partition the cipher space cleanly, that is, all ciphertexts of each plaintext forms a single connected "pile". To this end, we turn to Topological Data Analysis for help.

Fig. 1. The "Matrix Spectrum" : The colored parts indicate piles of distinct encryption, and the white parts indicate unused part of the space (Color figure online).

3.2 Methodology

In our experiments, for each set of parameters, we generate 200 encryptions each of 0 and 1 to form a single set of 400 ciphertexts. We then split this into a training set of 280 ciphertexts (70%) and a testing set of the remaining 120 ciphertexts (30%). We transform our sets according to a TDA pipeline, and finally train the model and score its accuracy against the testing set. Care was taken to ensure that the training data and the testing data do not overlap. That is, ciphertexts included in the testing set are never seen by the model during training (except if the encryption algorithm generated duplicates, which has an overwhelming probability to not happen.)

4 Results and Discussion

We ran our analysis over a range of parameters selected in accordance to the choice of parameters found in [4]. Note that the parameters ($n = 768$, $q = 2^{22}$, $m = 2 \times n \times \lfloor \log_2(q) \rfloor + 1$, $\chi = N_{Discrete}(\mu = \frac{q}{2^{3+\lambda} \times m}, \sigma^2 = 3.0)$) provide 129 bits of accuracy, according to [2]. We present the accuracy and time results of our attack below.

In both Tables 1, 2, n, q, and χ are the parameters to the LWE problem instance, and m is the size of the error vector drawn from χ.

Accuracy. Accuracy is measured as

$$\frac{\text{correct guesses of plaintext}}{\text{total number of ciphertexts}} \times 100$$

So, 100% accuracy would mean we can perfectly guess every ciphertext, while 50% accuracy means about half the time, we guess the ciphertext incorrectly. The exact analysis pipeline we used is included in [4]. In the following table, we note the accuracy of our attack scheme. While the classifiers have parameters one could tune in order to improve the score for each instance of our cryptosystem, the following table shows the accuracy of the globally "optimal" parameters, which produce good accuracy in most cases.

Table 1. Accuracy Results

n	q	m	χ	Accuracy
64	2^{18}	2305	$N_{Dis}(\mu=7.11,\sigma^2=1)$	79%
128	2^{18}	4609	$N_{Dis}(\mu=3.55,\sigma^2=1)$	98%
512	2^{22}	22529	$N_{Dis}(\mu=11.64,\sigma^2=1)$	86%
768	2^{22}	33793	$N_{Dis}(\mu=7.75,\sigma^2=1)$	81%

With these accuracy results in hand, we only have to worry about the feasibility of our attack.

Time Complexity. Recall, that no cryptosystem is completely unbreakable, as every scheme can be broken by the brute force attack, that is, simply trying every possible answer until we arrive at something correct. But it is not feasible to mount such an attack against a secure scheme, as it may take far too long to compute. A typically secure scheme has 128 bits of security, that is, in the worst case, it requires 2^{128} computations to complete the brute force attack. Assuming a reasonably fast computer can do around 3×10^9 computations per second, that will still take 10^{29} seconds, that is, $10^{11} \times$ **the age of the universe** to complete. We aim for an attack that performs better than this worst-case time of $\mathcal{O}(2^n)$.

Table 2. Time Complexity Results

n	q	m	χ	Time Estimate for Brute Force Attack	Time Taken
64	2^{18}	2305	$N_{Discrete}(\mu=7.11,\sigma^2=1)$	< 300 s	7200 s
128	2^{18}	4609	$N_{Discrete}(\mu=3.55,\sigma^2=1)$	300 s	15346 s
512	2^{22}	22529	$N_{Discrete}(\mu=11.64,\sigma^2=1)$	10^{12} Seconds	829898 s
768	2^{22}	33793	$N_{Discrete}(\mu=7.75,\sigma^2=1)$	10^{25} Seconds	2342399 s

In Table 2, we list the wall time of our attack on a single core of a relatively modern CPU (Intel(R) Xeon(R) E5-2697A), against the input parameter n. This data is then plotted in Figure 2. It shows that our approach (of complexity $\mathcal{O}(n^{2.21})$, polynomial in n) in single thread is much better than the worst case of $\mathcal{O}(2^n)$ (brute force). We refrain from multithreading our analysis as it currently results in a significant blowup in memory usage, and does not provide a proportional speedup. This restricts practical attacks against more secure parameters. One could consider spending effort trying to improve the efficiency of our attack, especially in leveraging a multiprocessing attack with the same idea. At the moment, we only use the n_jobs parameter of the functions provided by the various data science libraries. Right now, this provides only $\sim 30\%$ decrease in time when using 32 processors instead of just 1.

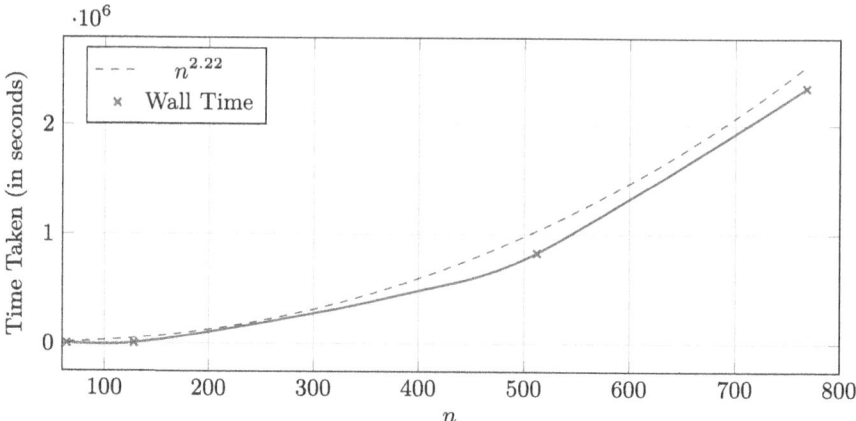

Fig. 2. Time Complexity Graph

5 Conclusions

We have constructed the cryptosystem described by [3] with parameters in accordance with the latest available literature. We use novel and previously unconsidered methods to construct our attack. Instead of classical cryptanalysis to recover the secret key, we turn to data science for help. We completely bypass the recovery of a secret key, and use inherent structure present in this homomorphic system to find patterns in the ciphertext, which a classical attack would perhaps never see. We were able to use a Topological Data Analysis pipeline to discern between encrypted texts of 0 and 1.

We employed this same strategy of attack against the base Learning With Errors instance on which the cryptosystem is based, to no success. However, the attack begins to find patterns in very early stages of the cryptosystem, and fails to satisfy the requirements of the security proof from [5] from being applied in this case. This is explained in detail in the full paper. As our attack is ineffective against the base LWE, and the security of this scheme depends on a proof from [5], it should be possible to check whether the proof is improperly adapted in its use.

References

1. Ajtai, M.: The shortest vector problem in L2 is NP-hard for randomized reductions (extended abstract). In: Proceedings of the Thirtieth Annual ACM Symposium on Theory of Computing, STOC '98, pp. 10–19. Association for Computing Machinery, New York, NY, USA (1998). https://doi.org/10.1145/276698.276705
2. Albrecht, M.R., Player, R., Scott, S.: On the concrete hardness of learning with errors. Cryptology ePrint Archive, Paper 2015/046 (2015). https://eprint.iacr.org/2015/046

3. Gentry, C., Sahai, A., Waters, B.: Homomorphic encryption from learning with errors: conceptually-simpler, asymptotically-faster, attribute-based. Cryptology ePrint Archive, Paper 2013/340 (2013). https://eprint.iacr.org/2013/340
4. Kaushik, A.: Investigating resilience of levelled fully homomorphic encryption system against data scientific attacks (2023). https://akaushik.edufor.me/resources/MasterThesis.pdf
5. Regev, O.: On lattices, learning with errors, random linear codes, and cryptography. J. ACM (2009). https://doi.org/10.1145/1568318.1568324

Novel Formalisations of Mathematics in Lean

Formalising Families of ℓ-adic Galois Representations in Lean 4

Ivan Farabella[✉]

Imperial College London, London, United Kingdom
ivan.farabella21@imperial.ac.uk

Abstract. Families of ℓ-adic Galois representations are an important tool in modern algebraic number theory. Andrew Wiles [1] used families of representations associated with elliptic curves to prove Fermat's Last Theorem and the Langlands philosophy conjectures a deep connection to the theory of automorphic forms [2]. We formalise the definition of families of ℓ-adic Galois representations as well as the definition of compatibility on them for the first time in an interactive theorem prover and discuss the formalisation process.

Keywords: formalisation · Lean · compatible systems · Galois representations

1 Introduction

1.1 Mathematical Definitions

Unless otherwise stated, in these definitions, K and E are number fields, $n \in \mathbb{N}$, ℓ is a prime number and $\overline{\mathbb{Q}}_\ell$ denotes the algebraic closure of the field of ℓ-adic numbers. We will endow $\mathrm{Gal}(\overline{K}/K)$ with the Krull topology.

Definition 1. A collection of maps for which every prime number ℓ and ring homomorphism
$$\lambda : E \to \overline{\mathbb{Q}}_\ell$$
is associated to some continuous homomorphism
$$\rho_{\ell,\lambda} : \mathrm{Gal}(\overline{K}/K) \to \mathrm{GL}_n(\overline{\mathbb{Q}}_\ell)$$
is called a **family of ℓ-adic Galois representations**.

Note that this definition does not specify how the ring homomorphisms λ are actually used in the corresponding representation $\rho_{\ell,\lambda}$. In fact, we can get a perfectly valid family of ℓ-adic Galois representations by allowing
$$\rho_{\ell,\lambda} : \mathrm{Gal}(\overline{K}/K) \to \mathrm{GL}_n(\overline{\mathbb{Q}}_\ell)$$
to be the trivial representation
$$g \mapsto 1$$
for all ℓ and λ.

Definition 2. Such a family of representations is called a **trivial family**.

Given a finite field extension L/K and a homomorphism

$$h : \text{Gal}(L/K) \to \text{GL}_n(E)$$

there is a family of ℓ-adic Galois representations given by the composition

$$\rho_{\ell,\lambda} = f_{\ell,\lambda} \circ h \circ g$$

$$\text{Gal}(\overline{K}/K) \xrightarrow{g} \text{Gal}(L/K) \xrightarrow{h} \text{GL}_n(E) \xrightarrow{f_{\ell,\lambda}} \text{GL}_n(\overline{\mathbb{Q}_\ell})$$

where g is the 'restriction' of $\text{Gal}(\overline{K}/K)$ to $\text{Gal}(L/K)$ - that is, the map that takes an element of the former and restricts it to an element of the latter. The map $f_{\ell,\lambda}$ comes from $\lambda : E \to \overline{\mathbb{Q}_\ell}$, by taking an element of $\text{GL}_n(E)$, which is a matrix, and applying λ to every entry.

Definition 3. We will refer to this family as the **family induced by** h.

We sometimes like to consider families of ℓ-adic Galois representations that have the following regularity condition.

Definition 4. A family of ℓ-adic Galois representations is said to be **compatible** if there exists a finite set S of prime ideals of $\mathcal{O}_K$ such that for all prime ideals $\mathfrak{p} \notin S$ of $\mathcal{O}_K$, there is a polynomial $H_\mathfrak{p}$ with the property that $H_\mathfrak{p}$ is equal to the characteristic polynomial of

$$\rho_{\ell,\lambda}(\text{Frob}_\mathfrak{p})$$

whenever ℓ is a prime number not in $\mathfrak{p}$, and $\text{Frob}_\mathfrak{p}$ is any Frobenius element of $\text{Gal}(\overline{K}/K)$ associated with $\mathfrak{p}$.

2 Formalisation

A danger in projects such as this where the aim is to formalise a definition as opposed to providing a proof is that it can be difficult to tell if the ensuing definition is mathematically useful. Towards justifying our formalisation of ℓ-adic Galois representations and compatibility, we prove that the trivial family and the family induced from a homomorphism satisfy our definitions, modulo some intermediary results, proofs of which were beyond the scope of this project.

We found great utility in utilising an 'AKLB' setup where A is an integrally closed domain, K the fraction field of A, L a finite dimensional extension of K and B the integral closure of A in L. An example of this is when K is a number field, L is a finite dimensional extension of K, A is taken to be the ring of integers of K and B the ring of integers of L. It would have been possible to work in this particular case throughout the entire project but the general AKLB approach has a few notable advantages. First of all, much of the material in Mathlib relevant to our project already makes use of an AKLB approach so utilising it here gives us access to a rich library of powerful results. Furthermore, it enables a greater level of generality since some results do not rely on K being a number field and instead just use the default assumptions of AKLB. In fact, class field theory provides analogues of results about number fields for other algebraic objects, like function fields and using an AKLB approach helps unify them.

2.1 Definitions and Initial Challenges

Extracting a suitable definition in Lean from our mathematical definition of families of ℓ-adic Galois representations was fairly straightforward, iterating over primes and suitable ring homomorphisms and providing a suitable continuous homomorphism.

In Lean 4, we can express a continuous homomorphism from A to B as a term of type `ContinuousMonoidHom A B`, where A and B are both topological monoids. In the context of families of ℓ-adic Galois representations, Mathlib has an instance of `TopologicalSpace` and `Monoid` for $\mathrm{Gal}(\overline{K}/K)$ but only an instance of `Monoid` for $\mathrm{GL}_n(\overline{\mathbb{Q}_\ell})$. In order to get the missing `TopologicalSpace` instance, we first note that $\overline{\mathbb{Q}_\ell}$ is a normed commutative ring. The proof of this in Lean 4 [3] is not yet in Mathlib so this was expressed as an instance of a `NormedCommRing` with a proof marked `sorry`. With this 'black box' in place, Lean's typeclass inference was able to infer that $\mathrm{GL}_n(\overline{\mathbb{Q}_\ell})$ is a topological space.

Coming next was the definition of compatibility, which required some preliminary definitions, notably a predicate on Frobenius elements of $\mathrm{Gal}(\overline{K}/K)$ `IsFrobenius'`. The predicate on Frobenius elements was done by first defining it on $\mathrm{Gal}(L/K)$ for finite dimensional extensions L/K then extending this definition to $\mathrm{Gal}(\overline{K}/K)$ by insisting the finite predicate was satisfied for all intermediate fields $\overline{K}/L/K$ where L/K is a finite dimensional extension.

2.2 Trivial Example

Recall the trivial family of ℓ-adic Galois representations, where every entry is associated with a constant representation, sending every element to the identity of $\mathrm{GL}_n(\overline{\mathbb{Q}_\ell})$. Proving that this satisfies our definition of a family in Lean amounts to a proof that this function is a continuous homomorphism. This can be done without issue using existing Mathlib content.

2.3 A Further Example

Given a homomorphism

$$h : \mathrm{Gal}(L/K) \to \mathrm{GL}_n(E),$$

recall the family induced by h where

$$\rho_{\ell,\lambda} = f_{\ell,\lambda} \circ h \circ g.$$

Our formalisation of this family will begin by proving that g and $f_{\ell,\lambda} \circ h$ are both continuous homomorphisms. Once this has been established, Mathlib has that compositions of continuous homomorphisms are continuous homomorphisms and so their composition, $\rho_{\ell,\lambda}$, is a `ContinuousMonoidHom` for all ℓ, λ and hence satisfies our definition of a family of ℓ-adic Galois representations.

g is a Continuous Homomorphism Recall that g is the restriction of $\mathrm{Gal}(\overline{K}/K)$ to $\mathrm{Gal}(L/K)$. This is in Mathlib as `AlgEquiv.restrictNormalHom L`, where it is already proved to be a homomorphism. What remained to be proved was continuity, which was split into two steps. First of all, proving that a homomorphism between topological groups is continuous if its kernel is open and then a proof that the kernel of g is open to deduce that g is continuous.

For the first of these, we write the pre-image of any open set as a union of cosets of the kernel of the map. Each of these cosets is open by `IsOpen.smul` in `Mathlib.Topology.Algebra.ConstMulAction` so we deduce that the pre-image of any open set is open and hence such a homomorphism is continuous.

Showing that the kernel of g is open utilises Mathlib's implementation of the Krull topology. In particular, the group filter basis of $\mathrm{Gal}(L/K)$ consists of the fixing subgroups of intermediate fields E finite dimensional over K. From here, we prove that the kernel of g is such a fixing subgroup so, as a member of this filter, it is an open set.

$f_{\ell,\lambda} \circ h$ is a continuous homomorphism Recall that we define
$$f_{\ell,\lambda} : \mathrm{GL}_n(E) \to \mathrm{GL}_n(\overline{\mathbb{Q}_\ell})$$
to be the function that applies the map $\lambda : E \to \overline{\mathbb{Q}_\ell}$ to every entry of the source matrix. We can write this in Mathlib as
`Units.map (RingHom.mapMatrix λ).toMonoidHom`, which immediately gives that it is a homomorphism. Since h is also a homomorphism, $f_{\ell,\lambda} \circ h$ is a homomorphism. Continuity of this composition comes from proving that the domain, $\mathrm{Gal}(L/K)$, has the discrete topology as Mathlib already has a proof that a map between topological spaces is continuous if the domain has the discrete topology.

Our proof that $\mathrm{Gal}(L/K)$ has the discrete topology also makes use of the group filter basis of the Krull topology. We first note that the fixing subgroup of L is the trivial subgroup. And then, because L/K is finite dimensional, the trivial group is a member of this filter basis and hence is open. Mathlib already has a proof that a topological group has the discrete topology if and only if the singleton set $\{1\}$ is open so we deduce that $\mathrm{Gal}(L/K)$ has the discrete topology and that $f_{l,\lambda} \circ h$ is continuous.

2.4 Compatibility

Both of the families discussed are also compatible, with the formalisation of the compatibility of the trivial family being particularly simple. Since the characteristic polynomial of the n-dimensional identity matrix is just $(x-1)^n$, the set of 'bad' primes to avoid is empty and the rest of the proof follows.

Compatibility of families induced by a homomorphism was much more difficult and was only partially completed. For example, we require that Frobenius elements are conjugate when the associated prime is unramified. A proof of this probably entails another project of similar scope so was left incomplete but the consequence is that the finite set of primes to be avoided are those which divide the discriminant of L since conjugate elements are mapped to similar matrices under $\rho_{\ell,\lambda}$. Similar matrices have the same characteristic polynomial, so we have outlined a 'proof' of compatibility. This property of the characteristic polynomial and even the finiteness of the set of primes discussed above were also not formalised. Completing these would probably be a modest task for an intermediate Lean user since the underlying mathematics is quite simple but there are some remaining proofs which continue to provide a significant challenge. These are mostly technical lemmas, for example that restricting Frobenius elements of $\text{Gal}(\overline{K}/K)$ to an intermediate field L gives a Frobenius element of $\text{Gal}(L/K)$. This result is probably more accessible to those with some familiarity with category theory since it effectively reduces to the showing that certain diagrams commute. We also use, without proof, the existence of Frobenius elements, although a proof can be found here.

It should be noted that using results without proof is against the spirit of formalisation and was done here only to demonstrate that our definition of compatibility can be used on a non-trivial family of ℓ-adic Galois representations.

3 Conclusion

This formalisation demonstrates that formalisation of modern algebraic number theory is accessible in Lean 4. In particular, the robustness of Mathlib and the utility of the 'AKLB' approach became evident over the course of the project.

In terms of further work, an obvious next step is to formalise the results used in the 'proof' of the compatibility of the family induced by a homomorphism, particularly the conjugacy of Frobenius elements in the unramified case. From here, other compatible families like the cyclotomic character can be formalised. The formalisation can be found here.

Acknowledgement. This paper is based on projects completed in by the author as part of the undergraduate course Formalising Mathematics at Imperial College London. I would like to thank Kevin Buzzard for running this course and providing guidance that made this possible. I would also like to thank the graduate teaching assistants on this course for providing much needed Lean support.

References

1. Wiles, A.: Modular elliptic curves and Fermat's last theorem. Ann. Math. **141**(3), 443–551 (1995). https://doi.org/10.2307/2118559
2. Buzzard, K., Gee, T.: The conjectural connections between automorphic representations and Galois representations. In: Diamond, F., Kassaei, P.L., Kim, M. (eds.) Automorphic Forms and Galois Representations, pp. 135–187. Cambridge University Press (2014). https://doi.org/10.1017/CBO9781107446335.006

3. de Frutos-Fernández, M.I.: Formalizing norm extensions and applications to number theory. In: 14th International Conference on Interactive Theorem Proving (ITP 2023), Leibniz International Proceedings in Informatics (LIPIcs), Volume 268, pp. 13:1–13:18, Schloss Dagstuhl - Leibniz-Zentrum für Informatik (2023). https://doi.org/10.4230/LIPIcs.ITP.2023.13

Formalization of the Existence of Frobenius Elements

Jou Glasheen[✉]

Department of Mathematics, Imperial College London, London, UK
jou.glasheen23@imperial.ac.uk

Abstract. We use Lean 4 to formalize a proof of the existence of Frobenius elements for finite Galois extensions of number fields.

Keywords: Frobenius · Lean · Proof assistant · $AKLB$

1 Introduction

We present a `sorry`-free formalization in Lean 4 of a proof of the existence of Frobenius elements for finite Galois extensions L / K of number fields. Our formalization is based on the "direct proof" in [9, p. 141], which translates[1] as:

```
theorem exists_frobenius : ∃ σ : L ≃ₐ[K] L,
    (σ ∈ decomposition_subgroup_Ideal' Q) ∧
    (∀ γ : B, ((γ ^ q) - (galRestrict A K L B σ) γ) ∈ Q) :=
  ⟨Frob, Frob_is_in_decompositionSubgroup P Q_ne_bot, fun γ =>
    for_all_gamma P Q_ne_bot γ⟩
```

For the full formalization, see [4].

This work was completed in Kevin Buzzard's Formalising Mathematics module, taught at Imperial in Spring 2024; and it contributes to Kevin Buzzard's project to formalize the proof of Fermat's Last Theorem. Amelia Livingston and Jujian Zhang, Kevin Buzzard's PhD students, taught me enough Lean to go from having no experience to accomplishing this formalization.

In this report, I will 1. introduce relevant 'mathematical background' material; 2. highlight the following components of `exists_frobenius`: (i) construction of the transitive action of the Galois group (L ≃ₐ[K] L) on the set of non-zero prime ideals of B which lie above a certain non-zero prime ideal P ⊂ A, where B, A are the integer rings of L, K, respectively; (ii) definition of the decomposition subgroup of a nonzero prime ideal Q ⊂ B lying over P; (iii) characterization of the orbit of Q under the action of (L ≃ₐ[K] L); (iv) proof that a polynomial (F : B[X]) which is a product of linear factors (X - C (galRestrict A K L B τ α)), (τ : L ≃ₐ[K] L), (α : B) algebraic over K, has coefficients in A; and 3. discuss relative advantages of using the A K L B setup for our formalization.

---

[1] Jujian Zhang helped to simplify the statement of the proof of `exists_frobenius`.

© The Author(s), under exclusive license to Springer Nature Switzerland AG 2024
K. Buzzard et al. (Eds.): ICMS 2024, LNCS 14749, pp. 63–71, 2024.
https://doi.org/10.1007/978-3-031-64529-7_7

2 Mathematical Background

2.1 A K L B Setup.

By A K L B setup,[2] we mean the assumptions that "A is an integrally closed domain; K is the fraction ring of A; L is a finite (separable) extension of K; B is the integral closure of A in L" [15, ll. 14–16].

In the A K L B setup, if Q ⊂ B is a non-zero prime ideal lying over a nonzero prime ideal P ⊂ A, then (B / Q) is a vector space of dimension [L : K] over (A / P) [1, p. 32][8, p. 45]; i.e., [IsScalarTower A (A / P) (B / Q)].

With the assumption [IsDedekindDomain A] (hence, so is B [14, Theorem 5.25.]), the A K L B setup exactly corresponds to a finite extension of number fields L / K with respective rings of integers B, A [3, p. 7][14, p. 43]. By explicitly assuming the A K L B setup; with the further assumptions that A is Dedekind and L / K is Galois; several authors write proofs of the existence of Frobenius elements for L / K [1, p. 66][14, p. 70]. Hence, there is ample justification for assuming the A K L B setup to formalize exists_frobenius.

2.2 Characteristic of Finite Fields

If F is a finite field, then F has characteristic p (CharP F p) for p some natural prime; the cardinality of F is a power of p [8, p. 191]; and every element of F is a pth power [2, p. 549]. Thus, the Frobenius endomorphism $\text{Frob}_p : a \mapsto a^p$ is an automorphism of F [10, pp. 9–10].

2.3 Galois Correspondence

Under the Galois correspondence [8, p. 188] for finite extensions, the entire Galois group corresponds to the base field [6, p. 262]:

theorem IsGalois.tfae...:...IntermediateField.fixedField ⊤ = ⊥

2.4 Integral Closure

As the integer ring of a number field is integrally closed [6, pp. 334–335] in its fraction ring [10, p. 30]; in addition to the assumption [IsIntegralClosure B A L] from the standard A K L B setup; we also have [IsIntegralClosure A A K]. So, any element (α : K) which is integral over A is actually in A [9, p. 28].

---

[2] variable (A K L B : Type*) [CommRing A] [CommRing B] [Algebra A B]
  [Field K] [Field L] [Algebra A K] [IsFractionRing A K] [Algebra B L]
  [Algebra K L] [Algebra A L] [IsScalarTower A B L] [IsScalarTower A K L]
  [IsIntegralClosure B A L] [FiniteDimensional K L] (See [15, ll. 28–31].).

3 Survey of Lean Code

3.1 gal_action_Ideal' and decomposition_subgroup_Ideal'

Given that a Frobenius element is defined "to be [an] element of [the decomposition subgroup of Q over K] that acts as the Frobenius automorphism on the residue field" [9, p. 141]; our first goal was to define the decomposition subgroup. Mathlib contains the definition ValuationSubring.decompositionSubgroup [5]; but it is according to the definition of primes of B as valuation subrings of L. Our definition decomposition_subgroup_Ideal' takes from ValuationSubring.decompositionSubgroup the idea of defining this subgroup as the stabilizer of the multiplicative action (gal_action_Ideal') of (L $\simeq_a$ [K] L) on the primes of B; but redefines 'primes of B' as prime ideals Q ⊂ B.

To ensure that it is well-defined on prime ideals of B, the multiplicative action is written as a composition involving preimages (Ideal.comap) of ideals:

1. galRestrict restricts an automorphism (σ : L $\simeq_a$ [K] L) to one (galRestrict A K L B σ : B $\simeq_a$ [A] B) of B;
2. AlgEquiv.symm (galRestrict A K L B _) is the inverse of (galRestrict A K L B _) as an A-algebra automorphism of B;
3. Ideal.comap ensures that the composition Ideal.comap (AlgEquiv.symm (galRestrict A K L B _)) maps prime ideals to prime ideals (this is due to the theorem Ideal.comap_isPrime).

In sum, Ideal.comap (AlgEquiv.symm (galRestrict A K L B _)) maps prime ideals of B to prime ideals of B, under automorphisms (σ : L $\simeq_a$ [K] L) restricted to (B $\simeq_a$ [A] B).

Note 1. As galRestrict restricts (σ : L $\simeq_a$ [K] L) to (B $\simeq_a$ [A] B), not to B alone; to avoid an error concerning metavariables; defining gal_action_Ideal' requires using, in place of Ideal B, the "type synonym"[3] Ideal' A K L B, whose definition comprises not only an instance of A, but also the relations amongst A, K, L and B prescribed by the A K L B setup.

Thence, analogously to ValuationSubring.decompositionSubgroup, the decomposition subgroup of a prime[4] ideal Q ⊂ B is defined as the stabilizer of Q under gal_action_Ideal':

```
def decomposition_subgroup_Ideal' (Q : Ideal' A K L B) :
    Subgroup (L ≃ₐ[K] L) := MulAction.stabilizer (L ≃ₐ[K] L) Q
```

---

[3] This type synonym was written by Amelia Livingston.
[4] In Lean, it is not necessary to assume that Q is prime in order to define its decomposition subgroup.

3.2 exists_generator

```
theorem exists_generator : ∃ (ρ : B)
   (h : IsUnit (Ideal.Quotient.mk Q ρ)),
   (∀ (x : (B / Q)ˣ), x ∈ Subgroup.zpowers h.unit)∧
   (∀ τ : L ≃ₐ[K] L,
   (τ ∉ decomposition_subgroup_Ideal' Q) → ρ ∈ (τ • Q)) :=
```

Note 2. In Sect. 3.3 of this paper, the variable α is an element of B which has the properties of the ρ distinguished by `exists_generator`.

As Milne states, `exists_generator` follows "[b]y the Chinese Remainder Theorem" (CRT) [9, p. 141]. In the context of coprime ideals in integer rings, the version of CRT which implies the existence of such a (ρ : B) is `IsDedekindDomain.exists_forall_sub_mem_ideal` [11]. This theorem tells us that, given a set of pairwise coprime ideals $P_1, \ldots, P_k$ of a Dedekind domain R, with corresponding ramification indices $e_i \in \mathbb{N}$; and given a tuple $(x_1, \ldots, x_k)$, where $x_i \in R, i = 1, \ldots, k$; it is possible to find a representative $y \in R$ such that $y \equiv x_i \mod P_i^{e_i}, i = 1, \ldots, k$.

The key to interpreting the fact that $\rho \in (\tau \bullet Q)$, for any τ not in the decomposition subgroup of Q over K; and the relation of this fact to the condition that ρ generates (B / Q)ˣ; is to hold two views of the statement (τ • Q) $\neq$ Q, simultaneously : on the one hand, (τ • Q) $\neq$ Q, because ($\tau \notin$ decomposition_subgroup_Ideal' Q) implies τ does not stabilize Q; on the other hand, (τ • Q) $\neq$ Q, Q $\neq \bot$ is the hypothesis of CRT, that Q, (τ • Q) are pairwise coprime ideals $\forall$ ($\tau \notin$ decomposition_subgroup_Ideal' Q). Thus, we understand (τ • Q) and Q, not as ideals *per se*, but as elements in the orbit of Q under `gal_action_Ideal'` [2, pp. 115–116]; i.e., (by the orbit-stabilizer theorem) as cosets[5] in the quotient

```
(MulAction.orbit  (L ≃ₐ[K] L) Q) ≃
   (L ≃ₐ[K] L) / (decomposition_subgroup_Ideal' Q)
```

The idea of `exists_generator` is to choose a generator g of (B / Q)ˣ (such a g exists by `residue_field_units_is_cyclic`, because (B / Q)ˣ is cyclic); then, given the distinct cosets $\{[1], [\tau_1], \ldots, [\tau_{d-1}]\}$, where d is the size of the orbit of Q; to use the isomorphism

$$B/\left(Q \times \prod_{i=1}^{d-1} (\tau_i \bullet Q)\right) \cong B/Q \times B/(\tau_1 \bullet Q) \times \cdots \times B/(\tau_{d-1} \bullet Q)$$

given by CRT, to choose an element (ρ : B) such that (ρ - g) $\in$ B / Q and (ρ - 0) $\in$ B / (τ_i • Q) for i = 1, ..., d - 1. This is where we apply `exists_forall_sub_mem_ideal`.[6] I compartmentalize this application in:

---

[5] Amelia Livingston led me to this insight.
[6] Jujian Zhang helped me to work out this application in Lean.

```
lemma crt_representative (b : B) : ∃ (y : B),
    ∀ (i : (L ≃ₐ[K] L) / decomposition_subgroup_Ideal' Q),
    if i = Quotient.mk'' 1 then y - b ∈ f Q i else y ∈ f Q i :=
  by ...
  have := IsDedekindDomain.exists_forall_sub_mem_ideal
    (s := Finset.univ) (f Q) (fun _ ↦ 1)
```

To supply `exists_forall_sub_mem_ideal` with its required arguments, I preface `crt_representative` with the following statements:

1. `noncomputable def f :`
   `((L ≃ₐ[K] L) / (decomposition_subgroup_Ideal' Q)) →`
   `(Ideal' A K L B)`
   corresponds to the map (f : ι → Ideal R) in
   `Ideal.quotientInfRingEquivPiQuotient` (generic CRT), with domain
   `Fintype ι`. We take ι to be the orbit of Q.
2. `noncomputable instance : Fintype ((L ≃ₐ[K] L) /`
   `decomposition_subgroup_Ideal' Q)`
   synthesizes an instance of the domain of `f` as a `Fintype`;[7] then, we set
   (s: Finset ι) to be the whole orbit, i.e., `Finset.univ'`.
3. `instance MapPrimestoPrimes (σ : L ≃ₐ[K] L)`
   `(Q : Ideal' A K L B) [Ideal.IsPrime Q] :`
   `Ideal.IsPrime (σ • Q)`
   implies every element in the orbit of Q is a prime ideal.
4. `lemma coprime_ideals_non_equal_prime (I J : Ideal' A K L B)`
   `[Imax : Ideal.IsMaximal I] [Jmax : Ideal.IsMaximal J]`
   `(h : I ≠ J) : IsCoprime I J`
   gives us that Q and (τ • Q) are coprime. To prove this, we assume Q and (τ • Q) are maximal ideals of B, which follows from the facts that they are nonzero, prime (Q by assumption; (τ • Q) by `MapPrimestoPrimes`); and that B is a Dedekind domain [14, Theorem 5.25.].

3.3 ∃ σ, αq ≡ σ α mod Q (pow_q_is_conjugate)

The lemma `pow_q_is_conjugate` is a consequence of the result that

$$F(\alpha^q) \equiv F(\alpha)^q \equiv 0 \mod Q \qquad (1)$$

Note 3. α is an element of B which is an instance of the ρ distinguished by `exists_generator` (see Sect. 3.2, *Note 2.*); "q" => `Fintype.card (A / P)`; F is local notation for F' : B[X] := $\prod$ τ : L ≃ₐ[K] L, (X - C (galRestrict A K L B τ α)).

---

[7] Amelia Livingston helped me to write this instance.

Note 4. In `exists_frobenius`, Frob denotes an element of (L $\simeq_a$ [K] L) which has the property of the σ distinguished by `pow_q_is_conjugate`.

Both the right-hand side $F(\alpha)^q \equiv 0 \mod Q$ of Eq. (1); and the fact that Eq. (1) implies `pow_q_is_conjugate`; depend on showing that `IsRoot F x` (i.e., eval x F = 0) if and only if x is a conjugate of α.[8]

To show the left-hand side of Eq. (1); i.e., that

```
lemma F_expand_eval_eq_eval_pow :
    (eval₂ (Ideal.Quotient.mk Q) (Ideal.Quotient.mk Q α) F) ^ q =
    (eval₂ (Ideal.Quotient.mk Q) (Ideal.Quotient.mk Q (α ^ q)) F)
```

we convene all of our 'mathematical background' material.

As we show in the `section FreshmansDream`,

1. `lemma q_is_p_pow_n` : p ^ n = q, where (n : ℕ) coincides with the degree of the finite field (A / P) over its prime subfield; and the prime (p : ℕ) coincides with the characteristic of (A / P): i.e., `CharP (A / P) p`.[9] This lemma is used to re-write:
2. `Polynomial.map_expand_pow_char` in the proof of:
   theorem pow_eq_expand (a : (A / P)[X]) :
   (a ^ q) = (expand _ q a)   :=

$$\text{i.e.,} \quad a(X^q) = a(X)^q$$

To show `pow_eq_expand`, we define $a(X)^q$ in terms of `Polynomial.expand`; thence, we rw [← map_expand_pow_char]:

$$a(X)^q \stackrel{\text{Freshman's Dream}}{=} a_n^q X^{qn} + a_{n-1}^q X^{q(n-1)} + \cdots + a_0^q$$

By applying the Frobenius automorphism `FiniteField.frobenius_pow` to the coefficients of $a_n^q X^{qn} + a_{n-1}^q X^{q(n-1)} + \cdots + a_0^q$, we obtain[10]

$$a_n^q X^{qn} + a_{n-1}^q X^{q(n-1)} + \cdots + a_0^q \equiv a_n X^{qn} + a_{n-1} X^{q(n-1)} + \cdots + a_0 = a(X^q)$$

To obtain `F_expand_eval_eq_eval_pow` from `pow_eq_expand`, we must recognize that, *a priori*, F has coefficients in B. Hence, we construct the intermediary polynomial (m : A[X])[11] whose lift to B coincides with F:

```
lemma ex_poly_in_A : ∃ m : A[X],
    Polynomial.map (algebraMap A B) m = F
```

---

[8] Apply lemmas which relate a polynomial to its (linear) factors: `Polynomial.eval_prod`, `Finset.prod_eq_zero_iff`, `Polynomial.IsRoot.def`, `Polynomial.root_X_sub_C`.

[9] *a priori*, we only know that q is the power of some prime p', but we do not know that p' is p, the characteristic of (A / P).

[10] The formalization closely resembles Amelia Livingston's handwritten proof.

[11] This is Amelia Livingston's idea.

By `pow_eq_expand`, we get that

$$m(X^q) \equiv m(X)^q \mod P$$

Thence, by applying `IsScalarTower.algebraMap_eq`,[12] we prove that, taken together, `ex_poly_in_A` and $m(X^q) \equiv m(X)^q \mod P$ imply

```
lemma pow_expand_A_B_scalar_tower_F :
   (Polynomial.map (algebraMap B (B / Q)) F) ^ q =
   (expand _ q (Polynomial.map (algebraMap B (B / Q)) F)) :=
```

Now, by the A K L B setting for prime ideals of B lying over prime ideals of A [1, p. 32][14, p. 46], we have an instance of [IsScalarTower A (A / P) (B / Q)]. Combining this scalar tower with [IsScalarTower A B L] (from the original A K L B setup), we have an instance of [IsScalarTower A B (B / Q)].

Thence, coordinating (I) the equivalence $m(X^q) \equiv m(X)^q \mod P$, lifted to (B / Q) along the scalar tower A (A / P) (B / Q); with (II) the equality `Polynomial.map (algebraMap A B) m = F`, lifted to (B / Q) along the scalar tower A B (B / Q); we obtain that

$$m(X^q) \equiv m(X)^q \mod P \implies F(X^q) \equiv F(X)^q \mod Q$$

Finally, evaluating F at (Ideal.Quotient.mk Q α) yields:

$$F(X^q) \equiv F(X)^q \mod Q \implies F(\alpha^q) \equiv F(\alpha)^q \equiv 0 \mod Q$$

Nevertheless, proving `ex_poly_in_A` does not come without effort. We construct m as a polynomial whose coefficients, when lifted from A to L, are the coefficients of F, lifted from B to L. Hence, to each coefficient of F, there must correspond an (a : A), such that

```
theorem coeff_lives_in_A (n : ℕ) : ∃ a : A,
   algebraMap B L (coeff F n) = (algebraMap A L a) :=
```

To obtain `coeff_lives_in_A`, we combine the 'mathematical background' material concerning both the Galois correspondence and integral closure; with instances of [IsScalarTower A B L] and [IsScalarTower A K L], both components of the classic A K L B setup. By the Galois correspondence, having shown (`gal_smul_F_eq_self`) that the group (L ≃$_a$ [K] L) fixes F, we have:

```
lemma coeff_lives_in_K (n : ℕ) : ∃ k : K,
   algebraMap B L (coeff F n) = (algebraMap K L k) :=
```

That is, the coefficients of F lie in the bottom subfield, since, by `coeff_lives_in_fixed_field`, they are in the fixed field of the top subgroup.

---

[12] Given `IsScalarTower R S A`, `IsScalarTower.algebraMap_eq` unpacks an `algebraMap R A` as a composition of `algebraMap`'s along the scalar tower.

Then, assuming A is the integer ring of the number field K, we have an instance of [IsIntegralClosure A A K] [9, p. 30]; to which we apply IsIntegralClosure.isIntegral_iff; to give us that:[13]

$$\text{coeff\_lives\_in\_K} \implies \text{coeff\_lives\_in\_A}$$

4 Discussion

One salient drawback of the A K L B setup is that it causes the number of variables to become unwieldy. For example, for the proof of pow_pow_gen_eq_pow, it was necessary to supply prop_γ_not_in_Q_is_pow_gen with eight variables:

```
lemma pow_pow_gen_eq_pow {γ : B} (h : γ ∉ Q) :
    ((α ^ ((i h) * q)) - (γ ^ q)) ∈ Q := by ...
    prop_γ_not_in_Q_is_pow_gen A K L B Q Q_ne_bot γ h
```

To address the problem of an excessive number of variables, two solutions we applied were: 1. to introduce local notation; 2. to make variables implicit.[14] Still, this disadvantage of the A K L B setup is outweighed by the absolute disadvantage of using the alternative definitions NumberField and NumberField.ringOfIntegers [12]; of which the second presents problems with deterministic timeouts. At the time of writing this report, Lean encounters deterministic timeouts upon attempting to synthesize each of the following instances: [Algebra (𝒪 K) (𝒪 L)]; [Algebra ((𝒪 K) / P) ((𝒪 L) / Q)]; [IsScalarTower (𝒪 K) (𝒪 L) L]; [IsScalarTower (𝒪 K) ((𝒪 K) / P) ((𝒪 L) / Q)].[15] By comparison, assuming A K L B permitted us to synthesize the equivalent instances [Algebra A B]; [Algebra (A / P) (B / Q)]; [IsScalarTower A B L]; [IsScalarTower A (A / P) (B / Q)]. All these instances were indispensable to the application of the Galois correspondence to show that $F(X^q) \equiv F(X)^q \mod Q$.

Therefore, in circumventing typeclass errors encountered by ringOfIntegers, while still accurately describing finite extensions of number fields; for the purpose of formalizing the existence of Frobenius elements, the A K L B setup proved to be a happy medium between too little and too much generality.[16]

Acknowledgement. I would like to thank Amelia Livingston, for intellectual mentorship; Jujian Zhang, for Lean wizardry; Paul Lezeau, for curiosity concerning my project; and Kevin Buzzard, for heartfelt encouragement.

---

[13] This idea is Kevin Buzzard's. Jujian Zhang provided the proofs of the generalizations of the lemmas necessary for showing coeff_lives_in_K.
[14] I thank Jujian Zhang and Paul Lezeau, for teaching me how to do this.
[15] failed to synthesize SMul ↑(𝒪 K) ↑(𝒪 L), SMul (↑(𝒪 K)) (↑(𝒪 K) / P) (deterministic) timeout at 'typeclass'.
[16] cf. [7, p. 22:3] [13, p. 1].

Disclosure of Interests. The author has no competing interests to declare that are relevant to the content of this article.

References

1. Ash, R.B.: A Course in Algebraic Number Theory. Dover, Mineola, New York (2010)
2. Dummit, D.S., Foote, R.M.: Abstract Algebra, 3rd edn. Wiley, Hoboken (2014)
3. Gardner, Z.: Local class field theory (2019). https://web.ma.utexas.edu/SMC/2019/Local%20Class%20Field%20Theory.pdf
4. Glasheen, J.: ExistsFrobenius, 30 April 2024. https://github.com/ImperialCollegeLondon/FLT/blob/2386cce66c9ba2eae446d1f48c79e7ba544a27a5/FLT/ExistsFrobenius.lean
5. Karatarakis, M.: Mathlib.ringtheory.valuation.ramificationgroup (2022). https://github.com/leanprover-community/mathlib4/blob/dc8cf8b25927b121ce49a85d620be120409d51a0/Mathlib/RingTheory/Valuation/RamificationGroup.lean#L28-L29
6. Lang, S.: Algebra, 3rd edn. Springer, New York (2002)
7. Livingston, A.: Group cohomology in the lean community library. In: Leibniz International Proceedings in Informatics (LIPIcs), vol. 268, pp. 22:1–22:17, 26 July 2023. https://doi.org/10.4230/LIPIcs.ITP.2023.22
8. Marcus, D.A.: Number Fields, 2 edn. Springer, Cham (2018). https://doi.org/10.1007/978-3-319-90233-3
9. Milne, J.S.: Algebraic Number Theory (v3.08) (2020). https://www.jmilne.org/math/CourseNotes/ANT.pdf. www.jmilne.org/math/
10. Milne, J.S.: Fields and Galois Theory (v5.10) (2022). https://www.jmilne.org/math/CourseNotes/FT.pdf. www.jmilne.org/math/nbsp;
11. Nakagawa, K., Baanen, A., Nuccio, F.A.E.: https://github.com/leanprover-community/mathlib4/blob/dc8cf8b25927b121ce49a85d620be120409d51a0/Mathlib/RingTheory/DedekindDomain/Ideal.lean#L64-L64
12. Narayanan, A., Baanen, A.: Mathlib.numbertheory.numberfield.basic (2021). https://github.com/leanprover-community/mathlib4/blob/dc8cf8b25927b121ce49a85d620be120409d51a0/Mathlib/NumberTheory/NumberField/Basic.lean#L37-L39
13. Nash, O.: Engel's theorem in mathlib. J. Autom. Reasoning **67**, 18 (2023). https://doi.org/10.1007/s10817-023-09668-0
14. Sutherland, A.: 18.785 number theory I (2021). https://ocw.mit.edu/courses/18-785-number-theory-i-fall-2021/pages/lecture-notes/. Lecture notes
15. Yang, A.: Mathlib.ringtheory.integralrestrict (2023). https://github.com/leanprover-community/mathlib4/blob/dc8cf8b25927b121ce49a85d620be120409d51a0/Mathlib/RingTheory/IntegralRestrict.lean#L38-L54

Formalising Analysis in Lean: Compactness and Dimensionality

Dawid Lipinski

Department of Mathematics, Imperial College London, London, UK
`dawid.lipinski23@imperial.ac.uk`

Abstract. The theorem that a closed unit ball is compact if, and only if, its vector space is finite dimensional showcases how unintuitive infinite dimensional spaces can be. Many proofs skip over what's considered obvious, leaving readers unaware of the underlying assumptions until they attempt formalization. We begin by proving a particular formulation of Riesz's lemma in Lean. We then use it to construct a sequence in the unit ball where the distance between all elements is 1 and show that such a sequence cannot contain a convergent subsequence. Subsequently, we'll establish that a closed unit ball is not sequentially compact. A significant challenge we encountered during this proof was the necessity to define the sequence through strong recursion, which posed some difficulties in its formalisation. We showcase Lean's role in enhancing understanding and generalization of proofs by prompting us to explore broader definitions and theorems within the Mathlib library.

Keywords: Lean · Analysis · Compactness · Dimensionality

1 Introduction

Analysis is a well-established branch of mathematics, heavily utilized across various fields like differential equations, topology, and probability theory. While the fundamental concepts in analysis are generally intuitive, the more complex notions often challenge our intuition. For instance, consider the Heine-Borel Theorem which characterizes compact sets in $\mathbb{R}$, yet fails to extend to infinite-dimensional spaces, where the closed unit ball is not necessarily compact.

This motivates us to formalise a theorem: in a normed vector space, the closed unit ball is compact if and only if the vector space is finite-dimensional. Recognising the counterintuitive nature of this theorem, formalisation becomes particularly important. By grounding our understanding in precise statements derived from fundamental principles rather than relying solely on intuition, we deepen our insights into the subject.

We formalise this theorem using a classical proof commonly taught in functional analysis courses. We use Riesz's lemma to construct a sequence recursively, which is crucial for demonstrating non-compactness in the infinite-dimensional case. In Sect. 2, we present a standard mathematical proof. In Sect. 3, we translate this proof into Lean 4 [4]. We discuss instances where deviation or expansion becomes necessary, shedding light on the complexities of formalising mathematical concepts.

While Mathlib [5] already contains versions of both the lemma [3] and the theorem [2], it's worth noting that the lemma's formulation lacks the guarantee that the elements given by the lemma have norm 1, as it's proven over a general nontrivially normed field. To align with conventional functional analysis literature, we assume the underlying space to be $\mathbb{R}$, simplifying the proof. Similarly, there is a result in Mathlib which assures one direction of the implication, and in this paper we aim to formalise the fact that when the field is $\mathbb{R}$, both directions are true. We believe our proof offers an intuitive and familiar perspective on Theorem 1. We also note that this theorem has been formalised in Lean 3 [1].

2 Mathematical Proof

Before outlining the proof in Lean, we present the mathematical statement and proof of Theorem 1 and the Riesz's lemma (Lemma 1). This initial presentation is intended to provide a fundamental understanding and serve as a reference point in further sections.

Theorem 1. *Let X be a normed vector space. Then, the following are equivalent:*

(i) $\dim X < \infty$.
(ii) $\overline{\mathcal{B}}_1 := \{x \in X : \|x\| \leq 1\}$ *is compact.*

Lemma 1 (Riesz's lemma). *Let X be a normed vector space, and $Y \subset X$ a closed subspace of X such that $Y \neq X$. Then for all $\epsilon \in (0,1)$ there exists $x \in X \setminus Y$ such that:*

(i) $\|x\| = 1$
(ii) $d(x, Y) := \inf_{y \in Y} \|x - y\| > 1 - \epsilon$

We begin by proving the Riesz's lemma (Lemma 1), which we will then use to prove Theorem 1:

Proof. Let $x' \in X \setminus Y$. Since Y is closed $d(x', Y) > 0$. By definition of infimum we can find y', such that $d(x', Y) \leq \|x' - y'\|$, and $\|x' - y'\| < \frac{d(x',Y)}{1-\epsilon}$. Let $x = \frac{x'-y'}{\|x'-y'\|}$. Then, by definition, $\|x\| = 1$ and for all $y \in Y$ we obtain

$$\|x - y\| = \left\| \frac{x' - y'}{\|x' - y'\|} - y \right\| = \frac{\|x' - y' - \|x' - y'\|y\|}{\|x' - y'\|} \geq \frac{\|d(x', Y)\|}{\|x' - y'\|} > 1 - \epsilon \quad (1)$$

This finishes the proof. □

Given Riesz's lemma, we prove Theorem 1, stating that in a normed vector space the closed unit ball is compact if and only if the vector space is finite dimensional.

Proof. The backward implication is trivial and available in the Mathlib library in Lean.

We prove the forward implication using the contrapositive of the statement. Assume the space is not finite dimensional. Let (y_n) be a sequence of linearly independent vectors in X. Then $Y_n = \text{span}\{y_k : 1 \leq k \leq n\}$ has finite dimension and so it is closed. Pick $x_1 = \frac{y_1}{\|y_1\|}$ and for all $n \geq 2$ we can use Riesz's lemma (with $X = Y_n$, $Y = Y_{n-1}$ and $\epsilon = \frac{1}{2}$) to obtain $x_n \in Y_n \setminus Y_{n-1}$ with $\|x_n\| = 1$ and $d(x_n, Y_{n-1}) > \frac{1}{2}$.

Hence, for all $m \geq n$ we have $Y_n \subset Y_m$ and we obtain

$$\|x_n - x_m\| \geq d(x_n, Y_m) > 1 - \epsilon = \frac{1}{2} \qquad (2)$$

Clearly, (x_n) has no convergent subsequence, but $(x_n) \subset \overline{\mathcal{B}}_1$. Therefore, $\overline{\mathcal{B}}_1$ is not sequentially compact, hence it cannot be compact in the normed space. □

3 Proof in Lean

The Mathlib library contains numerous lemmas and theorems allowing us to skip many trivial steps. However, we still begin the proof by stating some of the basic properties not available in Mathlib, such as $\forall z \in Y, \inf_{y \in Y} \|x - y\| \leq \|x - z\|$, where X is a normed vector space, and Y is a closed subspace. This allows us to simplify further proofs and prevent repetition of the same arguments. We omit such proofs from this paper for conciseness, as we do not consider them insightful.[1]

One of the first major lemmas that needs to be proven states that for all $x' \in X$ and $\epsilon \in (0,1)$ there exists a $y' \in Y$ such that $\|x' - y'\| \leq \inf_{y \in Y} \frac{\|x' - y\|}{1 - \epsilon}$. We formalize this in Source Code 1.

```
lemma norm_leq_iInf_div_eps
    {X : Type} [NormedAddCommGroup X] [NormedSpace ℝ X]
    {Y : Subspace ℝ X} (hFc : IsClosed (Y : Set X))
    (x' : X) (hF : x' ∉ Y)
    (ε : ℝ ) (hε : ε > 0) (hε2 : ε < 1):
    ∃ y' : Y, ‖x'-y'‖ ≤ ⊓ y : Y, ‖x'-y‖/(1-ε)
```

Source Code 1: Property of an infimum of a norm.

The proof of this statement follows from a rearrangement of terms and properties of the infimum. However, in Lean all assumptions required for those operations need to be proven. This results in a lengthy proof, despite using multiple theorems already available in the Mathlib library. For example, the proof that $\inf_{y \in Y}(\|x'-y\|*(1-\epsilon)) < \inf_{y \in Y}\|x'-y\|$ requires almost 20 lines of code as many simple assumptions such as $\inf_{y \in Y}(\|x'-y\|*(1-\epsilon)) = (\inf_{y \in Y}\|x'-y\|)*(1-\epsilon)$

---

[1] Our code is available at https://github.com/DawidLipin/-Formalising-Analysis-in-Lean-Compactness-and-Dimensionality.

need to be proven. We continue by proving the statement shown in Source Code 2.

```
lemma eps_leq_normal_diff
    {X : Type} [NormedAddCommGroup X] [NormedSpace ℝ X]
    {Y : Subspace ℝ X} (x' : X)
    (hF : x' ∉ Y) (y' : Y) (ε : ℝ ) (hε2 : ε < 1)
    (ht : ‖x'-y'‖ ≤ ⊓ y : Y, ‖x'-y‖/(1-ε)):
    ∀ y : Y, 1 - ε ≤ ‖(1/‖x'-y'‖) • (x' - y')-y‖
```

Source Code 2: Property of a norm.

The lemma follows from the assumption we refer to as ht in Source Code 2 and the fact that $(y' + \|x' - y'\| \cdot y) \in Y$ by rearranging terms. We note that, due to the vector spaces being implemented as an extension of modules in many of the proofs, we rely mostly on statements using modules and submodules rather than vector spaces directly. However, often the statements are easy to verify, especially given that in Lean subspaces are definitionally equal to the corresponding submodules. This showcases an important advantage of Lean: all statements in Lean can be easily generalized by examining the assumptions required for different theorems used. These observations allow us to significantly shorten the proof of Riesz's lemma in the form shown in Source Code 3.

```
theorem riesz_lemma_norm
    {X : Type} [NormedAddCommGroup X] [NormedSpace ℝ X]
    {Y : Subspace ℝ X}
    (hFc : IsClosed (Y : Set X)) (hF : ∃ z : X, z ∉ Y)
    (ε : ℝ) (hε : ε > 0) (hε2 : ε < 1):
    ∃ x : X, x ∉ Y ∧ ‖x‖=1 ∧ ⊓ y : Y, ‖x-y‖ ≥ 1-ε
```

Source Code 3: Statement of Riesz's lemma.

After obtaining $x' \notin Y$ using the hypothesis (hF : $\exists z : X, z \notin Y$), the statement can be proven by setting $x = (1/\|x' - y'\|) \cdot (x' - y')$, where y' is obtained from the statement shown in Source Code 1.

The next key elements of this proof are the two functions g_riesz_next (Source Code 4) and g_riesz (Source Code 5), which in combination construct a sequence defined using Riesz's lemma and retaining the properties given by the lemma.

```
noncomputable def g_riesz_next
    {X : Type} [NormedAddCommGroup X]
    [NormedSpace ℝ X] (h_inf : ¬FiniteDimensional ℝ X)
    (ε : ℝ) (hε : ε > 0) (hε2 : ε < 1)
    (k : ℕ) (g' : (m : ℕ) → m < k → X) : X :=
  let Y : Subspace ℝ X :=
    Submodule.span ℝ {x | ∃ i : {i : ℕ // i < k},
      g' i.val i.property = x}
  have _: FiniteDimensional ℝ Y := fin_dim_Y_span_riesz k g'
  (riesz_lemma_norm (Submodule.closed_of_finiteDimensional Y)
    (strict_sub_Y_span_riesz h_inf k g') ε hε hε2).choose
```

Source Code 4: The first function used to define the sequence using Riesz's lemma. Constructs a subspace given by the span of all previous elements and picks an element outside of it using Riesz's lemma.

```
noncomputable def g_riesz
    {X : Type} [NormedAddCommGroup X]
    [NormedSpace ℝ X] (h_inf : ¬FiniteDimensional ℝ X)
    (ε : ℝ) (hε : ε > 0) (hε2 : ε < 1) : ℕ → X :=
  fun n => Nat.strongRec (g_riesz_next h_inf ε hε hε2) n
```

Source Code 5: The second function used to define the sequence using Riesz's lemma and strong recursion.

The first function, g_riesz_next (Source Code 4), defines a sequence that constructs the subspaces given by Y_n in the proof presented in Sect. 2. This allows us to construct the second function, g_riesz (Source Code 5), representing the desired sequence using strong recursion. The use of strong recursion is necessary, since we require all the previous elements in a sequence to be able to use them to construct the needed subspace. It also enables us to retain all the properties of elements of this sequence, which we need to prove Theorem 1. Despite this part of the proof accounting for a significant proportion of the formalised proof, this is not the case in the mathematical proof, as we simply state that such a sequence can be defined by Riesz's lemma.

Similarly to the proof of Theorem 1 in Sect. 2, we state and prove the backward implication using its contrapositive shown in Source Code 6.

```
theorem dim_inf_implies_not_compact
    {X : Type} [NormedAddCommGroup X] [NormedSpace ℝ X]:
    ¬FiniteDimensional ℝ X → ¬IsCompact (Metric.closedBall (0 : X) 1)
```

Source Code 6: Contrapositive of the backward implication.

Although long, the proof very closely follows the proof from Sect. 2. It primarily uses lemmas proven above and most of the code is used to prove simple, but necessary, assumptions. We show that any subsequence cannot be Cauchy, and consequently, cannot be convergent. This proof is relatively straightforward, given that the construction of `g_riesz` enables us to preserve all the properties of elements given by Riesz's lemma. Therefore, by choosing $\epsilon = \frac{1}{2}$, it follows that all elements in the sequence must be at least $\frac{1}{2}$ apart from each other. Consequently, any subsequence formed under these conditions cannot be Cauchy.

4 Conclusions

We prove in Lean that in a normed vector space, a closed unit ball is compact if and only if the vector space is finite dimensional. We construct a specific version of Riesz's lemma to allow for construction of a sequence proving the contrapositive of the forward implication. The proof underscores the challenges associated with working with such constructed sequences, while also showcasing how Lean facilitates the straightforward generalization of mathematical statements. We demonstrate that, assuming adequate rigour in the initial proof, it can be accurately reconstructed in Lean, enabling the reader to effortlessly follow each step.

Acknowledgements. The code was developed as part of "Formalising Mathematics 2023–2024" course at Imperial College London ran by Professor Kevin Buzzard. As such, parts of the code have been written by or with the help of Professor Kevin Buzzard, Dr. Bhavik Mehta and David Angdinata. I would also like to thank Olaf Lipinski for reviewing this paper.

References

1. Lean 3 proof that a normed vector space with compact unit ball is finite dimensional. https://github.com/LAC1213/compact_unit_ball/tree/lean-3.4.2
2. Mathlib.Analysis.NormedSpace.FiniteDimension. https://leanprover-community.github.io/mathlib4_docs/Mathlib/Analysis/NormedSpace/FiniteDimension.html#FiniteDimensional.of_isCompact_closedBall
3. Mathlib.Analysis.NormedSpace.RieszLemma. https://leanprover-community.github.io/mathlib4_docs/Mathlib/Analysis/NormedSpace/RieszLemma.html#riesz_lemma
4. Moura, L., Ullrich, S.: The lean 4 theorem prover and programming language. In: Platzer, A., Sutcliffe, G. (eds.) CADE 2021. LNCS (LNAI), vol. 12699, pp. 625–635. Springer, Cham (2021). https://doi.org/10.1007/978-3-030-79876-5_37
5. The mathlib Community: The lean mathematical library. In: Proceedings of the 9th ACM SIGPLAN International Conference on Certified Programs and Proofs, pp. 367–381. CPP 2020, Association for Computing Machinery, New York, NY, USA (Jan 2020).https://doi.org/10.1145/3372885.3373824

Formalisation of the Category of Hopf Algebras in Lean4

Jujian Zhang[✉][iD], Yunzhou Xie[iD], Yichen Feng[iD], and Yanqiao Zhou[iD]

Imperial College London, London SW7 2AZ, UK
{jujian.zhang19,yunzhou.xie21,yichen.feng21,
yanqiao.zhou21}@imperial.ac.uk

Abstract. Hopf algebras are used in various areas of mathematics and theoretical physics, such as algebraic geometry and quantum groups. In this article, we formalise a few basic results about Hopf algebras. In particular, we prove that the set of R-algebra homomorphisms $\text{Hom}_R(A, L)$ from a commutative R-Hopf algebra A to a commutative R-algebra L can be endowed with a group structure under the convolution product. Furthermore, we also formalise an anti-equivalence between the category of commutative R-Hopf algebras and the category of affine group schemes over R, where the latter is defined as group objects of corepresented functors from commutative R-algebras to Set.

Keywords: Lean · Formalisation · Hopf Algebra · Affine Group Schemes · Anti-equivalence

1 Introduction

Hopf algebras are used in various areas of mathematics and theoretical physics, such as algebraic geometry and quantum groups. We formalised some basic theory about Hopf algebras at [1]. Throughout the article, let R be a commutative ring. We begin by defining coalgebras, bialgebras and Hopf algebras.

Definition 1. *An R-coalgebra A is an R-module with an R-linear map $\Delta : A \to A \otimes_R A$, which we call the comultiplication, and an R-linear map $\varepsilon : A \to R$, which we call the counit, such that the following diagrams commute:*

$$
\begin{array}{ccc}
A & \xrightarrow{\Delta} & A \otimes_R A \\
\Delta \downarrow & & \downarrow \Delta \otimes \text{id} \\
A \otimes_R A & \xrightarrow{\text{id} \otimes \Delta} & A \otimes_R A \otimes_R A
\end{array}
\qquad
\begin{array}{ccc}
A & \xrightarrow{\Delta} & A \otimes_R A \\
\Delta \downarrow & \searrow{\text{id}} & \downarrow \text{id} \otimes \varepsilon \\
A \otimes_R A & \xrightarrow{\varepsilon \otimes \text{id}} & R \otimes_R A \cong A \cong A \otimes_R R
\end{array}.
$$

Definition 2. *An R-bialgebra is an R-algebra that has an R-coalgebra structure where the comultiplication and counit maps are also R-algebra homomorphisms.*

Y. Xie, Y. Feng and Y. Zhou—The three coauthors contributed equally.

Definition 3. *An R-Hopf algebra A is an R-bialgebra with an R-linear map $S : A \to A$, which we call the antipodal map, such that the following diagram commutes*

$$\begin{array}{ccccc}
A \otimes_R A & \xleftarrow{\Delta} & A & \xrightarrow{\Delta} & A \otimes_R A \\
{\scriptstyle \mathrm{id} \otimes S} \downarrow & & \downarrow {\scriptstyle \rho \circ \varepsilon} & & \downarrow {\scriptstyle S \otimes \mathrm{id}} \\
A \otimes_R A & \xleftarrow{\Delta} & A & \xrightarrow{\Delta} & A \otimes_R A
\end{array}$$

where ρ is the structure map of A as R-algebra.

Since comultiplication plays an important role in co/bi/Hopf algebras, we often find it useful to expand $\Delta(x)$ as a sum of pure tensors $\sum x_1 \otimes x_2$, and manipulate expressions in terms of x_1's and x_2's. This technique is called Sweedler notation, and is used extensively throughout the project. For example, the diagram on the right in Definition 1 can be written as

$$\sum \varepsilon(x_1) \otimes x_2 = 1 \otimes x, \quad \sum x_1 \otimes \varepsilon(x_2) = x \otimes 1;$$

and the diagram in Definition 3 can be written as

$$\sum S(x_1) \cdot x_2 = \sum x_1 \cdot S(x_2) = \rho(\varepsilon(x)).$$

All the axioms of co/bi/Hopf algebras are defined in terms of commutative diagrams, i.e. equality of R-linear maps or R-algebra homomorphisms, both on paper and in Lean. Thus Sweedler notation on paper quickly turns equalities of morphisms to equality of elements; in Lean4, we implement this "notation" as series of lemmas such as

```
lemma antipode_repr_eq_smul {I : Type*} (x : A) (S : Finset I) (x1 x2 : I → A)
    (repr : Coalgebra.comul a = ∑ i in S, x1 i ⊗t[R] x2 i) :
    ∑ i in S, antipode (R := R) (x1 i) * x2 i =
    (Coalgebra.counit x : R) • (1 : A) :=
  ...
```

Here is a glossary of the formalised results:

- for any commutative R-Hopf algebra A and commutative R-algebra L, the set of R-algebra homomorphisms $\mathrm{Hom}_R(A, L)$ is a group under the convolution product;
- some computations of expressions involving comultiplication with Sweedler notation;
- the category of R-Hopf algebras is a symmetric monoidal category;
- the category of commutative R-Hopf algebras is anti-equivalent to the category affine group schemes over R.

We will explain the group structure in Sect. 2. We define affine group schemes in terms of *corepresented* functors instead of *corepresentable* functors. A corepresented functor F is a functor F together with an R-algebra A and an isomorphism

$F \cong \mathrm{Hom}_{\mathsf{CommAlg}_R}(A, -)$; meanwhile, a corepresentable functor G in mathlib4 is a functor G such that there exists some A where $G \cong \mathrm{Hom}_{\mathsf{CommAlg}_R}(A, -)$. We will explain why the constructive version is better during the formalisation of the anti-equivalence in Sect. 3.

2 Group Structure on the Hom Set

2.1 Mathematical Details

Let A be a commutative R-Hopf algebra and L a commutative R-algebra. In this section, we present a proof and a formalisation that $\mathrm{Hom}_R(A, L)$ forms a group under the convolution product following [2, pages 49-52].

Definition 4 (convolution product). *Let $f, g \in \mathrm{Hom}_R(A, L)$ be two R-algebra homomorphisms; the convolution of f and g is defined as*

$$f \star g : A \xrightarrow{\Delta} A \otimes_R A \xrightarrow{f \otimes g} L \otimes_R L \xrightarrow{m} L \ ,$$

where $\Delta : A \to A \otimes_R A$ is the comultiplication and $m : L \otimes L \to L$ is the multiplication map defined by $x \otimes y \mapsto x \cdot y$.

The name "convolution product" is due to the fact that $(f \star g)(x) = \sum f(x_1) \cdot g(x_2)$ in Sweedler notation.

Lemma 1 (group structure of the Hom set). $\mathrm{Hom}_R(A, L)$ *can be endowed with a group structure with binary operation $\star$, neutral element $1 : A \xrightarrow{\varepsilon} R \xrightarrow{\rho} L$ and inverse $f^{-1} := f \circ S$ of any $f \in \mathrm{Hom}_R(A, L)$, where ε is the counit of A and ρ the structure map of L as an R-algebra.*

Proof. The associativity of the convolution product is checked via the following commutative diagram:

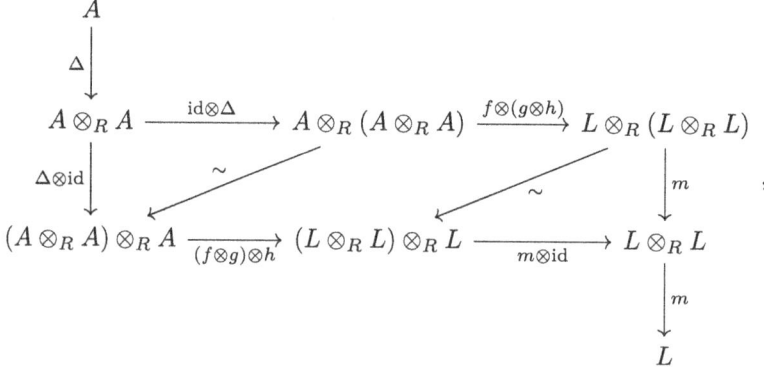

where the upper path is $f \star (g \star h)$, the lower path is $(f \star g) \star h$ and the triangles commute because of (co)associativity of (co)multiplication.

We use Sweedler notation to help us check that $f \star 1 = 1 \star f = f$ and that $f^{-1} \star f = f \star f^{-1} = 1$. For example, we can check $(f \star f^{-1})(x) = 1(x)$, where 1 is the neutral element under the convolution product and *not* the identity, by transforming the left hand side to

$$\sum f(x_1) \cdot f(S(x_2)) = f\left(\sum x_1 \cdot S(x_2)\right) = f(\rho(\varepsilon(x))) = \rho(\varepsilon(x)) = 1(x).$$

In Lean4, since Sweedler notation is implemented as series of lemmas, we perform `rewrite` and `simp` using these lemmas to finish the proof.

2.2 Formalisation

Since Hopf algebras in mathlib4 are defined as bialgebras with a *linear* antipodal map, we need to first prove that the antipodal map is in fact an algebra homomorphism in the commutative case. Thus, we divide our formalisation of the above construction into four stages:

1. We construct a monoid structure on the set of linear maps $\text{Hom}_R(A, L)$ with any ring R, R-coalgebra A and R-module L.
   [1, for_mathlib/Coalgebra/Monoid.lean > instMonoid]
2. We prove that the tensor product of coalgebras (*resp.* bialgebras) can be given a natural coalgebra (*resp.* bialgebra) structure.
   [1, for_mathlib/Coalgebra/TensorProduct.lean > tensorProduct]
3. Using the tensor product of bialgebras, we prove that the antipodal map is anti-commutative, i.e. $S(a \cdot b) = S(b) \cdot S(a)$, so that the linear antipodal map is in fact an R-algebra homomorphism in the commutative case.
   [1, for_mathlib/HopfAlgebra/Basic.lean > antipode_anticommute]
4. We prove that $f \circ S$ is indeed the inverse of f under the convolution product using Sweedler notation, and hence finish the construction of the group structure.
   [1, for_mathlib/HopfAlgebra/Basic.lean > instGroup]

Alongside the result in stage 2, we also formalised that the tensor product of two Hopf algebras is another Hopf algebra, so that the category of Hopf algebras is symmetric monoidal. The main difficulties encountered during the formalisation of this section were calculations with comultiplication on tensor products. For example, to prove that $A \otimes (B \otimes C) \cong (A \otimes B) \otimes C$ as R-bialgebras or R-Hopf algebras, we need to check the following computations

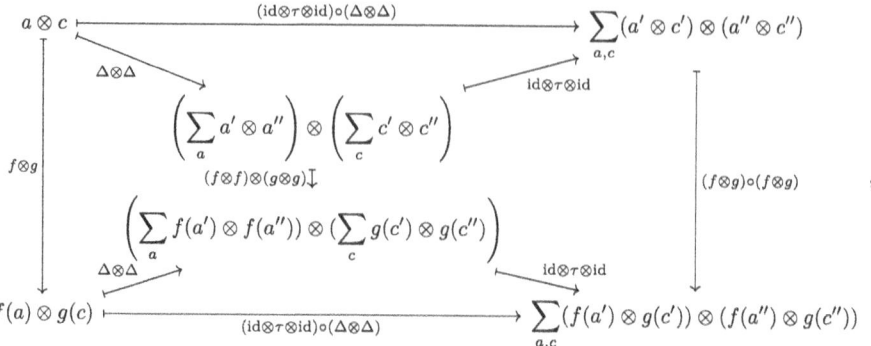

where τ is the map defined by $x \otimes y \mapsto y \otimes x$. However, with enough lemmas about Sweedler notation, the formalisation needed only four uses of `rewrite`, three uses of `simp` and can be finished off by `aesop`, a tactic in Lean4 to automatically find proofs for simple problems [1, `HopfCatMonoidal.lean > comul_comp_assoc`]. Although we did not use these results in Sect. 3, once the construction of the category of Hopf algebras was completed, the formalisation that the category of Hopf algebras is a symmetric monoidal category was only a few lemmas away, using that the forgetful functor to the symmetric monoidal category of algebras preserves tensors.

3 Affine Group Scheme

In this section, we give a brief account of the formalisation of the anti-equivalence between affine group schemes and Hopf algebras. The entire formalisation can be found in [1, `AffineFunctor.lean`].

3.1 Mathematical Details

In this section, we denote the functor $\mathrm{Hom}_{\mathsf{CommAlg}_R}(R, -)$ by $\star$.

Definition 5 (affine group schemes over R). *We say F is an affine group scheme if we have the following [3, Definition 3.1][4, pages 7–9]:*

- *F is corepresentable; in other words, $F \cong \mathrm{Hom}_{\mathsf{CommAlg}_R}(A, -)$ for some commutative R-algebra A;*
- *a natural transformation $m : F \times F \Longrightarrow F$ which we call the multiplication map;*
- *a natural transformation $e : \star \Longrightarrow F$ which we call the unit;*
- *a natural transformation $i : F \Longrightarrow F$ which we call the inverse;*
- *five commutative diagrams:*

$$(F \times F) \times F \xrightarrow{m \times \text{id}} F \times F \quad \star \times F \xrightarrow{e \times \text{id}} F \times F \quad F \times \star \xrightarrow{\text{id} \times e} F \times F$$
$$\downarrow \sim \qquad\qquad \downarrow m \qquad \downarrow \sim \quad \swarrow m \qquad\qquad \downarrow \sim \quad \swarrow m$$
$$F \times (F \times F) \xrightarrow{\text{id} \times m} F \qquad\qquad F \qquad\qquad\qquad\qquad F$$

$$F \xrightarrow{(\text{id},i)} F \times F \qquad\qquad F \xrightarrow{(i,\text{id})} F \times F$$
$$\downarrow \qquad\qquad \downarrow m \qquad\qquad \downarrow \qquad\qquad \downarrow m \quad ,$$
$$\text{Hom}(A,-) \xrightarrow{\text{Hom}(\rho,-)} \star \xrightarrow{e} F \qquad \text{Hom}(A,-) \xrightarrow{\text{Hom}(\rho,-)} \star \xrightarrow{e} F$$

where A represents F and $\rho : R \to A$ is the structure map of A as a R-algebra.

The five commutative diagrams correspond to associativity of multiplication, e being the left and right neutral element and i being the left and right inverse.

Definition 6 (morphism between affine group schemes). *Let F and G be two affine group schemes. A morphism between F and G is a natural transformation $\alpha : F \implies G$ such that*

- *the unit of G is equal to the composition of α and the unit of F: $\alpha \circ \varepsilon_F = \varepsilon_G$;*

- *α is compatible with multiplication maps of F and G:*
$$\begin{array}{ccc} F \times F & \xrightarrow{\alpha \times \alpha} & G \times G \\ m_F \downarrow & & \downarrow m_G \\ F & \xrightarrow{\alpha} & G \end{array}.$$

Our main theorem is the following:

Theorem 1. *The category of affine group schemes over R is anti-equivalent to the category of commutative R-Hopf algebras.*

$$\mathsf{AffineGroupScheme}_R^{\text{op}} \cong \mathsf{HopfAlg}_R \tag{1}$$

The intuition behind Theorem 1 is that the axioms of affine group schemes and Hopf algebras are the same except the direction of the arrows is reversed — we demonstrate this with (co)associativity of (co)multiplications:

$$(F \times F) \times F \xrightarrow{m \times \text{id}} F \times F \qquad (H \otimes H) \otimes H \xleftarrow{\Delta \otimes \text{id}} H \otimes H$$
$$\downarrow \sim \qquad\qquad \downarrow m \quad \leftrightsquigarrow \qquad \uparrow \sim \qquad\qquad \uparrow \Delta$$
$$F \times (F \times F) \xrightarrow{\text{id} \times m} F \qquad H \otimes (H \otimes H) \xleftarrow{\text{id} \otimes \Delta} H$$

Thus, the key to proving and formalising the antiequivalence in Theorem 1 is to "flip" the commutative diagram. To this end, we introduce the following lemma:

Lemma 2. *Let F and G be two corepresentable functors $\mathsf{CommAlg}_R \Longrightarrow \mathsf{Set}$ corepresented by A and B respectively. We have that*

$$\mathrm{NatTrans}(F, G) \cong \mathrm{Hom}_{\mathsf{CommAlg}_R}(B, A).$$

The bijection respects composition of natural transformations as well.

The lemma follows from the Yoneda lemma, but we give and formalise a direct proof instead. The reason for not using the Yoneda lemma is the distinction between corepresentable functors and corepresented functors and is postponed to Sect. 3.2.

Proof (of Lemma 2). Write α_F as the natural isomorphism between $F \cong \mathrm{Hom}(A, -)$ and similarly for G.

Given a natural transformation n, then we have $\alpha_G^{-1}(A) : G(A) \cong \mathrm{Hom}(B, A)$, $n(A) : F(A) \to G(A)$ and $\alpha_F(A) : \mathrm{Hom}(A, A) \cong F(A)$. Thus $\alpha_G^{-1}(n(\alpha_F(1_A)))$ is an R-algebra homomorphism from B to A.

Conversely, if f is an R-algebra homomorphism from B to A, then $\mathrm{Hom}(f, -)$ is a natural transformation from $\mathrm{Hom}(A, -)$ to $\mathrm{Hom}(B, -)$ by pre-composition. Thus $\alpha_G \circ \mathrm{Hom}(f, -) \circ \alpha_F^{-1}$ is a natural transformation $F \to G$.

Given Theorem 2, the proof of Theorem 1 is simply by following one's nose:

Proof (of Theorem 1). Let F be an affine group scheme corepresented by A. By Theorem 2 and the fact that $F \times F$ is corepresented by $A \otimes_R A$, m corresponds to a homomorphism $\Delta : A \to A \otimes A$, ε to $e : A \to R$ and i to $S : A \to A$. We check A is a Hopf algebra with comultiplication Δ, counit ε and antipodal map S.

Similarly if A is a Hopf algebra with comultiplication Δ, counit ε and antipodal map S, then under Theorem 2 again, Δ corresponds to $m : \mathrm{Hom}(A, -) \times \mathrm{Hom}(A, -) \to \mathrm{Hom}(A, -)$, e to $\varepsilon : \mathrm{Hom}(R, -) \to \mathrm{Hom}(A, -)$ and S to $i : \mathrm{Hom}(A, -) \to \mathrm{Hom}(A, -)$. We check that $\mathrm{Hom}(A, -)$ is an affine group scheme as well with multiplication m, unit e and inverse i.

Both of the procedures are functorial. Checking that all the axioms are satisfied is a matter of finding the correct naturality of the correct functor to use.

3.2 Formalisation

Corepresentable Versus Corepresented. In mathlib4, corepresentable functors are already defined non-constructively: F is copresentable if and only if there exists some object A such that $F \cong \mathrm{Hom}(A, -)$. We choose to use a constructive version instead:

```
structure CorepresentedFunctor (F : CommAlgebraCat R ⟹ Type v) :=
  coreprX : CommAlgebraCat R
  coreprW : coyoneda.obj (op coreprX) ≅ F
```

The reason is that if we were to use mathlib4's version, we would have no control over the object corepresenting F: for example, the object corepresenting

$\mathrm{Hom}(A, -)$ is not definitionally equal to A. With our more constructive version, we can have *definitional equality*: for example, $\mathrm{Hom}(A, -) \times \mathrm{Hom}(B, -)$ will be corepresented by $A \otimes_R B$ by definition. Compared to the non-constructive approach, we can only prove that the object corepresenting $\mathrm{Hom}(A, -) \times \mathrm{Hom}(B, -)$ is *isomorphic*, but not definitionally equal, to $A \otimes_R B$. Thus, by using corepresented functors, we can upgrade some isomorphisms to definitional equalities. Since we choose to deviate from the corepresentable functor definition in mathlib4, our proof of Theorem 2 had to start from scratch.

Naturality and Why is it Hard? Let us write α and β for the functors between affine group schemes and Hopf algebras that form the anti-equivalence in Theorem 1. We outlined the proof by saying that it is all about finding the correct naturality of the correct functor to use. Thus the proof is natural and therefore the formalisation is expected to be easy. However, the formalisation is harder than its pen-and-paper counterpart. Let $\Delta : A \to A \otimes_R A$ be an arbitrary linear map; then, applying Lemma 2 twice, we will have another map $A \to A \otimes_R A$ that is equal to Δ, but this equality is far from being definitionally true:

$$A \to A \otimes_R A \xrightarrow{\beta} \mathrm{Hom}(A, -) \times \mathrm{Hom}(A, -) \to \mathrm{Hom}(A, -) \xrightarrow{\alpha} A \to A \otimes_R A \ .$$

equal to but not definitionally equal to

The non-definitional equality will manifest when defining the counit $\mathbf{1}_{\mathsf{HopfAlg}_R} \cong \alpha \circ \beta$ of the anti-equivalence in Eq. (1). Given any R-Hopf algebra H with comultiplication Δ, counit ε and antipodal map S, after going through two functors $\beta \circ \alpha$, the resulting R-Hopf algebra H' has a comultiplication Δ', a counit ε' and an antipodal map S' that are equal to but not definitionally equal to Δ, ε and S respectively, and thus id : $H \to H'$ will not type check. On the other hand, when we start with $m : F \times F \to F$ where F is corepresented by A with iso : $F \cong \mathrm{Hom}(A, -)$ and we apply Lemma 2 twice, we don't have any equality without composing enough isomorphisms:

$$F \times F \to F \xrightarrow{\alpha} A \to A \otimes_R A \xrightarrow{\beta} \mathrm{Hom}(A, -) \times \mathrm{Hom}(A, -) \to \mathrm{Hom}(A, -) \ .$$

need to compose iso on domain and codomain.

The discrepancies cause difficulties when defining the unit $\mathbf{1}_{\mathsf{AffineGroupScheme}} \cong \beta \circ \alpha$ of the anti-equivalence in Eq. (1). For example, if F is an affine group scheme corepresented by X, we need a morphism $F \to \beta(\alpha(F))$. This means that for any commutative R-algebra A and any R-algebra homomorphism $f : X \to A$, we need to construct an element of $F(A)$. If F is *literally* $\mathrm{Hom}(X, -)$, then the element required would *literally* be f. For an arbitrary affine group scheme F, the element will have to be $\mathfrak{f}(A)(\mathrm{id}_A)$ where $\mathfrak{f} : \mathrm{Hom}(A, -) \to F$ is obtained by Theorem 2 again. Not surprisingly, the identification of f with $\mathfrak{f}(A)(\mathrm{id}_A)$ is rather delicate to work with. Note that we do not have to define $\mathfrak{f}$ with an extra isomorphism as $\mathrm{Hom}(A, -) \to \mathrm{Hom}(X, -) \cong F$ precisely because we have used *corepresented* functors instead of corepresentable functors!

4 Conclusion

We have formalised some basic results about Hopf algebras in **Lean4**, such as the group structure on $\mathrm{Hom}_{\mathsf{CommAlg}_R}(A, L)$ between a commutative R-Hopf algebra and a commutative R-algebra. We also set up a symmetric monoidal category structure on the category of Hopf algebras. We formalised the anti-equivalence between affine group schemes and commutative Hopf algebras which we hope to extend to an anti-equivalence between affine group schemes defined in terms of prime spectrums and commutative Hopf algebras in the future. There are still many basic results missing from our project, such as group-like elements, Hopf algebra structures on group algebras, etc. We would also like to invite contributors to explore other directions in the theory of Hopf algebras as well such as applications to quantum groups.

Acknowledgments. We would like to thank David Ang for his patience in proofreading the article, and Kevin Buzzard for coming up with and encouraging us to take on the project of formalising some basic results about Hopf algebras. The first author is funded by the Schrödinger Scholarship Scheme from Imperial College London.

References

1. All the contributors: Fermat's last theorem. https://github.com/ImperialCollegeLondon/FLT/tree/CrazyAffine/FLT/Proj3 (2024)
2. Kassel, C.: Quantum groups. In: Graduate Texts in Mathematics (GTM), vol. 155. Springer, Heidelberg (1995)
3. Milne, J.S.: Basic theory of affine group schemes (2012). www.jmilne.org/math/
4. Waterhouse, W.C.: Introduction to Affine Group Schemes, vol. 66. Springer, Heidelberg (2012). https://doi.org/10.1007/978-1-4612-6217-6

Software for the Applications of Group Theory to Combinatorics

Computing the Group of an Algebraic Variety over a Finite Field

Abdullah Alazemi[ID] and Anton Betten[(✉)][ID]

Faculty of Science, Department of Mathematics, Kuwait University, P.O. Box 5969, 13060 Safat, Kuwait
anton.betten@ku.edu.kw

Abstract. We present a practical algorithm to computing the automorphism group of a variety. Our algorithm is based on combinatorial tools from graph theory, in particular the notion of a canonical form. Performance issues and relations to other approaches will be discussed.

Keywords: Isomorphism · Algebraic Variety · Classification · Canonical Form

1 Introduction

The problem of computing the automorphism group of an algebraic variety is difficult. No polynomial time algorithm is known. Here, we consider the case of algebraic varieties over a finite field. Using tools from graph theory, we present a practical algorithm to computing the automorphism group of a variety. Experiments show that the algorithm performs better than previously known algorithms. Because of the exponential complexity, the algorithm is bound to small problem instances. This can still be helpful to researchers in the field. Until the P vs NP dilemma in complexity theory has been resolved, we do not see any other way to proceed.

We present our implementation of the algorithm, utilizing the Orbiter software [4]. A key part of this algorithm relies on Nauty [18], a software tool to compute the canonical form of a graph. The connection between geometries and graphs and groups make this approach feasible. An early form of this algorithm was applied in [1], where the problem of classifying arcs in projective planes was considered. The present algorithm is more elaborate that the predecessor, in that it can deal with projective spaces. It is also able to compute the stabilizer of a variety in case that it differs from the set stabilizer of the set of rational points.

Some general comments on the contribution are in order:

The varieties considered here are projective, though the algorithm can be adapted to other kinds of varieties. For affine varieties, one could embed the variety in a suitable projective variety and distinguish the hyperplane whose complement is the affine space.

The algorithm described here is relevant for classification, as canonical forms provide a tool for solving the classification problem. The classification problem can only be solved for one field at a time. However, the examples from one specific field may give sufficient information to say something about the problem in general. For instance, it is often possible to generalize objects arising in specific fields to general families. A family is a construction of a class of objects that works for infinitely many fields. Some comments are in order.

The algorithm described here does not make any assumptions about the variety, other than finiteness, being projective, and being reasonably small. Thus, we call it a general purpose isomorphism test for varieties. Classification is the process of determining all isomorphism classes of a particular type of objects. A classification algorithm of a type of variety may be faster than the isomorphism test between two objects based on the algorithm here. This is because for specific types of varieties, special algorithms exist that can be more efficient. This lies in the fact that the classification may rely on extra structural knowledge of the object. For instance, for cubic surfaces, we can use the theory of Schäfli's double sixes to provide faster solutions. In [5], this is utilized to classify cubic surfaces over small finite fields. In Sect. 7, we will look at an example of isomorphism testing and classification of cubic surfaces.

Our algorithm is coded in Orbiter [4]. However, export to GAP [11] and Fining [2] as well as Magma [8] is possible.

2 Groups and Geometries

Our notation is as follows: $PG(n, F)$ is the n-dimensional projective geometry over the field F. If $F = \mathbb{F}_q$, we write $PG(n, q)$. A point in $PG(n, F)$ is written as $\mathbf{P}(x_0, \ldots, x_n)$, not all x_i equal to zero. If $f(X_0, \ldots, X_n)$ is an equation in n variables, $\mathbf{v}(f)$ is the variety of f, i.e., the set of points in $PG(n, F)$ whose coordinates satisfy $f(x_0, \ldots, x_n) = 0$. We call these the F-rational points. The same equation may be considered over different fields.

The group of projective space acts on points, subspaces, and algebraic sets. The action on points is coupled with an action on equations (and subspaces) in such a way that for any group element, the image equation corresponds to the image of the set of rational points. The action on equations is called the contragredient action, and it is given by substituting the columns of the inverse matrix into the equation. There is a more general definition that extends the action to the semilinear group. Over a given finite field, a variety may have multiple defining equations.

Two varieties are isomorphic if there is a collineation that maps one to the other. This also means that the equations of one variety transform to the equations of the other. The big problem in the field are the classification of objects under the group. Because we consider varieties over finite fields, the problem has a combinatorial character. The objects can be studied by reducing to the associated set of rational points. From there, it is often easy to lift the solution to the variety, which is different because the action on the equation is different from the action on points.

3 The Algorithm

Our approach to computing the group of a variety is based on two steps. In the first step, the group of the set of rational points is computed. In the second step, the group of the variety is computed from the stabilizer of the set of rational points.

Computing the stabilizer of the set of rational points is done by utilizing a tool from graph theory, namely that of computing the canonical form of a graph. The graph is the associated decorated Levi graph of the geometry. This is a bipartite graph, and we will say more about it below. For background on canonical froms of graphs, we refer to [19]. The canonical form of the graph can be computed practically by Nauty [18], which is a tool for computing graph isomorphism, automorphisms, and canonical forms. When computing the canonical form of the graph, generators for the automorphism group are computed as well. Once the stabilizer of the graph is known, we translate it back to the level of the geometry and obtain the collineation stabilizer of the set of rational points. This completes step one of the algorithm.

Step two addresses the fact that the stabilizer of the variety may be different from the set of rational points. In fact, it is a subgroup. To this end, we consider the orbit of the equation under the stabilizer of the set of rational points. In many cases, the stabilizer of the set of rational points is small, and hence the orbit on the equations with the same set of rational points is even smaller. For this reason, we can apply Schreier's orbit algorithm, and compute the stabilizer using the algorithm of Schreier and Sims, see [12, 21]. There are cases when this algorithm degenerates. For instance, there are algebraic varieties which have no rational point over a given field. In that case, the second stage would run with the full collineation group, which may be inefficient.

Another way to treat the problem of computing isomorphisms and automorphisms of variety is in a purely permutation group theoretic setting. In this approach, one computes the canonical set on the orbit of the group on the set of rational points, disregarding the fact that the stabilizer of the set of rational points is different from the actual stabilizer of the variety. This means, the result is often false. But the stabilizer of the variety is a subgroup of the stabilizer of the set of rational points, so the answer is not too far off. An implementation of a least image algorithm for sets has been developed by Linton [17] and later by Jefferson et.al. [13]. An implementation in GAP [11] is available in form of a GAP package called least images [14].

4 Implementation in Orbiter

We will primarily use the computer algebra system Orbiter [3, 4]. There are some conventions about coding of algebraic objects. These codes help Orbiter translate mathematical data into computer friendly data structures. So, let us briefly review some of the basic concepts.

In Orbiter, the elements of the field $\mathbb{F}_q$ are coded as integers from the interval $[0, q-1]$. The elements of the prime field $\mathbb{F}_p$ are identified naturally with the elements in the interval $[0, p-1]$. For extension fields, the polynomial representation of a field element over the prime field is considered as integer, simply by inserting the characteristic p into the polynomial and evaluating over the integers.

Orbiter treats a semilinear matrix group such as $\mathrm{P\Gamma L}(n, q)$ as a semidirect product of $\mathrm{PGL}(n, q)$ and the automorphism group of the field. The elements of the linear group $\mathrm{PGL}(n, q)$ are given as matrices (to be considered modulo scalars). A subscript at the matrix represents the field automorphism in the semidirect product.

Orbiter establishes a fixed bijection between the points of a given projective space and a 0-based interval of integers. This way, a point with homogeneous coordinates $\mathbf{P}(x_0, \ldots, x_n)$ has a certain integer code. An integer code can be converted back to a projective point by going the bijection in the other way. Such codes are very useful to set up a permutation action of the group, as well as for long term data storage. The bijection is fixed, but depends on the choice of primitive polynomial in the case of an extension fields. Orbiter uses a built-in set of primitive polynomials, so the element codes are fixed, even for fields that are not prime fields. When importing and exporting data to other computer algebra systems, an isomorphism between the different fields is needed. Such an isomorphism can be established by identifying field elements with respect to the exponents according to the representation as powers of a primitive element.

Orbiter is a Unix/Linux command line program, which can be used using unix makefiles or shell scripts. Orbiter commands and Unix/Linux command line tools can be mixed. Makefiles offer the ability to store commands under keyword, and the make command would search through the makefile for that keyword and execute the associated command sequence. They also reduce file clutter, as multiple command sequences can be stored in one single makefile.

5 Tactical Decompositions

The combinatorial tools of tactical decompositions and partition refinement is important in the algorithmic approach to the problem of computing automorphism and isomorphisms of geometries. Tactical decompositions were introduce by E.H. Moore in the 19th century [20]. Their cousins are the equitable partitions of graphs. By doing the Levi graph reduction from a geometry to a graph, the tactical decompositions become the equitable partitions. The refinement is supposed to preserve the structure of the geometry, so it must follow the structural properties of the object.

An algorithm to refine a tactical decomposition based on combinatorial invariants has been presented in [7]. Once the automorphism group of a geometry is known, we can also consider the tactical decomposition by automorphisms (TDA). This is the decomposition whose classes are the orbits of the automorphism group on points and lines. The TDA may be assumed to be a refinement of the TDO. One can compute the TDO first, and then refine the classes of the TDO according to orbits of the automorphism group. The TDA and TDO may or may not be equal. Examples for both cases exist.

6 Example 1: The Edge Quartic

An undulation (or hyperflex) of a quartic curve is a point where the tangent line has a fourfold point of contact. Edge [10] describes a family of curves with at least 8 such points. In projective coordinates, the curve is given by

$$X^4 - Y^4 - Z^4 + 4fX^2YZ + 2f^2Y^2Z^2 = 0,$$

where f is a primitive element in the field $\mathbb{F}_q$. We decide to test out algorithm on this family of objects. Our results are summarized in Table 1.

Table 1. Performance Testing for the Edge Quartic

Q	F	PTS	AUT	NCPX	TORB	TGAP
17	3	12	24	9	30	10
29	2	20	8	7	170	349
49	7	44	8	9	1333	3411
53	2	68	8	7	1830	50406
59	2	72	4	7	3940	73042
61	2	96	8	7	3340	−1
127	3	116	8	7	128240	−1
137	3	108	24	9	97380	−1

Here, Q is the order of the finite field, F is the primitive root in orbiter numbering, PTS is the number of $\mathbb{F}_Q$-rational points, AUT is the order of the collineation stabilizer, NCPX is the Nauty complexity, TORB is the execution time of our algorithm in Orbiter, and TGAP is the execution time of the canonical image algorithm in GAP. The timings are given in milliseconds. The Nauty complexity is measuring the size of the search tree as follows. Nauty performs a backtrack search over equitable partitions of the graph. The implementation is spread over 4 recursive functions, called:

firstpathnode,
othernode,
processnode and
firstterminal.

The nauty complexity is the cumulative number of times that any one of these four functions is executed. This number measures the size of the search tree, which is somewhat related to the order of the automorphism group.

The practical performance depends on the implementation, which takes into account many other factors, such as speed of the code, efficiency of data structures etc. We observe that the GAP algorithms fails for fields of size 61 or above, indicated by an entry of –1 in the timing column. This was because the least image GAP program yields an error message "reached the pre-set memory limit." This seems to indicate that there is a memory barrier which is limiting the algorithm. For $Q = 59$, which is the largest order of a file for which both algorithms succeed, we see that the Orbiter-Nauty approach is 18 times faster than GAP. Regarding the timing of the algorithms, the Orbiter time includes the total time of the Orbiter run, while the GAP time only measures the time spent in the "least image" command. We wanted to eliminate unrelated startup costs from the measurement of the algorithm. The Orbiter startup time is relatively low, while the startup time for GAP is rather high. This is because Orbiter is a compiled program whereas GAP has a large library of code written in the GAP language, which is read during startup time.

7 Comparison with Classification

We will now compare the general purpose isomorphism test described here with the classification algorithm for Cubic surfaces from [5]. We will consider two cubic surfaces over the field $\mathbb{F}_{11}$. The surfaces are defined using the 4-parameter normal form $\mathcal{F}_{a,b,c,d}$ from [6], where a,b,c,d are field elements. For the sake of completeness, the equation is

$$\begin{aligned}
\mathcal{F}_{a,b,c,d} = &-(abc - abd - acd + bcd + ad - bc)(b - d)X_0^2 X_2 \\
&+(abc - abd - acd + bcd + ad - bc)(a + b - c - d)X_0 X_1 X_2 \\
&+(a^2 c - a^2 d - ac^2 + bc^2 + ad - bc)(b - d)X_0 X_1 X_3 \\
&-(ad - bc)(abc - abd - acd + bcd + ad - bc)X_0 X_2^2 \\
&-(a^2 cd - abc^2 - a^2 d + abd + bc^2 - bcd)(b - d)X_0 X_2 X_3 \\
&-(a - c)(abc - abd - acd + bcd + ad - bc)X_1^2 X_2 \\
&-(a - c)(abc - abd - acd + bcd + ad - bc)X_1^2 X_3 \\
&+(ad - bc)(abc - abd - acd + bcd + ad - bc)X_1 X_2^2 \\
&+(2a^2 bcd - a^2 bd^2 - 2a^2 cd^2 \\
&-2ab^2 c^2 + ab^2 cd + 2abc^2 d + abcd^2 \\
&-b^2 c^2 d - a^2 bc + a^2 cd + a^2 d^2 + ab^2 c + abc^2 \\
&-4abcd - ac^2 d + acd^2 + b^2 c^2)X_1 X_2 X_3 \\
&+ca(ad - bc - a + b + c - d)(b - d)X_1 X_3^2
\end{aligned}$$

We test whether $\mathcal{F}_{2,4,4,2}$ and $\mathcal{F}_{5,10,7,5}$ are isomorphic (they are!). The equations are

$$\mathcal{F}_{2,4,4,2} = X_0^2 X_2 + 10 X_1^2 X_2 + 10 X_1^2 X_3 + 5 X_0 X_2^2 +$$
$$6 X_1 X_2^2 + X_1 X_3^2 + 3 X_0 X_1 X_3 + 6 X_0 X_2 X_3 + 7 X_1 X_2 X_3$$
$$\mathcal{F}_{5,10,7,5} = X_0^2 X_2 + 4 X_1^2 X_2 + 4 X_1^2 X_3 + 2 X_0 X_2^2 +$$
$$9 X_1 X_2^2 + X_1 X_3^2 + 6 X_0 X_1 X_2 + 8 X_0 X_1 X_3 +$$
$$X_0 X_2 X_3 + 10 X_1 X_2 X_3$$

The algorithm from [5] takes about 2.6 s. In this time, the program classifies all cubic surfaces over the field $\mathbb{F}_{11}$ and then finds an isomorphism from the second to the first. The isomorphism is the projectivity associated with the matrix

$$\begin{bmatrix} 0 & 1 & 1 & 6 \\ 0 & 1 & 7 & 3 \\ 1 & 8 & 2 & 4 \\ 4 & 10 & 10 & 9 \end{bmatrix}.$$

Because we need the contragredient action, the matrix is in fact the inverse of the projectivity. By substituting the columns of this matrix into $\mathcal{F}_{2,4,4,2}$, we arrive at $5\mathcal{F}_{5,10,7,5}$:

$$\mathcal{F}_{2,4,4,2}(X_2 + 4X_3, X_0 + X_1 + 8X_2 + 10X_3, X_0 + 7X_1 + 2X_2 + 10X_3,$$
$$6X_0 + 3X_1 + 4X_2 + 9X_3) = 5\mathcal{F}_{5,10,7,5}.$$

Since scalars do not matter, this is the required isomorphism. For comparison, the Nauty based algorithm, using the canonical form of the Levi graph, takes 12.6 s to execute. It does not perform the classification, but it also finds that the two cubic surfaces are isomorphic. The time is dominated by two canonical form computations, one for each surface. By finding that the canonical forms are identical, the fact that the surfaces are isomorphic is established.

For larger surfaces, the algorithm based on Nauty canonical forms is increasingly slower than the algorithm based on Schläfli double sixes. In fact, the graphs get so large that the method becomes rather impractical. This is rather disappointing, but it shows that there is value in working on the classification problem for specific stuctures. Likewise, it is valuable to have standard tools for isomorphism testing that can be applied to wide classes of objects, just like the Nauty-based canonical form algorithm discussed here.

Acknowledgements. The authors thank Sajeeb Roy Chowdhury for his parser of algebraic expressions, which is used in Orbiter to read algebraic equations over finite fields.

References

1. Al-ogaidi, A., Betten, A.: Large arcs in small planes. Congr. Numer. **232**, 119–136 (2019)
2. Bamberg, J., Betten, A., Cara, P., De Beule, J., Lavrauw, M., Neunhöffer, M.: FinInG – Finite Incidence Geometry, Version 1.4.1 (2018)
3. Betten, A.: The orbiter ecosystem for combinatorial data. In: ISSAC'20—Proceedings of the 45th International Symposium on Symbolic and Algebraic Computation, pp. 30–37. ACM, New York (2020)
4. Betten, A.: Orbiter – a program to classify discrete objects (2022). https://github.com/abetten/orbiter
5. Betten, A., Karaoglu, F.: Cubic surfaces over small finite fields. Des. Codes Cryptogr. **87**(4), 931–953 (2019)
6. Betten, A., Karaoglu, F.: The Eckardt point configuration of cubic surfaces revisited. Des. Codes Cryptogr. **90**(9), 2159–2180 (2022)
7. Betten, D., Braun, M.: A tactical decomposition for incidence structures. Ann. Disc. Math. **52**, 37–43 (1992)
8. Bosma, W., Cannon, J., Playoust, C.: The Magma algebra system. I. The user language. J. Symb. Comput. **24**(3-4), 235–265 (1997)
9. Conway, J.H., Curtis, R.T., Norton, S.P., Parker, R.A., Wilson, R.A.: Atlas of Finite Groups. Oxford University Press, Eynsham (1985)
10. Edge, W.L.: A plane quartic with eight undulations. Proc. Edinburgh Math. Soc. **2**(8), 147–162 (1950)
11. The GAP Group. GAP – Groups, Algorithms, and Programming, Version 4.11.1 (2021). http://www.gap-system.org
12. Holt, D.F., Eick, B., O'Brien, E.A.: Handbook of computational group theory. In: Discrete Mathematics and its Applications. Chapman & Hall/CRC, Boca Raton (2005)
13. Jefferson, C., Jonauskyte, E., Pfeiffer, M., Waldecker, R.: Minimal and canonical images. J. Algebra **521**, 481–506 (2019)
14. Jefferson, C., Pfeiffer, M., Waldecker, R., Jonauskyte, E.: GAP package images (2019)
15. Leon, J.S.: Permutation group algorithms based on partitions. I. Theory and algorithms. J. Symb. Comput. **12**(4–5), 533–583 (1991)
16. Leon, J.S.: Partitions, refinements, and permutation group computation. In: Groups and computation, II (New Brunswick, NJ, 1995), vol. 28 of DIMACS Series Discrete Mathematical Theoretical Computer Science, pp. 123–158. American Mathematical Society, Providence (1997)
17. Linton, S.: Finding the smallest image of a set. In: ISSAC 2004, pp. 229–234. ACM, New York (2004)
18. McKay, B.: Nauty and Traces (Version 2.7r1), Australian National University (2020)
19. McKay, B.D.: Isomorph-free exhaustive generation. J. Algor. **26**(2), 306–324 (1998)
20. Moore, E.H.: Tactical Memoranda I–III. Amer. J. Math. **18**(3), 264–290 (1896)
21. Seress, Á.: Permutation group algorithms. In: Cambridge Tracts in Mathematics, vol. 152. Cambridge University Press, Cambridge (2003)

Computer Classification of Linear Codes Based on Lattice Point Enumeration

Sascha Kurz[1,2]

[1] Friedrich-Alexander-University Erlangen-Nuremberg, Erlangen, Germany
sascha.kurz@uni-bayreuth.de
[2] University of Bayreuth, Bayreuth, Germany

Abstract. Linear codes related to applications in Galois Geometry often require a certain divisibility of the occurring weights. In this paper we present an algorithmic framework for the classification of linear codes over finite fields with restricted sets of weights. The underlying algorithms are based on lattice point enumeration and integer linear programming. We present new enumeration and non-existence results for projective two-weight codes, divisible codes, and additive $\mathbb{F}_4$-codes.

Keywords: linear codes · classification · enumeration · lattice point enumeration · integer linear programming · two-weight codes

1 Introduction

A linear code of length n is a k-dimensional linear subspace C of the vector space $\mathbb{F}_q^n$, where $\mathbb{F}_q$ is the finite field with q elements. The vectors in C are called codewords. The weight $\operatorname{wt}(c)$ of a codeword $c = (c_1, \ldots, c_n) \in C$ is the number of nonzero positions $\#\{1 \le i \le n : c_i \ne 0\}$. For two codewords $c', c'' \in C$ the Hamming distance is given by $d(c', c'') := \operatorname{wt}(c' - c'')$. With this, the minimum distance $d(C)$ is given by the minimum occurring Hamming distance between two different codewords, i.e., $d(C) := \min\{d(c', c'') : c', c'' \in C, c' \ne c''\}$. An $[n, k, d]_q$-code is a k-dimensional subspace C of $\mathbb{F}_q^n$ with minimum distance at least d. We also speak of an $[n, k]_q$-code if we do not want to specify the minimum distance. Having applications in error correction in mind, a main problem in coding theory is the maximization of d, k, or $-n$ fixing the other two parameters.

A famous problem in Galois Geometry is the maximum size of a partial t-spread in $\mathbb{F}_q^k$, where a partial t-spread is a set $\mathcal{T}$ of t-dimensional subspaces whose pairwise intersection is the zero vector. To such a set $\mathcal{T}$ we can associate an $[n, k]_q$-code C whose weights are divisible by $\Delta := q^{t-1}$, where $n = (q^k - 1)/(q - 1) - \#\mathcal{T} \cdot (q^t - 1)/(q - 1)$. So, non-existence results for Δ-divisible codes imply upper bounds on the size of partial spreads. Using the fact that the relevant codes have to be projective (defined later) we remark that indeed all currently known upper bounds for partial spreads can be deduced from non-existence results for projective Δ-divisible codes, see e.g [11] for details. Linear codes with few weights have applications in e.g. cryptography, designs, and secret sharing schemes. To

© The Author(s), under exclusive license to Springer Nature Switzerland AG 2024
K. Buzzard et al. (Eds.): ICMS 2024, LNCS 14749, pp. 97–105, 2024.
https://doi.org/10.1007/978-3-031-64529-7_11

sum up, there is a wide interest in the enumeration of linear $[n, k]_q$-codes with certain restrictions on the occurring weights.

Algorithms for the computer classification of linear codes date back at least to 1960 [18], see also [3,5,12,13,15] for more recent literature. Here we want to focus on the approach from [5] using lattice point enumeration algorithms, see e.g. [1,2,16]. We refine some of the algorithmic techniques and apply integer linear programming, see e.g. [17].

The remaining part of this paper is structured as follows. We start introducing the preliminaries in Sect. 2. Since the geometric representation of linear codes as multisets of points in projective spaces plays a major role we briefly introduce this concept in Subsect. 2.1. The main idea of extending linear codes using lattice point enumeration from [5] is briefly outlined in Subsect. 2.2. The extension of each linear code is performed in two phases: In **Phase 1** lattice points of a certain Diophantine equation system are enumerated and in **Phase 2** additional checks are executed in order to reduce the number of extension candidates. In Sect. 3 we state integer linear programming (ILP) formulations for some of those checks from **Phase 2** that are moved to **Phase 0** that is executed prior to **Phase 1**. Since we are interested in the practical performance of our proposed algorithm we discuss computational results in Sect. 4.

2 Preliminaries

A common representation of an $[n, k]_q$-code is as the row-span of a $k \times n$ matrix over $\mathbb{F}_q$ – called the generator matrix. As an example consider the $[56, 6]_3$-code C spanned by

$$\begin{pmatrix} 10000022110100110202111100101201021211112200120020012211 \\ 01000011101210101121120010211222111210000210222200222010 \\ 00100022220221020011200101120020202002111221211222001112 \\ 00010010112222022102002210010101002222100222112122221200 \\ 00001020121022112112001021102211121000021202220212201001 \\ 00000112202002201012122002011020121222212002100202211222 \end{pmatrix}.$$

It is the code of the famous Hill cap [10]. We say that a linear code is Δ-divisible if the weights of all codewords are divisible by Δ and we speak of a t-weight code if t different non-zero weights occur. Our example is a 9-divisible 2-weight code with weights 36 and 45. If a partial 3-spread of $\mathbb{F}_3^8$ of size 248 exists, then the set of uncovered points need to form a Hill cap, see [11] for the details.

2.1 Geometric Representation of Linear Codes

Permuting the columns of the stated generator matrix or multiplying arbitrary columns by arbitrary elements in $\mathbb{F}_q^* := \mathbb{F}_q \backslash \{0\}$ yields an isomorphic linear code. We can factor out those symmetries by using the geometric representation of linear codes as multisets of points in a projective space $\mathrm{PG}(k-1, q)$. Here we just give a brief sketch of the most important facts and refer to e.g. [8] for details. The 1-dimensional subspaces of $\mathbb{F}_q^k$ are the points of $\mathrm{PG}(k-1, q)$ which we denote by $\mathcal{P}_k$. A multiset of points $\mathcal{M}$ in $\mathrm{PG}(k-1, q)$ is a mapping from $\mathcal{P}_k$ to $\mathbb{N}$, i.e., to each point $P \in \mathcal{P}_k$ we assign a point multiplicity $\mathcal{M}(P)$. Starting from a generator matrix of a linear code we obtain a multiset of points by considering the multiset of one-dimensional subspaces spanned by the respective columns. In the other direction, we can take $\mathcal{M}(P)$ arbitrary generators for a point P and place them at arbitrary positions in a generator matrix. Two linear codes are ismorphic iff their associated multisets of points are ismorphic. We call 2-dimensional subspaces lines. Furthermore, $(k-1)$-dimensional subspaces in $\mathrm{PG}(k-1,q)$ are called hyperplanes and their set is denoted by $\mathcal{H}_k$. For each arbitrary non-zero codeword $c \in C$ also $\alpha \cdot c$ is a codeword for all $\alpha \in \mathbb{F}_q^*$ and we can associate $\mathbb{F}_q^* \cdot c$ with a hyperplane $H \in \mathcal{H}_k$. The weight $\mathrm{wt}(c)$ of a codeword c equals $\#\mathcal{M} - \mathcal{M}(H)$ for the associated hyperplane H, where $\#\mathcal{M} := \sum_{P \in \mathcal{P}_k} \mathcal{M}(P)$ and $\mathcal{M}(H) := \sum_{P \in \mathcal{P}_k : P \in H} \mathcal{M}(P)$. The residual code of a non-zero codeword c is the restriction $\mathcal{M}|_H$ of $\mathcal{M}$ to the corresponding hyperplane H. We say that a linear $[n,k]_q$-code C is projective iff we have $\mathcal{M}(P) \in \{0,1\}$ for all $P \in \mathcal{P}_k$ for the corresponding multiset of points $\mathcal{M}$.

2.2 Extending Linear Codes Using Lattice Point Enumeration

We say that a generator matrix $G \in \mathbb{F}_q^{k \times n}$ of an $[n,k]_q$-code C is systematic if it is of the form $G = (I_k | R)$, where I_k is the $k \times k$ unit matrix and $R \in \mathbb{F}_q^{k \times (n-k)}$. Our general strategy to enumerate linear codes is to start from a systematic generator matrix G of a code and to extend G to a systematic generator matrix G' of a "larger" code C', c.f. [5, Section III]. Here we assume the form

$$G' = \begin{pmatrix} I_k & 0 \ldots 0 & R \\ 0 & \underbrace{1 \ldots 1}_{r} & \star \end{pmatrix} \quad (1)$$

where $G = (I_k | R)$ and $r \geq 1$, i.e., C' is an $[n+r, k+1]_q$-code. Let $\mathcal{M}$ and $\mathcal{M}'$ be multisets of points corresponding to C and C', respectively. Geometrically, $\mathcal{M}$ arises from $\mathcal{M}'$ by projection through a point $P \in \mathcal{P}_{k+1}$, i.e., for each line L through P we define $\mathcal{M}(L/P) = \mathcal{M}'(L) - \mathcal{M}'(P) = \sum_{Q \in L : Q \neq P} \mathcal{M}'(Q)$ and use $\mathrm{PG}(k,q)/P \cong \mathrm{PG}(k-1,q)$. Given the assumed shape of G' we have $P = \langle e_{k+1} \rangle$ for the $(k+1)$th unit vector e_{k+1} with $\mathcal{M}'(P) = r$. However, we may choose any point $P \in \mathcal{P}_{k+1}$ with $\mathcal{M}'(P) \geq 1$ to construct $\mathcal{M}$. While the linear code C corresponding to $\mathcal{M}$ may not admit a systematic generator matrix, there always is an isomorphic linear code which does. So, in general there are a lot of extensions ending up in a given code C'. In order to reduce the number of possible paths in [5] the authors speak of "canonical length extension" if

$$\min \{\mathcal{M}'(Q) : \mathcal{M}'(Q) > 0, Q \in \mathcal{P}_{k+1}\} = r \quad (2)$$

is satisfied, c.f. [5, Corollary 9], i.e., the smallest possible value of r is chosen.

The workhorse for the algorithmic approach based on lattice point enumeration is [5, Lemma 7]:

Lemma 1. *Let G be a systematic generator matrix of an $[n,k]_q$ code C whose non-zero weights are contained in $\{i\Delta : a \leq i \leq b\} \subseteq \mathbb{N}_{\geq 1}$. By $c(P)$ we denote* the number of columns of G whose row span equals P for all points $P \in \mathcal{P}_k$ and set $c(\mathbf{0}) = r$ for some integer $r \geq 1$. Let $\mathcal{S}(G)$ be the set of feasible solutions of

$$\Delta y_H + \sum_{P \in \mathcal{P}_{k+1}: P \leq H} x_P = n - a\Delta \forall H \in \mathcal{H}_{k+1} \tag{3}$$

$$\sum_{q \in \mathbb{F}_q} x_{\langle(u|q)\rangle} = c(\langle u \rangle) \forall \langle u \rangle \in \mathcal{P}_k \cup \{\mathbf{0}\} \tag{4}$$

$$x_{\langle e_i \rangle} \geq 1 \forall 1 \leq i \leq k+1 \tag{5}$$

$$x_P \in \mathbb{N} \forall P \in \mathcal{P}_{k+1} \tag{6}$$

$$y_H \in \{0, ..., b-a\} \forall H \in \mathcal{H}_{k+1}, \tag{7}$$

where e_i denotes the ith unit vector in $\mathbb{F}_q^{k+1}$. Then, for every systematic generator matrix G' of an $[n+r, k+1]_q$ code C' whose first k rows coincide with G and whose weights of its non-zero codewords are contained in $\{i\Delta : a \leq i \leq b\}$, we have a solution $(x, y) \in \mathcal{S}(G)$ such that G' has exactly x_P columns whose row span is equal to P for each $P \in \mathcal{P}_{k+1}$.

Our algorithmic strategy is to enumerate all lattice points satisfying constraints (3)-(7) in **Phase 1** and consider them as extension candidates C' for a given linear $[n, k]_q$-code C, where additional checks may be applied, in **Phase 2**. Of course we have to deal with the problem of eliminating isomorphic copies. On the other hand, there are some theoretic insights that allow to directly reject some of the lattice points as candidates in **Phase 2**, see [5] for details.

For our purpose, a few remarks are in order. Equations (3) ensure that C' is Δ-divisible with minimum weight at least $a\Delta$ and maximum weight at most $b\Delta$. Of course we may always choose $\Delta = 1$, but the larger we choose Δ and the tighter we choose a, b the less lattice points will satisfy constraints (3)-(7). Inequalities (4) and (5) model the assumed shape of (1). (Technically, Inequalities (5) are removed in a preprocessing step before invoking software for the enumeration of lattice points.) The variables x_P model $\mathcal{M}'(P)$ and the variables y_H parameterize $\mathcal{M}'(H)$ as detailed in (3). Actually, constraints (6) and (7) are just saying that we are only interested in lattice points, i.e., integral solutions.

Due to the availability of practically fast lattice point enumeration algorithms, the algorithmic strategy to generate many extension candidates in **Phase 1** and filter out suitable candidates afterwards in **Phase 2** turned out to be quite efficient if the number of constraints and variables is not too large, see [5]. It was also observed that considering just a subset of the constraints (3) can reduce computation times in many situations, i.e., generating more candidates in **Phase 1** can pay off if a simpler system allows faster generation of lattice points and the checks in **Phase 2** can be implemented efficiently.

In the implementation described in [5], the check of condition (2) as well as checks based on possible gaps in the assumed weight spectrum $\{i\Delta : a \leq i \leq b\}$ are moved to **Phase 2**. The idea of this paper is to demonstrate that it sometimes can pay off to move such checks to integer linear programming computations in a **Phase 0** prior to **Phase 1**.

3 Enhancing the Algorithmic Approach via Integer Linear Programming Computations

Of course one can check the feasibility or infeasibility of an ILP using lattice point enumeration, after possibly transforming inequalities into equalities. In the other direction, many ILP solvers can also enumerate all lattice points of a polytope. However, in many situations ILP solvers can find a single feasible solution or show infeasibility faster than the full enumeration by a lattice point enumeration algorithm. However, the situation changes if one wants to enumerate a larger set of solutions exhaustively.

So, a first idea is to check feasibility of constraints (3)-(7) using an ILP solver in a Phase 0 prior to Phase 1. This pays off in those situations where some extension problems don't have a solution. We remark that several calls of ILP solvers for different random target functions can also be used in a heuristic approach if it is not necessary to classify all possible codes with certain parameters but just to find some examples.

The second idea is to move some of the checks of Phase 2 into an initial feasibility test based on ILP computations. Starting with Inequality (2) we observe that it can be rewritten as $x_P = 0 \ \lor \ x_P \geq r$ for every point $P \in \mathcal{P}_{k+1}$. Having an upper bound $x_P \leq \Lambda_P$ at hand, which we have in most applications, we can linearize to

$$x_P \leq \Lambda_P \cdot u_P \quad \land \quad x_P \geq r \cdot u_P \tag{8}$$

using an additional binary variable u_P.

Instead of solving a larger ILP some cases can also be eliminated in a simple preprocessing step. Suppose that an instance of Inequality (4) reads $\sum_{i=1}^{q} x_{P_i} = c$ and we have $x_{P_i} \leq \Lambda$ as well as $x_{P_i} = 0 \ \lor \ x_{P_i} \geq r$ for all $1 \leq i \leq r$. If $\lfloor \frac{c}{r} \rfloor < \lceil \frac{c}{\Lambda} \rceil$, then no solution exists since at most $\lfloor \frac{c}{r} \rfloor$ variables x_{P_i} have to be non-zero and at least $\lceil \frac{c}{\Lambda} \rceil$ variables x_{P_i} have to be non-zero. We have observed the applicability of this criterion in practice for parameters $(q, r, \Lambda, c) = (2, 3, 4, 5)$.

With respect to gaps in the weight spectrum we assume that the possible non-zero weights are contained in

$$\{a_1 \Delta_1, \ldots, b_1 \Delta_1, a_2 \Delta_2, \ldots, b_2 \Delta_2, \ldots, a_l \Delta_l, \ldots, b_l \Delta_l\}, \tag{9}$$

where $l \geq 2$, $a_i < b_i$ for $1 \leq i \leq l$, and $b_{i-1}\Delta_{i-1} < a_i \Delta_i$ for $2 \leq i \leq l$. With this, we can replace Inequalities (3) and Inequalities (7) by

$$\sum_{i=1}^{l} \Delta_i y_H^i + \sum_{P \in \mathcal{P}_{k+1} : P \leq H} x_P = n - \sum_{i=1}^{l} a_i \Delta_i z_H^i \Delta \forall H \in \mathcal{H}_{k+1}, \tag{10}$$

$$y_H^i \leq (b_i - a_i) z_H^i \forall H \in \mathcal{H}_{k+1}, \forall 1 \leq i \leq l, \tag{11}$$

$$\sum_{i=1}^{l} z_H^i = 1 \forall H \in \mathcal{H}_{k+1}, \tag{12}$$

$$z_H^i \in \{0, 1\} \forall H \in \mathcal{H}_{k+1}, \forall 1 \leq i \leq l, \tag{13}$$

$$y_H^i \in \mathbb{N} \forall H \in \mathcal{H}_{k+1}, \forall 1 \leq i \leq l. \tag{14}$$

The ILP for **Phase** 0 consists of inequalities (4)-(6) and (10)-(14) if we have a possible weight spectrum as in (9) with $l \geq 2$ or, alternatively, of inequalities (3)-(7). If $r \geq 2$, see (1), then we additionally add the constraints (8) and additional variables $u_P \in \{0, 1\}$ for all $P \in \mathcal{P}_{k+1}$.

4 Computational Results

In this section we want to present a few computational results that have been obtained with our refined algorithmic approach of computer classification of linear codes based on lattice point enumeration and integer linear programming. We used a customary laptop and non-free software packages for ILP computations (IBM ILOG CPLEX 12.4) as well as lattice point enumeration (solvediophant [20]). Computation times are below a few hours in all cases.

We start with non-existence results for projective two-weight codes, see e.g. [6,7] for surveys. To this end, we slightly modify the notion of an $[n, k, d]_q$-code by replacing d with a set of occurring non-zero weights and writing $\leq n$ if the length is at most n.

Proposition 1. *No projective* $[66, 5, \{48, 56\}]_4$*-code exists.*

Proof. By exhaustive enumeration we have determined all three non-isomorphic $[\leq 65, 3, \{48, 56\}]_4$-codes. They have lengths 63, 64, 65 and orders 362880, 1728, 36 of their automorphism groups, respectively. None of them can be extended to an $[65, 4, \{48, 56\}]_4$-code, so that no projective $[66, 5, \{48, 56\}]_4$-code exists. □

Proposition 2. *No projective* $[35, 4, \{28, 32\}]_8$*-code exists.*

Proof. We have computationally determined the unique $[\leq 34, 3, \{28, 32\}]_8$-code. It has length 34 and an automorphism group of order 43008. We have computationally checked that no extension exists. The corresponding ILP computation took 5.6 h of computation time and checked 1 633 887 B&B-nodes. □

We remark that we have also enumerated all $[\leq 122, 3, \{108, 117\}]_9$-codes (with maximum point multiplicity 9). All of these 1147 non-isomorphic codes have length $n = 122$. It is an interesting open question whether one of these can be extended to a projective $[123, 4, \{108, 117\}]_9$-code, which is currently unknown. Applying the so-called subfield construction, see e.g. [7], would also yield a projective $[492, 8, \{324, 351\}]_3$-code – again currently unknown.

Next we consider projective Δ-divisible codes.

Proposition 3. *No projective 5-divisible* $[40, 4]_5$*-code exists.*

Proof. By exhaustive enumeration we have determined all 371 non-isomorphic 5-divisible $[39, 3]_5$-codes with maximum point multiplicity 4. None of them can be extended to a projective 5-divisible $[40, 4]_5$-code. Note that extending a $[39, 3]_5$-code with maximum point multiplicity to a projective $[40, 4]_5$-code would imply that the resulting code contains a two-dimensional simplex code, in geometrical terms a line, in its support. However, no projective 5-divisible $[34, k]_5$-code exists [14, Lemma 7.12]. □

As a last example we consider additive codes over $\mathbb{F}_4$, i.e. $\mathbb{F}_2$-linear subspaces of $\mathbb{F}_4^k$ where $k \in \mathbb{N}/2$. In general, each k-dimensional additive code of length n with minimum Hamming distance d over $\mathbb{F}_{q^2}$ geometrically corresponds to a multiset of lines in $\mathrm{PG}(2k-1, q)$ such that each hyperplane contains at most $n - d$ lines, see e.g. [4]. For $\mathbb{F}_4$ the case $k = 3.5$ was partially studied in [9]. Our aim is to construct examples for $(n, d) = (51, 38)$ which are $[153, 7, 76]_2$-codes.

Lemma 2. *Let $\mathcal{M}$ be the multiset of points in $\mathrm{PG}(6, 2)$ formed by 51 lines such that every hyperplane contains at most 13 lines and C denote the corresponding binary code. Then, C is $[153, 7, 76]_2$-code with $\mathrm{wt}(c) \le 102$ for every $c \in C$.*

Proof. Replacing the 51 lines by their 3 points yields a multiset $\mathcal{M}$ in $\mathrm{PG}(6, 2)$ with cardinality 153. Since each hyperplane H contains between 0 and 13 lines, we have $51 \le \mathcal{M}(H) \le 77$, so that $76 \le \mathrm{wt}(c) \le 102$ for all $c \in C \backslash \{\mathbf{0}\}$. □

Lemma 3. *No projective 2-divisible $[65, 6, 32]_2$-code with maximum point multiplicity exists.*

Proof. By exhaustive enumeration. □

Lemma 4. *Let C be a $[153, 7, 76]_2$-code with $\mathrm{wt}(c) \le 102$ for every $c \in C$. Then, the occurring non-zero weights are contained in $\{76, 80, 92, 96, 100\}$.*

Proof. Note that C is a Griesmer code and that 4 divides the minimum distance 76, so that we can conclude that C is 4-divisible [19]. The residual code of a codeword of weight 84 would be a $[69, 6, 34]_2$-code which does not exist. Now let C' be the residual code of a codeword of weight 88, so that C' is a $[65, 6, 32]_2$-code. Since C is 4-divisible and has maximum point multiplicity 2, C' has to be 2-divisible with maximum point multiplicity 2. However, we have just shown that such a code does not exist. □

Proposition 4. *There are exactly two non-isomorphic $[153, 7, 76]_2$-codes C with $\mathrm{wt}(c) \le 102$ for every $c \in C$.*

Proof. We enumerated all $[151, 6]_2$-codes with weight restrictions as in Lemma 4. Noting that C has maximum point multiplicity 2, it has to be an extension of such a $[151, 6]_2$-code and we have computationally verified the stated results. Those two codes have weight distributions $76^{107} 80^{15} 92^5$ and $76^{108} 80^{14} 92^4 96^1$. □

A generator matrix of the $[153, 7, \{76, 80, 92\}]_2$-code is given by the concatenation of

$$\begin{pmatrix} 111111111111111111111111111111111111111111111111111111111111111111111111111111111111111111111111111111111111111111111111111111111111111111111111111111111 \\ 000000000000000000000000000000000000000000000001111111111111111111111111111111111111111111111111111111111 \\ 000000000000000000000000000001111111111111111111111111000000000000000000000000001111111111111111111111111 \\ 000000000000111111110000000000000000011111111100000000000000011111111000000000000001111111100000000000011111 \\ 000001111110000111100000001111110000000111100000001111000001111000000011110000000111100000011110000001111100001 \\ 001110011001100110011000110011001100110011001100110011001100110011001100110011001100110011001100110011001100110 \\ 010010010101101010100011011011001010101110110010110110110101001011001001010000 \end{pmatrix}$$

and

$$\begin{pmatrix}11100000000000000000000000000000000000000000000000000000000010000000\\11111111111111111111111111111111111110000000000000000000000001000000\\11100000000000000000011111111111111111111111111111111110000000000100000\\11100000011111111111100000000011111111111100000001111111111111000010000\\11100011110000001111110000111100000011111100011110000001111110001111000100010\\01101100110001110001100110011000110001101100110001100011011001110000100\\10110100010010010110010101101011011001010010101010100100100110101111100000011\end{pmatrix}.$$

The corresponding multisets of points can be partitioned into 51 lines.

Disclosure Statement. The authors have no competing interests to declare that are relevant to the content of this article.

References

1. Aardal, K., Hurkens, C.A., Lenstra, A.K.: Solving a system of linear Diophantine equations with lower and upper bounds on the variables. Math. Oper. Res. **25**(3), 427–442 (2000)
2. Aardal, K., Wolsey, L.A.: Lattice based extended formulations for integer linear equality systems. Math. Program. **121**(2), 337–352 (2010)
3. Betten, A., Braun, M., Fripertinger, H., Kerber, A., Kohnert, A., Wassermann, A.: Error-correcting linear codes: Classification by isometry and applications. In: Algorithms and Computation in Mathematics, (Vol. 18), Springer (2006)
4. Blokhuis, A., Brouwer, A.E.: Small additive quaternary codes. Eur. J. Comb. **25**(2), 161–167 (2004)
5. Bouyukliev, I., Bouyuklieva, S., Kurz, S.: Computer classification of linear codes. IEEE Trans. Inf. Theory **67**(12), 7807–7814 (2021)
6. Brouwer, A.E.: Two-weight codes. In: Concise Encyclopedia of Coding Theory, pp. 449–462. Chapman and Hall/CRC (2021)
7. Calderbank, R., Kantor, W.M.: The geometry of two-weight codes. Bull. Lond. Math. Soc. **18**(2), 97–122 (1986)
8. Dodunekov, S., Simonis, J.: Codes and projective multisets. Electron. J. Comb. **5**, 1–23 (1998)
9. Guan, C., Li, R., Liu, Y., Ma, Z.: Some quaternary additive codes outperform linear counterparts. IEEE Trans. Inf. Theory **69**(11), 7122–7131 (2023)
10. Hill, R.: On the largest size of cap in $S_{5,3}$. Atti della Accademia Nazionale dei Lincei. Classe di Scienze Fisiche, Matematiche e Naturali. Rendiconti **54**(3), 378–384 (1973)
11. Honold, T., Kiermaier, M., Kurz, S.: Partial spreads and vector space partitions. In: Greferath, M., Pavčević, M.O., Silberstein, N., Vázquez-Castro, M.Á. (eds.) Network Coding and Subspace Designs, pp. 131–170. Springer International Publishing, Cham (2018). https://doi.org/10.1007/978-3-319-70293-3_7
12. Jaffe, D.B.: Optimal binary linear codes of length ≤ 30. Discret. Math. **223**(1–3), 135–155 (2000)
13. Kaski, P., Östergård, P.R.: Classification algorithms for codes and designs. In: Algorithms and Computation in Mathematics, (Vol. 15) Springer (2006)
14. Kurz, S.: Divisible codes. arXiv preprint **2112**, 11763 (2021)
15. Östergård, P.R.: Classifying subspaces of Hamming spaces. Des. Codes Crypt. **27**, 297–305 (2002)
16. Schnorr, C.P., Euchner, M.: Lattice basis reduction: Improved practical algorithms and solving subset sum problems. Math. Program. **66**, 181–199 (1994)

17. Schrijver, A.: Theory of linear and integer programming. John Wiley & Sons (1998)
18. Slepian, D.: Some further theory of group codes. Bell Syst. Tech. J. **39**(5), 1219–1252 (1960)
19. Ward, H.N.: Divisibility of codes meeting the Griesmer bound. J. Combinat. Theory, Series A **83**(1), 79–93 (1998)
20. Wassermann, A.: Attacking the market split problem with lattice point enumeration. J. Comb. Optim. **6**, 5–16 (2002)

Software for Proper Vertex-Colouring Exploiting Graph Symmetry

Leonard H. Soicher

School of Mathematical Sciences, Queen Mary University of London, Mile End Road, London E1 4NS, UK
L.H.Soicher@qmul.ac.uk

Abstract. We describe the methods used in the GAP package GRAPE for proper vertex-colouring a graph, including the determination of a minimum vertex-colouring and hence the chromatic number. These methods are designed to exploit the automorphism group of the graph.

Keywords: proper vertex-colouring · minimum vertex-colouring · chromatic number · graph symmetry · GRAPE · GAP

1 Introduction

The GAP system [2] is a freely available open-source computer system for algebra and discrete mathematics, with an emphasis on computational group theory. See [3] for a tutorial introduction to GAP.

The GRAPE package [12] for GAP provides extensive functionality for graphs, and is designed primarily for applications in algebraic graph theory, permutation group theory, design theory, and finite geometry. See [13] for a tutorial introduction to GRAPE, including the use of much of the functionality described in this article.

In GRAPE, a graph *gamma* always comes together with an associated group *gamma*.group of automorphisms. This group is set (automatically or by the user) when the graph is constructed, and is used by GRAPE to store the graph compactly and to speed up computations with the graph. Often, but not always, this group is the full automorphism group of the graph.

GRAPE includes functionality for constructing graphs, determining their regularity properties, and classifying their cliques. GRAPE also provides seamless interfaces to both the nauty [11] and bliss [7] computer packages for computing the automorphism group of a graph and testing graph isomorphism.

GRAPE now also has machinery for properly vertex-colouring a graph, which exploits the automorphism group of that graph. This functionality includes the calculation of a minimum vertex-colouring and hence the determination of the chromatic number of the graph. We shall describe the main ideas and methods used for this functionality. We expect these to be of broader interest and application.

Our proper vertex-colouring software in GRAPE is meant to be of practical use, especially for graphs with large automorphism groups. This software was used to compute many of the chromatic numbers given in [1, Chapter 10] for specific interesting strongly regular graphs. We present two further examples in the last section.

Throughout this article, all graphs are *simple*, meaning they are finite, undirected, and have no loops and no multiple edges.

2 Proper Vertex-Colouring and Cliques

Let Γ be a graph. A *proper vertex-colouring* of Γ is a labelling of its vertices by elements from a set of *colours*, such that adjacent vertices are labelled with different colours. Where k is a non-negative integer, a *vertex k-colouring* of Γ is a proper vertex-colouring using at most k colours. A *minimum vertex-colouring* of Γ is a vertex k-colouring with k as small as possible, and the *chromatic number* $\chi(\Gamma)$ of Γ is the number of colours used in a minimum vertex-colouring of Γ.

The problem of whether a given graph has a vertex k-colouring for a given k is a well-known NP-complete problem. Indeed, the problem is still NP-complete even for fixed $k = 3$ [4, Chapter 8]. Moreover, the problem of determining whether a graph has a vertex k-colouring for a given k appears to be very difficult in practice. As users of GRAPE are usually interested in graphs with non-trivial, and often large, automorphism groups, it is important to be able to exploit any symmetry a graph may have when determining a vertex k-colouring for a given k or showing that no such vertex-colouring exists.

A *clique* of Γ is a set of pairwise adjacent vertices. Where t is a non-negative integer, a *t-clique* is a clique of size t. A *maximal clique* of Γ is a clique which is contained in no larger clique, while a *maximum clique* of Γ is a t-clique with t as large as possible. The *clique number* $\omega(\Gamma)$ of Γ is the number of vertices in a maximum clique of Γ. Clearly, $\omega(\Gamma) \leq \chi(\Gamma)$.

The problem of whether a given graph has a t-clique for a given t is a well-known NP-complete problem [4, Chapter 8]. However, this problem appears to be less difficult in practice than determining whether the graph has a vertex k-colouring for a given k. GRAPE contains powerful machinery for clique classification and for the determination of a maximum clique.

Now note that, up to the naming of the colours, the vertex k-colourings of Γ are in one-to-one correspondence with the partitions of the vertex set of the complement $\bar{\Gamma}$ of Γ into at most k cliques of $\bar{\Gamma}$ (the parts in such a partition are the colour classes of a vertex k-colouring of Γ). Also note that if g is an automorphism of Γ (and hence of $\bar{\Gamma}$) and $\mathcal{C}$ is an ordered partition of $V(\bar{\Gamma})$ into m cliques, then the g-image of $\mathcal{C}$ is also an ordered partition of $V(\bar{\Gamma})$ into m cliques.

The first step in GRAPE to try to find a vertex k-colouring of Γ is to perform a relatively inexpensive heuristic proper vertex-colouring (see [9]), but if this does not result in a vertex k-colouring, then a backtrack search is performed to find the "least" (defined later) ordered partition $(C_1, C_2, \ldots, C_m)$ of the vertices of

$\bar{\Gamma}$ into cliques, such that $m \leq k$. This search will either find a vertex k-colouring of Γ or prove that such a colouring does not exist.

The method used at present in GRAPE for the determination of a minimum vertex-colouring of Γ (and hence $\chi(\Gamma)$) is a binary search for the least k for which a vertex k-colouring of Γ exists, together with the determination of a vertex k-colouring for this least k. To start with, a lower bound for k can be taken to be $\omega(\Gamma)$ or $\lceil |V(\Gamma)|/\omega(\bar{\Gamma}) \rceil$ and an upper bound for k can be obtained by a proper vertex-colouring heuristic, many of which are described in [9], the simplest of which is "greedy colouring".

In the GRAPE package for GAP, the user functions for proper vertex-colouring are VertexColouring (for vertex k-colouring), MinimumVertexColouring, and ChromaticNumber. In GRAPE, a proper vertex-colouring of a graph is output as a sequence of positive integers indexed by the vertices of the graph, with the i-th element of the sequence being the colour of vertex i.

3 The Main Tools

We now describe the three main tools from GRAPE that we use in our vertex k-colouring algorithm.

The first tool is used to classify the maximal cliques of given size in a graph, up to the action of a given group of automorphisms of that graph. In GRAPE, where *gamma* is a (simple) graph and t is a non-negative integer, the function call

CompleteSubgraphsOfGivenSize(*gamma*, t, 2, true)

returns a set of *gamma*.group orbit-representatives of all the maximal cliques of size t in *gamma*.

The second tool is Steve Linton's function SmallestImageSet, which is included in GRAPE. Where G is a permutation group on $X := \{1, \ldots, n\}$ and S is a subset of X, the function call

SmallestImageSet(G, S)

returns the lexicographically least set in the G-orbit of S with respect to the natural action of G on subsets of X, without explicitly computing this (possibly huge) orbit. We use the SmallestImageSet function to determine the lexicographically least clique in a group orbit, given an arbitrary clique in that orbit.

The algorithm for SmallestImageSet is given in [10]. Further developments in the computation of minimal and canonical images with respect to a group action are given in [5,6].

Our third tool again employs CompleteSubgraphsOfGivenSize, this time for the calculation of exact set covers, exploiting symmetry, as detailed in Fig. 1. See the GRAPE manual [12] for full documentation of the very flexible function CompleteSubgraphsOfGivenSize, as well as other GRAPE functions used.

```
ExactSetCover := function(G,blocks,n)
#
# Let  n  be a positive integer, let  G  be a permutation group on
# [1..n],   let   blocks   be a set of non-empty subsets of  [1..n],
# and suppose  S  is the union of the G-orbits of the sets in  blocks.
# Then this function returns an exact set cover of  [1..n]  by elements
# from  S,  if such a cover exists, and returns     'fail'   otherwise.
#
local gamma,i,j,wts,K;
gamma:=Graph(G,blocks,OnSets,
    function(x,y) return Intersection(x,y)=[]; end);
wts:=[];
for i in [1..OrderGraph(gamma)] do
    wts[i]:=ListWithIdenticalEntries(n,0);
    for j in VertexName(gamma,i) do
        wts[i][j]:=1;
    od;
od;
K:=CompleteSubgraphsOfGivenSize(gamma,
    ListWithIdenticalEntries(n,1),0,true,true,wts);
if K=[] then
    return fail;
else
    return Set(VertexNames(gamma){K[1]});
fi;
end;
```

Fig. 1. Exact set cover using GRAPE

4 A Total Ordering of Finite Sequences of Subsets of $\{1, \ldots, n\}$

Let n be a non-negative integer, and let $A := \{a_1, \ldots, a_r\}$ and $B := \{b_1, \ldots, b_s\}$ be subsets of $\{1, \ldots, n\}$, with $a_1 < \cdots < a_r$ and $b_1 < \cdots < b_s$. We define

$$A \preceq B$$

to mean either $r > s$, or $r = s$ and $(a_1, \ldots, a_r) \leq (b_1, \ldots, b_r)$ in lexicographic order (w.r.t. the usual $\leq$ on the integers). For example, $\{3,5,6\} \preceq \{2,3\}$, but $\{3,4,7\} \preceq \{3,5,6\}$.

Let n be a non-negative integer and let $\mathcal{A} := (A_1, \ldots, A_t)$ and $\mathcal{B} := (B_1, \ldots, B_u)$ be finite sequences of subsets of $\{1, \ldots, n\}$. We define

$$\mathcal{A} \preceq \mathcal{B}$$

to mean that $(A_1, \ldots, A_t)$ is less than or equal to $(B_1, \ldots, B_u)$ in lexicographic order, with respect to the order $\preceq$ on subsets of $\{1, \ldots, n\}$. For example,

$$(\{3,4,7\}, \{3,5,6\}, \{7,8\}) \preceq (\{3,4,7\}, \{2,3\}) \preceq (\{3,4,7\}, \{2,3\}, \{1,2,3,4\}).$$

5 The Backtrack Search

Now let Γ be a graph with non-empty vertex set $V(\Gamma) := \{1,\ldots,n\}$, and let Δ be the complement graph $\bar{\Gamma}$ of Γ, such that the vertices of Δ can be partitioned into at most k cliques of Δ, for a given positive integer k (that is, Γ has a vertex k-colouring). Then there is a unique least ordered such partition $(C_1,\ldots,C_m)$ with respect to $\preceq$, and we shall consider some properties of this least ordered partition. These properties give us very useful constraints on partial solutions for our backtrack search, which either proves that this least ordered partition does not exist or finds a vertex k-colouring of Γ (which need not correspond to $(C_1,\ldots,C_m)$).

We follow the general structure of backtrack search as described in [8, Section 4.1.2]. Full details of our backtrack search can be found in the open-source code of GRAPE [12]. The degree of difficulty for this backtrack search depends heavily on k and the clique structure of Δ.

5.1 Constraints on Partial Solutions

Let $(C_1,\ldots,C_m)$ be the least ordered partition of $V(\Delta)$ into cliques of Δ, with respect to $\preceq$, such that $m \leq k$. Let $\Delta_1 := \Delta$, let $G_1 := \mathrm{Aut}(\Delta)$, and for $i = 2,\ldots,m$, let Δ_i be the subgraph of Δ_{i-1} induced on $V(\Delta_{i-1}) \setminus C_{i-1}$ and let G_i be the image of the action on $V(\Delta_i)$ of the (setwise) stabilizer in G_{i-1} of C_{i-1} (G_i is the group we associate to Δ_i in GRAPE). Then it is easy to see that the following must hold:

- C_i is a maximal clique of Δ_i.
- C_i is the lexicographically least set in its G_i-orbit.
- If $i > 1$ then $C_{i-1} \prec C_i$.
- $(k-i+1)|C_i| \geq |V(\Delta_i)|$.

Given $(C_1,\ldots,C_{i-1})$, with $1 \leq i \leq m$, the possible C_i satisfying these properties can be generated (in increasing order w.r.t. $\preceq$) making use of the functions CompleteSubgraphsOfGivenSize and SmallestImageSet. Of course, if we are given $(C_1,\ldots,C_m)$ then we return this solution and stop.

We now note the following useful result.

Lemma 1. *Let $1 \leq j < i \leq m$, let D be a clique of Δ_j containing C_i and let $g \in G_j$. Then $C_j \preceq D^g$, the image of D under g.*

Proof. Suppose $D^g \prec C_j$. Then we could make an ordered partition $(D_1,\ldots,D_m)$ of $V(\Delta)$ into cliques, such that $(D_1,\ldots,D_m) \prec (C_1,\ldots,C_m)$, as follows. For $1 \leq \ell \leq j-1$, $D_\ell := C_\ell$; $D_j := D^g$; $D_i := (C_j \setminus D)^g$; for $j+1 \leq \ell \leq m$, $\ell \neq i$, $D_\ell := (C_\ell \setminus D)^g$. □

Now, for $1 \leq j < i \leq m$, let $C_{i,j}$ be the largest clique that can be obtained by adding (zero or more) elements of C_j to C_i. Then, applying Lemma 1, we obtain the following further constraints that we apply on partial solutions in our backtrack search.

- If $i > 1$ then the least G_{i-1}-image of $C_{i,i-1}$ is $\succeq C_{i-1}$.
- Suppose $(k - i + 1)|C_i| = |V(\Delta_i)|$. Then $m = k$ and $\{C_i, C_{i+1}, \ldots, C_k\}$ is a partition of $V(\Delta_i)$ into maximal cliques of Δ_i, each of size $|C_i|$.
  Now additionally suppose that G_i is small, say G_i has order at most 24 (so that G_i-orbits of cliques are also small). Then, after we classify the maximal cliques of size $|C_i|$ in Δ_i, up to the action of G_i, we determine which G_i-orbits of these cliques could possibly contain elements of $\{C_i, \ldots, C_k\}$. We use the requirement that if $i > 1$, then for $\ell = i, \ldots, k$, the least G_{i-1} image of $C_{\ell,i-1}$ is $\succeq C_{i-1}$ (note that this requirement need only be checked for one representative from each G_i-orbit). We then make use of (an inline version of) the function ExactSetCover either to complete the partial solution $(C_1, \ldots, C_{i-1})$ to an ordered partition of $V(\Delta)$ into k cliques or to show that no least (w.r.t. $\preceq$) such completion exists.

6 Examples

In the example given in Fig. 2, we load the GRAPE package (suppressing the banner), construct the M_{23}-graph on 253 vertices (see [1, Section 10.56]), and determine some of its properties. In particular, we show that this graph has chromatic number 15, a fact which appears to have been previously unknown. The whole calculation took about 15 min of CPU-time on an i5 laptop, with the calculation of the chromatic number taking almost all of the time.

```
gap> LoadPackage("grape",false);
true
gap> G:=PrimitiveGroup(253,5);
M(23)
gap> M23graph:=First(GeneralizedOrbitalGraphs(G,1),
>     x->VertexDegrees(x)=[112]);;
gap> GlobalParameters(M23graph);
[ [ 0, 0, 112 ], [ 1, 36, 75 ], [ 60, 52, 0 ] ]
gap> A:=AutomorphismGroup(M23graph);;
gap> Size(A);
10200960
gap> RankAction(A,[1..253]);
3
gap> CliqueNumber(M23graph);
4
gap> CliqueNumber(ComplementGraph(M23graph));
21
gap> ChromaticNumber(M23graph);
15
```

Fig. 2. Example calculation

Similarly, we have constructed the M_{22}-graph on 176 vertices (see [1, Section 10.51]), and have determined that this graph has chromatic number 12, which also appears to have been previously unknown. This calculation took about one-half minute of CPU-time on an i5 laptop.

References

1. Brouwer, A.E., Van Maldeghem, H.: Strongly Regular Graphs. Cambridge University Press, Cambridge (2022)
2. The GAP Group: GAP — Groups, Algorithms, and Programming. Version 4.13.0 (2024). https://www.gap-system.org
3. The GAP Group: GAP — A Tutorial. Release 4.13.0 (2024). https://www.gap-system.org/Manuals/doc/tut/manual.pdf
4. Gibbons, A.: Algorithmic Graph Theory. Cambridge University Press, Cambridge (1985)
5. Jefferson, C., Jonauskyte, E., Pfeiffer, M., Waldecker, R.: Minimal and canonical images. J. Algebra **521**, 481–506 (2019)
6. Jefferson, C., Pfeiffer, M., Waldecker, R., Jonauskyte, E.: The images package for GAP, minimal and canonical images. Version 1.3.2 (2024). https://gap-packages.github.io/images/
7. Junttila, T., Kaski, P.: Engineering an efficient canonical labeling tool for large and sparse graphs. In: Applegate, D. et al. (eds.) Proceedings of the Ninth Workshop on Algorithm Engineering and Experiments and the Fourth Workshop on Analytic Algorithmics and Combinatorics, pp. 135–149. SIAM, Philadelphia (2007). bliss homepage: http://www.tcs.hut.fi/Software/bliss/
8. Kaski, P., Östergård, P.R.J.: Classification Algorithms for Codes and Designs. Springer, Berlin (2006)
9. Lewis, R.M.R.: A Guide to Graph Colouring: Algorithms and Applications, 2nd edn. Springer International Publishing, Switzerland (2021)
10. Linton, S.: Finding the smallest image of a set. In: J. Gutierrez (ed.) ISSAC '04: Proceedings of the 2004 International Symposium on Symbolic and Algebraic Computation, pp. 229–234. ACM Press, New York (2004)
11. McKay, B. D., Piperno, A.: Practical graph isomorphism, II. J. Symbol. Comput. **60**, 94–112 (2014). nauty and Traces homepage: https://pallini.di.uniroma1.it
12. Soicher, L.H.: The GRAPE package for GAP. Version 4.9.0 (2022). https://gap-packages.github.io/grape/
13. Soicher, L.H.: Using GAP packages for research in graph theory, design theory, and finite geometry. In: Ivanov, A.A. (ed.) Algebraic Combinatorics and the Monster Group, London Mathematical Society Lecture Note Series 487, pp. 527–566. Cambridge University Press, Cambridge (2024)

Classical Algebraic Geometry
and Modern Computer Algebra:
Innovative Software Design and Its
Applications

Localization in Gromov—Witten Theory of Toric Varieties in a Computer Algebra System

Giosuè Muratore[✉]

CMAFcIO, Faculdade de Ciências da ULisboa, Campo Grande 1749-016
Lisbon, Portugal
muratore.g.e@gmail.com
https://sites.google.com/view/giosue-muratore

Abstract. The Atiyah–Bott localization formula is a powerful tool for calculating the degree of equivariant classes of the moduli space of rational stable maps $\overline{M}_{0,m}(X,\beta)$, where X denotes a smooth toric variety, m is a non-negative integer, and β is an effective 1-cycle. Implementation of the formula entails intricate computational challenges, involving graph theory, colorings, partitions, and other discrete objects. Furthermore, the computed solution is a large summation of rational numbers, underscoring the imperative nature of computational efficiency. This formula has been applied in very specific cases for computing Gromov–Witten invariants, addressing enumerative problems, and determining the small quantum ring of X, among other applications. A comprehensive implementation as a Julia package has been recently presented by the author. We show the features of the package with a particular emphasis to the noteworthy contribution of the package `Oscar.jl`. Finally, we delve into the fundamental prerequisites for extending the implementation to encompass algebraic GKM manifolds.

Keywords: Localization · Toric variety · stable map

1 Introduction

One of the most studied object in Algebraic Geometry is the moduli space $\overline{M}_{g,m}(X,\beta)$ of genus g, m-marked stable maps of class β and target X.

A Gromov-Witten invariant is the degree of some 0-cycle of $\overline{M}_{g,m}(X,\beta)$, and often a solution of an enumerative problem reduces to the computation of such invariants. In order to enumerate rational curves in the quintic threefold, Kontsevich applied the Atiyah–Bott formula to $\overline{M}_{0,m}(\mathbb{P}^n,\beta)$. His approach has been extended to other varieties X and also to positive genus.

The author, motivated by the enumeration of rational contact curves in $\mathbb{P}^{2n+1}$ (see [15,16]), created with Csaba Schneider a package to compute any

Gromov–Witten invariant of projective spaces in [18], using Kontsevich's approach. In [17], he created the package `ToricAtiyahBott.jl` for similar computations for smooth projective toric varieties using `Oscar.jl` [3,19]. In the present paper, we show more details of this last implementation, which is available at:

https://github.com/mgemath/ToricAtiyahBott.jl

Finally, we discuss a possible extension to GKM manifolds.

2 Atiyah–Bott Formula

A toric variety X is an algebraic variety containing a torus $T = (\mathbb{C}^*)^n$ as a dense open subset, in such a way that the action of T on itself extends to the whole X. There exists a correspondence between fans of cones in $\mathbb{Z}^n$ and normal separated toric varieties of dimension n, see [5].

Example 1. The following list includes some of the most common toric varieties:

- projective spaces $\mathbb{P}^n$,
- projectivization $\mathbb{P}(E)$ where E is the direct sum of line bundles on a toric variety,
- product of toric varieties,
- blow-up of a toric variety along a torus-invariant subvariety.

From now on, for any toric variety X with normal fan Σ, we use the following notation.

- $\Sigma^{(k)}$ is the subset of k-dimensional cones of Σ. So, $\Sigma^{(n)}$ are the maximal cones and they are usually denoted by σ.
- For each $\sigma \in \Sigma^{(n)}$, we denote by σ^* the set of all $\sigma' \in \Sigma^{(n)}$ such that $\sigma \cap \sigma' \in \Sigma^{(n-1)}$.
- $\Sigma^{(1)} = \{\rho_1, \ldots, \rho_r\}$ are the 1-dimensional cones (rays) of Σ.
- For each cone $\tau \in \Sigma^{(k)}$, $V(\tau)$ is the T-invariant subvariety of dimension $(n-k)$ defined by τ.

We denote by $\overline{M}_{0,m}(X, \beta)$ the coarse moduli space of genus 0 stable maps to X of class β with m marked points. When $m > 0$, $\mathrm{ev}_j \colon \overline{M}_{0,m}(X, \beta) \to X$ denotes the evaluation map at the jth marked point. This is the space of all tuples $(C, f, p_1, \ldots, p_m)$ where:

1. C is a connected projective complex scheme of dimension 1 and arithmetic genus 0,
2. p_i are smooth points of C called marks,
3. $f \colon C \to X$ is a morphism such that $f_*([C]) = \beta \in H_2(X, \mathbb{Z})$,
4. If $E \subseteq C$ is an irreducible component that is contracted by f, then E contains at least three points among marks and singular points.

We denote by ψ_i the first Chern class of the line bundle whose fiber at $(C, f, p_1, \ldots, p_m)$ is the cotangent bundle of C at p_i. If Ω_X denotes the cotangent bundle of X, the virtual dimension of $\overline{M}_{0,m}(X, \beta)$ is

$$\mathrm{virdim}(\overline{M}_{0,m}(X, \beta)) = \dim(X) - \deg(c_1(\Omega_X) \cdot \beta) + m - 3. \tag{1}$$

Given a vector bundle $\mathcal{E}$ on $\overline{M}_{0,m}(X, \beta)$ of rank r, classes $\gamma_1, \ldots, \gamma_m \in H^*(X, \mathbb{C})$, and non-negative integers $a_1, \ldots, a_m$, we define a Gromov–Witten invariant to be the following degree of a cohomology class of maximal codimension:

$$\int_{\overline{M}_{0,m}(X,\beta)} \mathrm{ev}_1^*(\gamma_1) \cdots \mathrm{ev}_m^*(\gamma_m) \cdot \psi_1^{a_1} \cdots \psi_m^{a_m} \cdot c_r(\mathcal{E}). \tag{2}$$

The action of T on X lifts to $\overline{M}_{0,m}(X, \beta)$ by composition: Any $t \in T$ is a map $t \colon X \to X$ that we compose with $f \colon C \to X$, hence

$$t \cdot (C, f, p_1, \ldots, p_m) = (C, t \circ f, p_1, \ldots, p_m).$$

The fixed point locus of this new action is the disjoint union of subvarieties of $\overline{M}_{0,m}(X, \beta)$ parameterized by all possible decorated graphs of $\overline{M}_{0,m}(X, \beta)$.

Definition 1. *A decorated graph of $\overline{M}_{0,m}(X, \beta)$ is the isomorphism class of the tuple $\Gamma = (g, \mathbf{c}, \mathbf{w}, \nu)$, such that*

1. *g is a tree; that is g is a simple undirected, connected graph without cycles. We denote by V_g and E_g the sets of vertices and edges of g.*
2. *The vertices of g are colored by the maximal cones of Σ; the coloring is given by a map $\mathbf{c} \colon V_g \to \Sigma^{(n)}$ such that: if v and v' are two vertices in the same edge, then $\mathbf{c}(v) \cap \mathbf{c}(v') \in \Sigma^{(n-1)}$.*
3. *The edges are weighted by a map $\mathbf{w} \colon E_g \to \mathbb{Z}$, such that if e is an edge with vertices (v_e, v_e'), then $\mathbf{w}(e) > 0$ and $\sum_{e \in E_g} \mathbf{w}(e) \cdot [V(\mathbf{c}(v_e) \cap \mathbf{c}(v_e'))] = \beta$.*
4. *The marks of the vertices of g are given by a map $\nu \colon A \to V_g$, where the set A is $A = \{1, \ldots, m\}$ if $m > 0$, and $A = \emptyset$ if $m = 0$.*

An automorphism of Γ is an automorphism of g compatible with $\mathbf{c}, \mathbf{w}, \nu$. We denote by a_Γ the integer

$$a_\Gamma := |\mathrm{Aut}(\Gamma)| \prod_{e \in E_g} \mathbf{w}(e),$$

and by F_Γ the fixed point locus relative to Γ.

Following [2], given a cohomology class $\alpha \in H^*(\overline{M}_{0,m}(X, \beta), \mathbb{C})$, we denote by $\alpha^T(\Gamma)$ the T-equivariant cohomology class induced by α. The following is known as the Atiyah–Bott localization formula (see [8,10,11,20]).

Theorem 1. *Let P be a symmetric polynomial in Chern classes of equivariant vector bundles of $\overline{M}_{0,m}(X, \beta)$. Then*

$$\int_{\overline{M}_{0,m}(X,\beta)} P = \sum_\Gamma \frac{1}{a_\Gamma} \frac{P^T(\Gamma)}{c_{\mathrm{top}}^T(N_\Gamma)(\Gamma)}, \tag{3}$$

where N_Γ is the normal bundle of F_Γ.

3 The Algorithm

3.1 Generation of the Decorated Graphs

In this section we describe how we generate the decorated graphs of Definition 1. We used the algorithm of [21] in order to generate all trees. Roughly speaking, it works in the following way. Let g be a tree, and z be a center. That is, z is a vertex such that the following equation is satisfied

$$\max_{v \in V_g}\{d(z,v)\} = \min_{u \in V_g}\{\max_{v \in V_g}\{d(u,v)\}\},$$

where $d(u, v)$ is the distance between the vertices u and v. Note that any tree has either one or two centers. Let us consider a bijective map $l\colon V_g \to \{1,\ldots,|V_g|\}$ such that

1. $l(z) = 1$,
2. if w is a descendant of u, then $l(u) < l(w)$,
3. if two vertices u, v are children of the same vertex, $l(u)$ is less than $l(v)$, and w is a descendant of u, then $l(u) < l(w) < l(v)$.

We call a level sequence of g with respect to some map l the sequence of integers

$$[s_1, s_2, \ldots, s_{|V_g|}],$$

such that $s_i = 1 + d(u, z)$ where $l(u) = i$. Among all possible level sequences, we call canonical the greatest in lexicographic order. See Fig. 1 for an example.

Fig. 1. A tree with two different labellings and relative level sequences. The labelling on the left defines the canonical one.

Two trees are isomorphic if and only if their canonical level sequences are the same. The algorithm we implemented generates all canonical level sequences for each fixed number of vertices. Working with level sequences simplified the implementation of an algorithm to compute the colorings **c**, the marks ν (both modulo isomorphism), and the number of automorphisms of a tree g.

Indeed, if g has one center, all automorphisms fix the vertex z. Moreover, any automorphism permutes two subtrees of g if and only if they produce the same integers in the canonical level sequence. If g has two centers, we may have automorphisms that do not fix z, but the number of automorphisms is still easy to compute.

So, we generate all trees, with relative colorings and marks, having at most p edges. The number p corresponds to the maximal number of T-invariant curves $\{C_i\}_{i=1}^p$ such that

$$\beta = [C_1] + [C_2] + \cdots + [C_p].$$

Using the fact that the Mori cone is generated by T-invariant curves and it is dual to the cone of nef divisors [5, §6.3], we have that $p = \beta \cdot (M_1 + \cdots + M_k)$ where M_i are the classes of the generators of the nef cone.

Finally, in order to generate the functions $\mathbf{w} \colon E_g \to \mathbb{Z}$, we proceed in the following way. Let us fix a tree g and a coloring $\mathbf{c} \colon V_g \to \Sigma^{(n)}$. By definition, for any map $\mathbf{w} \colon E_g \to \mathbb{Z}$ we have

$$\beta = \sum_{e \in E_g} \mathbf{w}(e) \cdot [V(\mathbf{c}(v_e) \cap \mathbf{c}(v'_e))]. \tag{4}$$

Thus, if $\bar{e} \in E_g$ is a fixed edge, since $\mathbf{w}(e) > 0$ we have that

$$\beta - \mathbf{w}(\bar{e}) \cdot [V(\mathbf{c}(v_{\bar{e}}) \cap \mathbf{c}(v'_{\bar{e}}))] - \sum_{e \in E_g \setminus \{\bar{e}\}} [V(\mathbf{c}(v_e) \cap \mathbf{c}(v'_e))] \tag{5}$$

is an effective 1-cycle. Thus, an upper bound for $\mathbf{w}(\bar{e})$ is given by the number $b_{\bar{e}}$ that is the maximal number that, in place of $\mathbf{w}(\bar{e})$, makes (5) an effective cycle. The package `Oscar.jl` provides all divisors that span the extremal rays of the nef cone, so (5) is effective if and only if it intersects non-negatively those nef divisors [5, Theorem 6.3.20]. Hence we can find $b_{\bar{e}}$. Finally, in order to find all maps $\mathbf{w}$, we adopt the following strategy: First we generate all possible functions $\mathbf{w}$ such that $1 \leq \mathbf{w}(e) \leq b_e$ for all $e \in E_g$, then we keep those satisfying Equation (4).

In this way, we generate all decorated graphs.

3.2 Equivariant Classes

The equivariant cycles are implemented as Julia functions that take a decorated graph as input and return a rational number. In order to implement (3), we sum all these rational numbers to a variable `ans`. Eventually, the value of `ans` will be the value of the desired Gromow–Witten invariant. Consider the following.

Lemma 1. *Let $\sigma_1, \sigma_2 \in \Sigma^{(n)}$ be such that $\tau = \sigma_1 \cap \sigma_2 \in \Sigma^{(n-1)}$.*

Let $\{\rho_{i_1}, \ldots, \rho_{i_{n-1}}\}$ be the rays of τ. Moreover, let ρ' be the ray of σ_1 that is not in τ. Fix $\omega_1, \ldots, \omega_r$ weights of the action of $(\mathbb{C}^*)^r$ on $\mathbb{C}^r$. The induced $\mathbb{C}^*$-action on the subvariety $V(\tau)$ has weight $\omega_{\sigma_2}^{\sigma_1}$ at the point $V(\sigma_1)$ given by:

$$\omega_{\sigma_2}^{\sigma_1} := \sum_{i=1}^{r} \langle \rho_i, u_n \rangle \omega_i$$

where $\{u_1, \ldots, u_n\}$ is the $\mathbb{Z}$-dual basis of $\{\rho_{i_1}, \ldots, \rho_{i_{n-1}}, \rho'\} \subset \mathbb{Z}^r$.

Let us give an example of a equivariant class. For any maximal cone $\sigma \in \Sigma^{(n)}$, for any ray $\rho \in \Sigma^{(1)}$, and for any non-negative integer k, we define the following

$$\lambda(\sigma, \rho, k) := \begin{cases} 0 & \text{if } \rho \text{ is not a ray of } \sigma, \text{ and } k > 0, \\ 1 & \text{if } \rho \text{ is not a ray of } \sigma, \text{ and } k = 0, \\ (\omega_\gamma^\sigma)^k & \text{if } \gamma \in \sigma^*, \text{ and } \rho \text{ is a ray of } \sigma \text{ but not of } \gamma. \end{cases}$$

The cone γ with the property of the last case exists and it is unique, so the definition makes sense. Let Z be the following cohomology class:

$$Z = [V(\rho_1)]^{k_1} \cdot [V(\rho_2)]^{k_2} \cdots [V(\rho_r)]^{k_r}. \tag{6}$$

For any maximal cone σ, let us define

$$\Lambda(\sigma, Z) := \prod_{j=1}^{r} \lambda(\sigma, \rho_j, k_j). \tag{7}$$

Since the cohomology ring of X is generated by invariant divisors [5, Chapter 12], we may extend (7) by linearity to any homogeneous class $Z \in H^*(X, \mathbb{Q})$.

For any $i = 1, \ldots, m$ the contribution of $\mathrm{ev}_i^*(Z)$ is

$$\mathrm{ev}_i^*(Z)^T(\Gamma) = \Lambda(\mathbf{c}(\nu(i)), Z). \tag{8}$$

Let M be a line bundle of X. We extend (7) and (8) to M by taking $Z = c_1(M)$. We refer to [17] for the equivariant classes of $\psi_1, \ldots, \psi_n$ and $c_{\mathrm{top}}^T(N_\Gamma)$.

Example 2. Consider the normal fan of $\mathbb{P}^1$, generated by the rays $\{\rho_1, \rho_2\} \subset \mathbb{Z}^1$, with maximal cones $\sigma_1 = \langle \rho_1 \rangle$ and $\sigma_2 = \langle \rho_2 \rangle$. Let $Z = [V(\rho_1)]$, and β be the class of a line. Using Theorem 1, let us compute

$$\int_{\overline{M}_{0,1}(\mathbb{P}^1, \beta)} \mathrm{ev}_1^*(Z). \tag{9}$$

Let g be the path tree with only one edge $e = (v_e, v_e')$. Let $\mathbf{c}(v_e) = \sigma_1$, $\mathbf{c}(v_e') = \sigma_2$, and $\mathbf{w}(e) = 1$. Finally, let $\nu_1(1) = v_e$, and $\nu_2(1) = v_e'$. There are only two decorated graphs: $\Gamma_1 = (g, \mathbf{c}, \mathbf{w}, \nu_1)$ and $\Gamma_2 = (g, \mathbf{c}, \mathbf{w}, \nu_2)$. By Equation (8),

$$\mathrm{ev}_1^*(Z)^T(\Gamma_1) = \Lambda(\mathbf{c}(\nu_1(1)), Z)$$
$$= \Lambda(\sigma_1, Z)$$
$$= \lambda(\sigma_1, \rho_1, 1) \cdot \lambda(\sigma_1, \rho_2, 0)$$
$$= \omega_{\sigma_2}^{\sigma_1} \cdot 1.$$

On the other hand, $\mathrm{ev}_1^*(Z)^T(\Gamma_2) = 0$ because $\lambda(\sigma_2, \rho_1, 1) = 0$. Now, we need to find $c_{\mathrm{top}}^T(N_{\Gamma_1})$. Using the notation of [17, Equation (3.4)], and considering that there are no ψ-classes in Equation (9), we have

$$\Delta(e) := \frac{(-1)}{(\omega_{\sigma_2}^{\sigma_1})^2}$$

$$\Xi(\Gamma_1) := \prod_{v \in V_g} \left(\prod_{\gamma \in \mathbf{c}(v)^*} \omega_\gamma^{\mathbf{c}(v)} \right)^{|F_v|-1} \Delta(e)$$

$$\frac{1}{c_{\mathrm{top}}^T(N_{\Gamma_1})(\Gamma_1)} = \Xi(\Gamma_1) \prod_{v \in V_g} \prod_{F \in F_v} \omega_F^{-1} \left(\sum_{F \in F_v} \omega_F^{-1} \right)^{|F_v|+|S_v|-3}.$$

F_v denotes the set of flags of v, and $|S_v|$ is the number of marks of v. If F is the flag (v_e, e) (resp., (v'_e, e)), then $w_F = w_{\sigma_2}^{\sigma_1}$ (resp., $w_F = w_{\sigma_1}^{\sigma_2}$). Hence

$$\frac{1}{c_{\text{top}}^T(N_{\Gamma_1})(\Gamma_1)} = \frac{(-1)}{(w_{\sigma_2}^{\sigma_1})^2}(w_{\sigma_2}^{\sigma_1})^{-1}((w_{\sigma_2}^{\sigma_1})^{-1})^{2-3}(w_{\sigma_1}^{\sigma_2})^{-1}((w_{\sigma_1}^{\sigma_2})^{-1})^{1-3}$$

$$= \frac{(-1)}{(w_{\sigma_2}^{\sigma_1})^2} w_{\sigma_1}^{\sigma_2}.$$

It is easy to see that $w_{\sigma_2}^{\sigma_1} = -w_{\sigma_1}^{\sigma_2} = w_1 - w_2$. Finally

$$\int_{\overline{M}_{0,1}(\mathbb{P}^1,\beta)} \text{ev}_1^*(Z) = \frac{1}{a_{\Gamma_1}} \frac{\text{ev}_1^*(Z)^T(\Gamma_1)}{c_{\text{top}}^T(N_{\Gamma_1})(\Gamma_1)} + \frac{1}{a_{\Gamma_2}} \frac{\text{ev}_1^*(Z)^T(\Gamma_2)}{c_{\text{top}}^T(N_{\Gamma_2})(\Gamma_2)}$$

$$= \frac{1}{1} \frac{w_1 - w_2}{w_1 - w_2} + \frac{1}{a_{\Gamma_2}} \frac{0}{c_{\text{top}}^T(N_{\Gamma_2})(\Gamma_2)}$$

$$= 1.$$

Let us give two examples of computations of Gromov–Witten invariants using our code. Let $\pi \colon X \to \mathbb{P}^3$ be the blow-up at two toric invariant points. The GW invariants of X can be used to count the number of curves in $\mathbb{P}^3$ with conditions on the singularities of two points, see [6, Example 8.3]. Let us denote by H the pull-back of a general line, and by E_1 and E_2 two lines inside two different exceptional divisors. Let [pt] be the cohomological class of a point. If $\beta = 2H - E_1 - E_2$, we expect that

$$\int_{\overline{M}_{0,3}(X,\beta)} \text{ev}_1^*(H)\text{ev}_2^*(H)\text{ev}_3^*([\text{pt}])$$

equals the number of conics in $\mathbb{P}^3$ through 3 points and two lines. This can be obtained in the following way.

```
P3 = projective_space(NormalToricVariety, 3);# the space P^3.
Y = domain(blow_up(P3, [1,1,1]; coordinate_name="Ex1"));
X = domain(blow_up(Y, [-1,0,0]; coordinate_name="Ex2"));
mg = moment_graph(X);
(H, E1, E2) = (mg[7,8], mg[4,5], mg[1,2]);
(d, e1, e2) = (2, -1, -1);
beta = d*H + e1*E1 + e2*E2;
P = ev(1, H)*ev(2, H)*ev(3, a_point(X));
IntegrateAB(X, beta, 3, P);
Result: 1
```

The second example finds the number of lines in $\mathbb{P}^3$ passing through a general point of a quintic surface with multiplicity 3. It is computed by the following invariant

$$\int_{\overline{M}_{0,1}(\mathbb{P}^3,1)} \frac{1}{5} \left(\text{ev}_1^*(5h)^3 + 3\text{ev}_1^*(5h)^2 \psi_1 + 2\text{ev}_1^*(5h)\psi_1^2 \right) \text{ev}_1^*(h)^2 = 2,$$

where h is the class of a line, see [4, 13, 14] and references therein. We compute it using the following code.

```
P3 = projective_space(NormalToricVariety, 3);
h = cohomology_class(toric_line_bundle(P3, [1]));
l = 5*h;
P = 1//5*(ev(1,1)^3+3*ev(1,1)^2*Psi(1)+2*ev(1,1)*Psi(2))*ev(1,h)^2
IntegrateAB(P3, h^2, l, P);
Result: 2
```

4 Generalization

A GKM manifold X is a smooth algebraic variety equipped with an algebraic action of a complex algebraic torus with only finitely many fixed points and finitely many 1-dimensional orbits. Such varieties have been introduced in [7].

A toric variety X is GKM as the number of fixed points is $|\Sigma^{(n)}|$, and the number of 1-orbits is $|\Sigma^{(n-1)}|$.

The Grassmannian variety $G(k,n)$ of k-planes in $\mathbb{C}^n$ is defined as a quotient of the group of complex invertible matrices of order n. The subgroup of diagonal matrices is a torus $(\mathbb{C}^*)^n$ acting on $G(k,n)$. It can be shown that it has a finite number of fixed points and 1-orbits. Thus $G(k,n)$ is GKM, but it is a toric variety if and only if $k=1$ or $k=n-1$ by [5, Exercise 7.3.10].

Theorem 1 is still valid when X is a GKM manifold. If X is projective, and β the cohomology class of a curve, then the fixed loci of the action on $\overline{M}_{0,m}(X,\beta)$ are in bijective correspondence with the same decorated graphs of Definition 1. In this case, vertices are colored by the fixed points of the action.

Edges $e=(v_e,v'_e)$ are in correspondences with 1-orbits, which are smooth rational curves passing through two fixed points. Explicit combinatorial formulas for computing Gromov–Witten invariants are known, see [12].

In order to extend the algorithm to GKM manifolds, one needs to compute the weights of the action of the torus on a 1-orbit (that is, a generalization of Lemma 1). Moreover, one needs also to know the generators of the nef cone and their intersection with the invariant curves. These methods are not available in Oscar.jl yet.

The Atiyah–Bott formula has been implemented numerous times. For instance, alongside the previously mentioned [10,20], the papers [1,9] compute interesting enumerative invariants. Additionally, there exists an unpublished implementation in Julia for 0-marked rational maps to $\mathbb{P}^n$ created by Jieao Song (personal communication, August 28, 2021). To the author's knowledge, all published implementations are tailored exclusively for a few specific computations. The author developed the packages in [17,18] to provide tools that are both easy to use and effective across a broad spectrum of cases. While the Atiyah–Bott formula may not always be the most efficient method for computing a GW invariant, it is generally the easiest to apply. We intend to enhance these packages by expanding the list of supported target varieties to GKM manifolds and reducing computation time.

Disclosure of Interests. The author has no competing interests to declare that are relevant to the content of this article.

Acknowledgement. I thank Csaba Schneider for his contribution to the package, and Martin Bies, Lars Kastner and Matthias Zach for their support with `Oscar.jl`. I also thank the reviewers for their diligent examination of the article.

This project is supported by FCT - Fundação para a Ciência e a Tecnologia, under the project: UIDP/04561/2020 (https://doi.org/10.54499/UIDP/04561/2020). The author is a member of GNSAGA (INdAM).

References

1. Amorim, É.: Curvas de contato no espaço projetivo. Ph.D. thesis, UFMG (2014), in English: *Contact curves in the projective space*, arXiv:1907.03973
2. Atiyah, M.F., Bott, R.: The moment map and equivariant cohomology. Topology **23**(1), 1–28 (1984). https://doi.org/10.1016/0040-9383(84)90021-1
3. Bies, M., Kastner, L.: Toric Geometry in OSCAR (2023). https://doi.org/10.48550/arXiv.2303.08110
4. Cieliebak, K., Mohnke, K.: Punctured holomorphic curves and Lagrangian embeddings. Invent. Math. **212**(1), 213–295 (2018). https://doi.org/10.1007/s00222-017-0767-8
5. Cox, D.A., Little, J.B., Schenck, H.K.: Toric varieties, Graduate Studies in Mathematics, vol. 124. American Mathematical Society, Providence, RI (2011). https://doi.org/10.1090/gsm/124
6. Gathmann, A.: Gromov-Witten invariants of blow-ups. J. Algebraic Geom. **10**(3), 399–432 (2001)
7. Goresky, M., Kottwitz, R., MacPherson, R.: Equivariant cohomology, Koszul duality, and the localization theorem. Invent. Math. **131**(1), 25–83 (1998). https://doi.org/10.1007/s002220050197
8. Graber, T., Pandharipande, R.: Localization of virtual classes. Invent. Math. **135**(2), 487–518 (1999). https://doi.org/10.1007/s002220050293
9. Hiep, D.T.: Rational curves on Calabi-Yau threefolds: verifying mirror symmetry predictions. J. Symbolic Comput. **76**, 65–83 (2016). https://doi.org/10.1016/j.jsc.2015.12.003
10. Kontsevich, M.: Enumeration of rational curves via torus actions. In: The moduli space of curves (Texel Island, 1994), Progr. Math., vol. 129, pp. 335–368. Birkhäuser Boston, Boston, MA (1995). https://doi.org/10.1007/978-1-4612-4264-2_12
11. Liu, C.C.M.: Localization in Gromov–Witten theory and orbifold Gromov–Witten theory. In: Handbook of moduli. Vol. II, Adv. Lect. Math. (ALM), vol. 25, pp. 353–425. Int. Press, Somerville, MA (2013)
12. Liu, C.C.M., Sheshmani, A.: Equivariant Gromov–Witten invariants of algebraic GKM manifolds. SIGMA Symmetry Integrability Geom. Methods Appl. **13**, Paper No. 048, 21 (2017). https://doi.org/10.3842/SIGMA.2017.048
13. McDuff, D., Siegel, K.: Counting curves with local tangency constraints. J. Topol. **14**(4), 1176–1242 (2021). https://doi.org/10.1112/topo.12204
14. Muratore, G.: A recursive formula for osculating curves. Ark. Mat. **59**(1), 195–211 (2021). https://doi.org/10.4310/arkiv.2021.v59.n1.a7
15. Muratore, G.: Enumeration of Rational Contact Curves via Torus Actions. Michigan Math. J. **73**(4), 875–894 (2023). https://doi.org/10.1307/mmj/20216025
16. Muratore, G.: Irreducible contact curves via graph stratification. Bull. Sci. Math. **186**, Paper No. 103273 (2023). https://doi.org/10.1016/j.bulsci.2023.103273

17. Muratore, G.: Computations of Gromov-Witten invariants of toric varieties. J. Symbolic Comput. **125**, 102330 (2024). https://doi.org/10.1016/j.jsc.2024.102330
18. Muratore, G., Schneider, C.: Effective computations of the Atiyah-Bott formula. J. Symbolic Comput. **112**, 164–181 (2022). https://doi.org/10.1016/j.jsc.2022.01.005
19. OSCAR – Open Source Computer Algebra Research system, Version 1.0.2 (2024), https://www.oscar-system.org
20. Spielberg, H.: A formula for the Gromov–Witten invariants of toric varieties. Ph.D. thesis, Université Louis Pasteur, Département de Mathématique, Institut de Recherche Mathématique Avancée, Strasbourg (1999)
21. Wright, R.A., Richmond, B., Odlyzko, A., McKay, B.D.: Constant time generation of free trees. SIAM J. Comput. **15**(2), 540–548 (1986). https://doi.org/10.1137/0215039

Advancing Computer Algebra
with Massively Parallel Methods

Massively Parallel Methods for Free Resolutions

Santosh Gnawali[✉]

Fachbereich Mathematik, RPTU, 67653 Kaiserslautern, Germany
gewalis55@gmail.com

Abstract. This paper describes work towards an approach to using massively parallel methods for computing syzygies and free resolutions of finitely generated modules over polynomial rings over fields. Our primary focus here is Schreyer's resolution. Our method exploits the inherent parallelism of the algorithm, primarily utilizing Petri nets, within the GPI-SPACE [10] framework as our language for parallel workflows. GPI-SPACE is a task-based workflow management system that employs Petri nets as its coordination layer, while the computation is carried out by the computer algebra system SINGULAR [9]. We outline how the algorithm is modeled through a Petri net, explaining the coordination of tasks and data structures within the parallel computing environment.

Keywords: Free resolutions · Schreyer's resolution · Petri net · massively parallel computations · Singular/GPI-Space framework · Singular

1 Introduction

In computational algebraic geometry, efficiently computing free resolutions of ideals and modules is crucial for many tasks, e.g., the computation of Ext modules or Betti numbers [8]. An algorithm for computing resolutions, starting from a Gröbner basis of a submodule of a free module, for example, of the image of the presentation matrix $F_1 \xrightarrow{\varphi_1} F_0 \to M \to 0$ of a finitely presented module, and hence of M, was given by Schreyer [8]. If M is a graded module over the polynomial ring $R = K[x_1, \ldots, x_n]$ with its standard grading, then there exists a minimal free resolution which is uniquely determined up to isomorphism [6, Chapter 20.1]. Non-minimal-free resolutions are more feasible and cheaper to compute. Despite their non-uniqueness, they are often sufficient for computational purposes, such as the computation of minimal Betti numbers. We propose a novel massively parallel method building on Schreyer's refined algorithm **liftTree** for computing syzygies using the SINGULAR/GPI-SPACE Framework. The **liftTree** algorithm is based on an idea from [13], which allows it to ignore so-called lower order terms in the computation of the generators of the syzygy module. The paper is structured as follows: In Sect. 2 we introduce the fundamental terminology, Subsect. 2.1 covers monomial orderings of free modules and syzygies, while Subsect. 2.2 recalls Schreyer's Theorem and the corresponding algorithm; Sect. 3 introduces Petri nets as a graphical coordination language, and how they integrate with the workflow management system GPI-SPACE; finally, in Sect. 4 we outline the Petri net model for our strategy for computing syzygies, and give an example.

2 Gröbner Bases and Syzygies

A Gröbner basis is a collection of multivariate polynomials that extends two well-known techniques: Gaussian elimination for solving linear systems and the Euclidean algorithm for finding the greatest common divisor of univariate polynomials; see [7] [4,12]. Let K be an arbitrary field, and let $R = K[x_1, \ldots, x_n]$ denote the polynomial ring in n variables over K. We denote by $\mathrm{Mon}(x_1, \ldots, x_n)$ the semigroup of monomials in R. A term order on R is called a global monomial ordering if it is a total order $>$ on the set of all monomials $x^\alpha := x^{\alpha_1} \cdots x^{\alpha_n}$ which has the following two properties: It respects multiplication, i.e., $x^\alpha > x^\beta$ implies $x^{\alpha+\gamma} > x^{\beta+\gamma}$ for all $\alpha, \beta, \gamma \in \mathbb{N}^n$, and $x^\alpha > 1$ for all $\alpha \neq (0, \ldots, 0)$. If I is an ideal of R, then a finite subset $G := \{f_1, \ldots, f_r\}$ of I with $0 \notin G$ is a Gröbner basis if $L(G) := \langle L(f_1), \ldots, L(f_r) \rangle = L(I)$, where $L(G)$ is the leading ideal generated by the lead monomials with respect to given monomial ordering $>$. It is well known that every ideal of R admits a Gröbner basis with respect to a global monomial ordering.

2.1 Finitely Generated Modules and Their Syzygies

These notions are extended to free R-modules as follows. Let F be a free R-module of rank r and let $\{e_i\}_{i=1,\ldots,r}$ be a basis of F. A monomial in F is the product of an element in $\mathrm{Mon}(x_1, \ldots, x_n)$ with a basis element e_i. A term in F is the product of a monomial in F with a scalar in K. A monomial ordering $>$ on F is a total ordering $>$ on $\mathrm{Mon}(F)$, such that if $m_1 e_i$ and $m_2 e_j$ are monomials in F and m is a monomial in R, then $m_1 e_i > m_2 e_j$ implies $(m \cdot m_1) e_i > (m \cdot m_2) e_j$.

Any $f \in F - \{0\}$ has a unique decomposition $f = cme_i + f^*$, where: $c \in K^\times$, $me_i \in \mathrm{Mon}(F)$, and $me_i > m^* e_j$ for any non-zero term $c^* m^* e_j$ of f^*. With this decomposition, we define the leading monomial, the leading coefficient, the leading term, and the tail of f as $\mathrm{LM}(f) := me_i$, $\mathrm{LC}(f) := c$, $\mathrm{LT}(f) := cme_i$, $\mathrm{tail}(f) := f - \mathrm{LT}(f)$, respectively. For any subset $G \subset F$, we call $L(G) := \langle \mathrm{LM}(f) | f \in G - \{0\} \rangle_R \subset F$ the leading module of G. The notion of a Gröbner basis is then extended accordingly for submodules of F. Let M be a R-module, let $G := \{f_1, \ldots, f_r\}$ be a finite subset of M and let $F := R^r$ be a free R-module of rank r. Consider the homomorphism $\psi_G : F \to M, e_i \mapsto f_i$. The (first) syzygy module of G is defined as $\mathrm{Syz}(G) := \ker \psi_G$. An element of $\mathrm{Syz}(G)$ is called a syzygy (of G).

Theorem 1. *(Hilbert Syzygy Theorem) [6, Corollary 19.8] Let M be a finitely generated module over the polynomial ring $R = K[x_0, \ldots, x_n]$. Then M has a finite free resolution*

$$0 \longrightarrow F_l \xrightarrow{\varphi_l} F_{l-1} \xrightarrow{\varphi_{l-1}} \cdots \longrightarrow F_1 \xrightarrow{\varphi_1} F_0 \xrightarrow{\varphi_0} M \longrightarrow 0,$$

where $l \leq n$.

2.2 Induced Ordering and Schreyer's Algorithms

Let $F_0 := R^s$, and let $G := \{f_1, \ldots, f_r\} \subset F_0 - \{0\}$, and let $F_1 := R^r$ with ψ_G as before. Following [8], from G and a monomial ordering $>$ on F_0 we define the induced ordering $\succ$ on F_1 as follows: $m_1 e_i \succ m_2 e_j$ if and only if $\mathrm{LM}(m_1 f_i) > \mathrm{LM}(m_2 f_j)$ or ($\mathrm{LM}(m_1 f_i) = \mathrm{LM}(m_2 f_j)$ and $i > j$). Denote $m_{ji} := \frac{\mathrm{LCM}(\mathrm{LM}(f_j), \mathrm{LM}(f_i))}{\mathrm{LT}(f_i)}$. With this terminology, we formulate at Schreyer's Theorem.

Theorem 2 (Schreyer). *[6, Theorem 15.10] Let $G = \{f_1, \ldots, f_r\} \subset F_0 := R^s$ be a Gröbner basis w.r.t. a monomial ordering $>$ on F_0. For each pair (f_i, f_j) with $i, j \in \{1, \ldots, r\}$, let $S(f_i, f_j) = m_{ji} f_i - m_{ij} f_j = g_1^{(ij)} f_1 + \ldots + g_r^{(ij)} f_r$ be a standard representation of the corresponding S-vector. Then the relations $m_{ji} e_i - m_{ij} e_j - (g_1^{(ij)} e_1 + \ldots + g_r^{(ij)} e_r) \in F_1 := R^r$ form a Gröbner basis of $\mathrm{Syz}(G)$ w.r.t. the monomial ordering on F_1 induced by $>$ and G. In particular, these relations generate $\mathrm{Syz}(G)$.*

By means of this theorem, we can compute a free resolution of a finitely generated free module as in Theorem 1. To refine this algorithm, it is necessary to introduce the notion of lifting.

Definition 1. *[8] Let F_0, G and F_1 be as above, and let $s \in \mathrm{L}_\succ(\mathrm{Syz}(G)) \subset F_1$ be a leading syzygy term. We call $\bar{s} \in F_1$ a lifting of s w.r.t. G and $\succ$, if the following conditions hold:*

(1) $\mathrm{LT}_\succ(\bar{s}) = s$, and
(2) $\bar{s} \in \mathrm{Syz}(G)$.

The refined Schreyer algorithm from [8] is described in Algorithm 1 and requires computation of these liftings. The computation of liftings is described in Algorithms 2 and 3. Lower order terms can be ignored in the computation. Following [8], we define LOT as follows: Let $S \subset F_0$ be a set of vectors and let $t \in F_0$ be a term. Then t is called a lower order term (LOT) w.r.t. S if $\mathrm{LM}(f) \nmid t$ for all $f \in S - \{0\}$. For an element $g \in F_0$, we define $\mathrm{LOT}(g|S)$ to be the sum of those terms occurring in g which are of lower order w.r.t. S. We apply the massive parallelism in two instances: First, in Algorithm 1 for the different leading syzygies; and second, in Algorithm 2 for the iterative computation of the liftings, which amounts to expanding a tree.

3 SINGULAR/GPI-SPACE Framework and Petri Nets

The SINGULAR/GPI-SPACE Framework utilizes GPI-SPACE, a workflow management system from Fraunhofer ITWM, designed for the automated parallel

Algorithm 1 SyzLift

Input: A Gröbner basis $G = \{f_1, \ldots, f_r\} \subset F_0 := R^s$ w.r.t. $>$ and an algorithm Lift to compute for a leading syzygy term $s \in L_\succ(G)$, a lifting w.r.t. G and $\succ$
Output: A Gröbner basis of $\text{Syz}(G) \subset F_1 := R^r$ w.r.t. $\succ$
1: $\mathcal{L} := \text{LeadSyz}(G)$
2: **for** $s \in \mathcal{L}$ **do**
3:     $\bar{s} := \text{Lift}(s)$
4:     $S := S \cup \{s\}$
5: **return** S

Algorithm 2 LiftTree

Input: A Gröbner basis $G = \{f_1, \ldots, f_r\} \subset F_0 := R^s$ w.r.t. $>$ and a leading syzygy term $s \in L_\succ(\text{Syz}(G))$
Output: A lifting $\bar{s} \in \text{Syz}(G) \subset F_1 := R^r$ of s w.r.t. G and $\succ$
1: $g := \psi(s)$
2: $T :=$ set of terms in $(g - \text{LOT}(g|G))$
3: $\bar{s} := s$
4: **for all** $t \in T$ **do**
5:     choose a term $me_i \in F_1$ with $m\,\text{LT}(f_i) = t$ and $s \succ me_i$
6:     $\bar{s} := \bar{s} - \text{LiftSubTree}(me_i)$
7: **return** $\bar{s}$

execution of algorithms. It ensures resource management, allows for the addition and removal of nodes during a running computation, and facilitates data sharing with its virtual memory manager and asynchronous data transfers to minimize latencies [10]. The main idea is to separate coordination and computation, as proposed by Gelernter in [11]. Its Petri net-based workflow engine provides a coordination layer with automated parallelization and dependency tracking. For the computation layer, SINGULAR is employed, which is a computer algebra system specializing in polynomial computations, whose core functionality is its Gröbner basis engine. There are multiple success stories of the SINGULAR/GPI-SPACE Framework; e.g., in algebraic geometry [5] (certifying smoothness) tropical geometry [3] (computing of tropicalizations with finite symmetries), geometric invariant theory (computing GIT-fans with finite symmetries), high energy physics [2] (computing integration-by-parts identities for Feynman integrals), and modular methods [1]. Our Petri net model will be discussed in Sect. 4.

Petri nets are bipartite graphs featuring two types of vertices: places (depicted as circles) and transitions (illustrated as rectangular boxes). Transitions represent the elementary functional units, while places can contain marked tokens, which symbolize pieces of data. Transitions can fire, consume one token from each input place and deposit one token onto each output place. Therefore, a transition is only enabled if all input places contain a token. A Petri net is executed by randomly firing enabled transitions. Figure 1 is a simple Petri that has two places; Input and Result, and one transition Sum. To fire transition Sum, there must be at least one token on both Input and Result.

Algorithm 3 LiftSubTree

Input: A Gröbner basis $G = \{f_1, \ldots, f_r\} \subset F_0 := R^s$ w.r.t. $>$ and a term $s \in F_1$
Output: A subtree lifting $\hat{s} \in \text{Syz}(G) \subset F_1 := R^r$ of s w.r.t. G and $\succ$
1: $g := \psi(s) - \text{LT}(\psi(s))$
2: $T :=$ set of terms in $(g - \text{LOT}(g|G))$
3: $\hat{s} := s$
4: **for all** $t \in T$ **do**
5:    choose a term $me_i \in F_1$ with $m \, \text{LT}(f_i) = t$
6:    $\hat{s} := \hat{s} - \text{LiftSubTree}(me_i)$
7: **return** $\hat{s}$

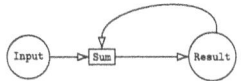

Fig. 1. Petri net for a sequential sum.

4 Modelling Free Resolutions

Our project's objective is to create a massively parallel algorithm for computing free resolutions by means of the SINGULAR/GPI-SPACE Framework. We achieve parallelism through the recursive tail-lifting approach of the enhanced Schreyer algorithm. We have reformulated the algorithm with an iterative rather than recursive structure to function as a Petri net, see Fig. 3. Place Input contains a token of type GroebnerBasis, which contains a reference pointing to the Gröbner basis of the given module. Place Lead has tokens of type LeadingSyz, containing the number of syzygies and a reference pointing to a leading syzygy vector. Places Tau and Tau' contain tokens of type Lifting. A Lifting consists of a LeadingSyz s, and a reference pointing to a term t. We initially place a token on Ctrl. Place ST holds a token of type Liftings. We introduce intermediate places named counter1, counter2, and counter3, along with transitions ti+1 and ti-1 to facilitate the termination of the Petri net. We initialize a list of counters stored in a single token in the place counter2, where the i-th entry, denoted as t_i, represents the counter for the number of terms of the i-th syzygy. Note that a double arrow produces multiple tokens. A dotted arrow stands for read-only access. We discuss step-by-step how the network works:

1. Place Start is initialized with a control token. LeadSyz produces multiple tokens on Lead corresponding to the lead syyzgies.
2. The transition Lift produces the first iteration of terms for the given leading syzygy.
3. SubLift determines new terms of the syzygy and produces corresponding tokens on Tau. It also produces a copy of its input on Tau', as well as a token containing the number i on the counter1 if it is computing towards the i-th syzygy. We introduce an alternating sign for the terms so that ADD can directly perform a summation.

4. LHS consumes the control token from Ctrl and a term from Tau' and produces a copy of its input on ST and sub.
5. RHS consumes a token (s, T') from Tau' and a token (s', T) from ST under the condition that $s = s'$. It produces a control token on Ctrl and copies the token from Tau' to rep.
6. ADD adds the terms (from the same syzygy) obtained from sub and rep, and places the sum on Tau', also producing a token containing the number i on counter3 in case ADD has added terms coming from the i-th syzygy.
7. Transitions ti+1 and ti-1 update the i-th entries of the counter on counter2; the former increases by 1 and the later decreases by 1. Finally, the transition FinalToken fires if all terms of all syzygies have been added (realized by the condition $t_i = 1$ for all i), and extracts all tokens from place sub.

Example 1. Let $F_0 := R := \mathbb{Q}[w, x, y, z]$ be endowed with the degrevlex ordering, denoted by $>$. We compute the first syzygy module of $G := (f_1, f_2, f_3, f_4, f_5) \subset R$ where $f_1 = w^2 - xz, f_2 = wx - yz, f_3 = x^2 - wy, f_4 = xy - z^2, f_5 = y^2 - wz$. Note that G is a Gröbner basis w.r.t. $>$. Let $\succ$ be the Schreyer ordering on $F_1 := R^5$. Define $\psi : F_1 \to F_0, e_i \mapsto f_i$. We have a minimal generating set of the leading syzygy module of G w.r.t. $\succ$ as $\mathcal{L} = \{we_2, we_3, we_4, xe_4, w^2e_5, xe_5\}$. We fix $s = we_3$. Table 1 summarizes the computation of the corresponding syzygy.

1. Lift finds $T = \{wx^2, -w^2y\}$, and computes the terms $(s, -xe_2)$, and (s, ye_1).
2. Two instances of SubLift can now fire simultaneously, as illustrated in the tree in Fig. 2. Note that there is no deterministic order of computation.
3. First sublift: Produces a token with the sublift $(s, -xe_2)$, and we have $T = \{xyz\}$. In the next iteration, sublift produces $(s, -ze_4)$. As now $T = \emptyset$, SubLift produces no new tokens on Tau.
4. Second sublift: Produces a token with the sublift (s, ye_1), and we have $T = \{-xyz\}$. In the next iteration, sublift finds (s, ze_4), and now $T = \emptyset$.

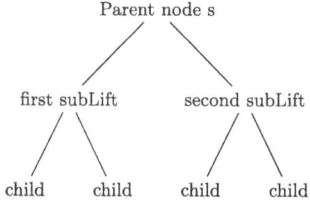

Fig. 2. Lifting tree for the example

Table 1. SubLift and sum in parallel

-	t	T	T'
First SubLift	$-xe_2$	$T = \{xyz\}$	-
Child SubLift	$-ze_4$	$T = \emptyset$	$T = \emptyset$
Second SubLift	ye_1	$T = \{-xyz\}$	
Child SubLift	ze_4	$T = \emptyset$	$T = \emptyset$
ADD	a	b	a+b
ADD	$-xe_2$	$-ze_4$	$-xe_2 - ze_4$
ADD	ye_1	ze_4	$ye_1 + ze_4$
ADD	$-xe_2 - ze_4$	$ye_1 + ze_4$	$we_3 - xe_2 + ye_1$

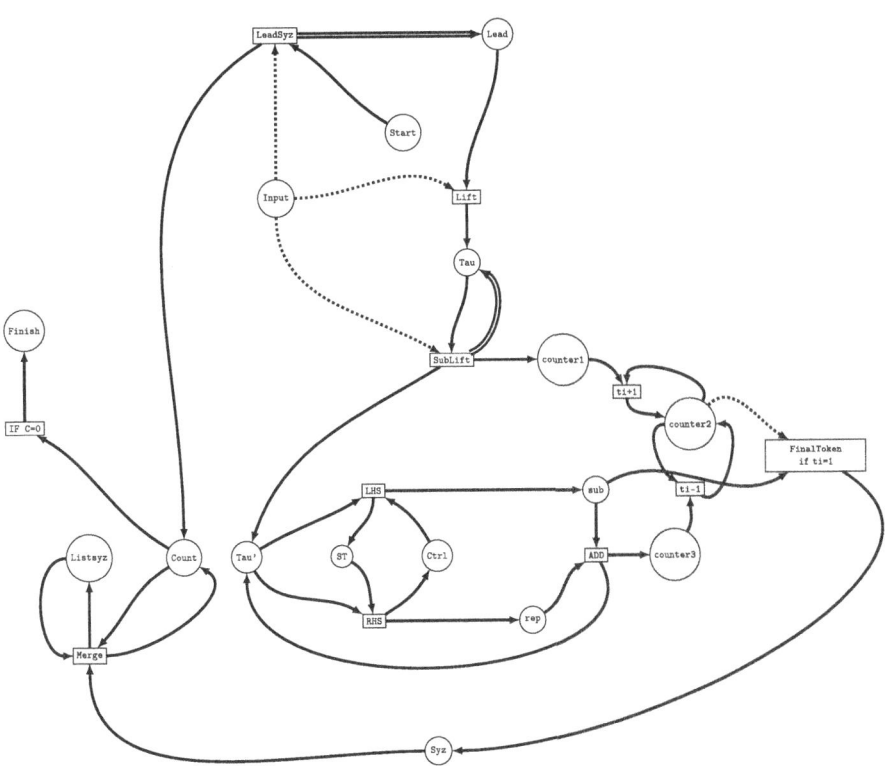

Fig. 3. Petri net for modelling the enhanced Schreyer resolution

Acknowledgement. The author would like to extend sincere gratitude to Janko Böhm, Wolfram Decker and Mirko Rahn for their continuous encouragement throughout the development of this work, as well as Hobihasina P. Rakotoarisoa for assistance on technical aspects. Gefördert durch die Deutsche Forschungsgemeinschaft (DFG) - Projektnummer 286237555 - TRR 195 [Funded by the Deutsche Forschungsgemeinschaft (DFG, German Research Foundation) - Project- ID 286237555 - TRR 195]. The work was supported by the German Academic Exchange Service (DAAD) through the Mathematics in Industry and Commerce (MIC) program.

References

1. Basson, D., Böhm, J., Marais, M.S., Rahn, M., Rakotoarisoa, H.P.: Massively Parallel Modular Methods in Commutative Algebra and Algebraic Geometry, 56 p. (2024, Preprint). arXiv:2401.11606
2. Bendle, D., et al.: Integration-by-parts reductions of Feynman integrals using Singular and GPI-Space. J. High Energ. Phys. **79**, 1–34 (2020)
3. Bendle, D., Böhm, J., Ren, Y., Schröter, B.: Massively parallel computation of tropical varieties, their positive part, and tropical Grassmannians. J. Symb. Comput. **120**, 102224 (2024)
4. Janko Böhm *Computer Algebra*, Lecture Notes (2024). https://www.mathematik.uni-kl.de/~boehm/lehre/24_CA/ca.pdf
5. Böhm, J., Decker, W., Frühbis-Krüger, A., Pfreundt, F.-J., Rahn, M., Ristau, L.: Towards massively parallel computations in algebraic geometry. Found. Comput. Math. **21**, 767–806 (2021)
6. Eisenbud, D.: Commutative Algebra: with a View Toward Algebraic Geometry. Springer Science & Business Media, New York (2013). https://doi.org/10.1007/978-1-4612-5350-1
7. Cox, D., Little, J., O'Shea, D., Sweedler, M.: Ideals, Varieties, and Algorithms. Springer, Cham (1997). https://doi.org/10.1007/978-3-319-16721-3
8. Eröcal, B., Motsak, O., Schreyer, F.-O., Steenpaß, A.: Refined algorithms to compute syzygies. J. Symb. Comput. **74**, 308–327 (2016)
9. Decker, W., Greuel, G.-M., Pfister, G., Schönemann, H.: Singular — A computer algebra system for polynomial computations (2024). http://www.singular.uni-kl.de
10. Fraunhofer ITWM, Competence Center High Performance Computing, GPI-Space (2024). https://www.gpi-space.de
11. Gelernter, D., Carriero, N.: Coordination languages and their significance. Commun. ACM **35**(2), 97–107 (1992)
12. Greuel, G.-M., Pfister, G., Bachmann, O., Lossen, C., Schönemann, H.: A Singular Introduction to Commutative Algebra. Springer, Cham (2008). https://doi.org/10.1007/978-3-540-73542-7
13. Schreyer, F.-O.: A standard basis approach to syzygies of canonical curves. J. Reine Angew. Math. **421**, 83–123 (1991)

Towards Parallel Methods in Birational Geometry

Benjamin Mirgain[1,2](✉)

[1] Fachbereich Mathematik, Universität des Saarlandes, Saarbrücken, Germany
[2] Division High Performance Computing, Fraunhofer ITWM, Kaiserslautern, Germany
benmirgain@hotmail.com

Abstract. Computational birational geometry is one of the key playing fields in an algorithmic approach to algebraic geometry, since birational maps are the fundamental way to relate algebraic varieties (or schemes). An important application is an algorithmic approach to the Minimal Model Program (MMP), which aims to classify algebraic varieties with mild singularities by finding simple birational models of such varieties in their birational equivalence class. This note presents work towards parallel methods to solve problems in birational geometry. Making use of a representation of algebraic schemes in terms of charts allows for a parallel computational approach for handling both the varieties and rational maps between them. In this note, we illustrate this approach on examples.

Keywords: Birational geometry · schemes · covered schemes · parallel algorithms · MMP

1 Introduction

In this note, with a view towards applications in the Mori Minimal Model Program, we present work towards parallel methods for algebraic schemes in terms of charts, which can be of use for computing blowups. We exemplify our approach at the computation of the image of a rational map, and the covering of the smooth scheme by complete intersection charts.

Ongoing work will make use of Petri nets as a coordination language to exploit the potential for parallelism in our algorithms. By presenting our schemes in terms of charts we hope to be able to run our computations locally on each individual chart, as well as for individual transition maps. Our implementation relies on the SINGULAR/GPI-SPACE framework [1]. We envision an integration with OSCAR to facilitate heavy computations.

2 Example: Determining Images of Rational Maps

If we are given a rational map $\phi : X \dashrightarrow Y$ between two schemes X and Y and X has an affine open cover $(U_i)_i$ and Y has an affine open cover $(V_j)_j$. Then our rational map induces rational maps $\phi_{ij} : U_i \longrightarrow V_j$ for each i, j by restricting our map ϕ to the set $U_i \cap \phi^{-1}(V_j)$. Therefore we get a collection of rational maps (ϕ_{ij}) over the affine open charts of X and Y, which define the map ϕ.

If we now want to compute the scheme theoretical image of the map ϕ as the Zariski closure of the Set theoretical image, one way to do it is to compute for each j the image of all the ϕ_{ij} in V_j. Then by taking the scheme theoretical union of these images, we get $V_j' := \mathrm{Im}(\phi) \cap V_j$. Thus we get an affine open cover $(V_j')_j$ of $\mathrm{Im}(\phi)$.

Of course this algorithm only works if we know how to compute the image of a rational map between two affine schemes $U := \mathrm{Spec}\ R[x_1,\ldots,x_n]/I$ and $V := \mathrm{Spec}\ R[y_1,\ldots,y_m]/J$. So assume we have a rational map $\phi : U \longrightarrow V$ and assume for now that the domain of ϕ is principal open, so $\mathrm{dom}(\phi) = D(f)$ for some $f \in R[x_1,\ldots,x_n]$. Then our map ϕ is given by a homomorphism $\psi : R[y_1,\ldots,y_m]/J \longrightarrow (R[x_1,\ldots,x_n]/I)_f$, which maps each y_j to some $\frac{g_j}{f^{e_j}}$. By Rabinowitsch's trick we have that $D(f) = \mathrm{Spec}\ R[x_1,\ldots,x_n,t]/(I + \langle t \cdot f - 1 \rangle)$. Then we can compute the image by going to

$$D(f) \times V = \mathrm{Spec}\ R[x_1,\ldots,x_n,y_1,\ldots,y_m,t]/(I + J + \langle t \cdot f - 1 \rangle)$$

and computing the graph $\Gamma(\phi)$ of ϕ, which is just given by $\Gamma(\phi) = V(y_j - t^{e_j} g_j \mid j = 1,\ldots,m)$. Then if $\pi_V : D(f) \times V \longrightarrow V$ is the canonical projection, we have $\mathrm{Im}(\phi) = \overline{\pi_V(\Gamma(\phi))}$. We have $\overline{\pi_V(\Gamma(\phi))} = R[y_1,\ldots,y_m]/K$ where $K = (I + J + \langle t \cdot f - 1 \rangle + \langle y_j - t^{e_j} g_j \mid j = 1,\ldots,m \rangle) \cap R[y_1,\ldots,y_m]$, which can be computed by choosing a suitable elimination order on $R[x_1,\ldots,x_n,y_1,\ldots,y_m,t]$.

Now if we assume that $\mathrm{dom}(\phi)$ is not principal open, then we still have that $\mathrm{dom}(\phi) = D(f_1) \cup \ldots \cup D(f_l)$ and then we get morphisms $\phi_k : D(f_k) \longrightarrow V$ by restricting ϕ to $D(f_k)$. Then we can just compute the image of ϕ by computing the images of the ϕ_k and taking the scheme theoretical union of them.

So we see that we can compute the image of a rational map $\phi : X \longrightarrow Y$ between two schemes X and Y by computing the images of a lot of smaller maps and stitching them together at the end. Each of the images of the smaller maps can be computed independently of each other so this computation lends itself naturally to being parallelized.

3 Covering a Scheme with Complete Intersection Charts

For the rest of this section, unless stated otherwise, let $X \subset \mathbb{A}_k^n$ be an equidimensional finite type affine scheme, with vanishing ideal $I_X \subset k[x_1,\ldots,x_n]$, where k is an algebraically closed field.

Definition 1. *Let $(R, \mathfrak{m})$ be a local Noetherian ring and $0 \neq f \in R$ any element. Then the order of f is defined to be*

$$\mathrm{ord}(f) = \max\{k \in \mathbb{N} \mid f \in \mathfrak{m}^k\}$$

Definition 2 ([4,5]). *Let $p \in X$. If $f_1,\ldots,f_s$ form a minimal standard basis of the extended ideal $I_X \mathcal{O}_{\mathbb{A}_k^n,p}$ with respect to a local degree ordering and the f_i are sorted by increasing order, set $\nu^*(X,p) = (\mathrm{ord}(f_1),\ldots,\mathrm{ord}(f_s))$*

Lemma 1 ([4,5]). *The sequence $\nu^*(X,p)$ depends only on X and p.*

Lemma 2 ([4], Chapter III). *The scheme X is singular at $p \in X$ iff*

$$\nu^*(X,p) >_{\text{lex}} (1,\ldots,1) \in \mathbb{N}^{\operatorname{codim}(X)}$$

where $>_{\text{lex}}$ denotes the lexicographical ordering.

The length of $\nu^*(X,p)$ can be larger than $\operatorname{codim}(X)$ if X is singular, but in that case at least one of the first $\operatorname{codim}(X)$ entries will be > 1.

For the rest of this section we suppose that we are given an embedding $X \subset W$, where W is a smooth complete intersection in $\mathbb{A}_k^n$ of codimension r. In particular W is equidimensional of dimension $d = n - r$. We assume that X has positive codimension in W, since otherwise X is necessarily smooth.

Definition 3. *If $(R, \mathfrak{m})$ is any local Noetherian ring, and $\langle 0 \rangle \neq J = \langle h_1, \ldots, h_t \rangle \subset R$ is any ideal, then the order of J is defined by setting*

$$\operatorname{ord}(J) = \max\{k \in \mathbb{N} \mid J \subset \mathfrak{m}^k\} = \min\{\operatorname{ord}(h_i) \mid i = 1, \ldots, t\}$$

For an ideal $\langle 0 \rangle \neq I \subset k[W]$ and a point $p \in W$, the order $\operatorname{ord}_p(I)$ of I at p is defined to be the order of the extended ideal $I\mathcal{O}_{W,p}$. For $0 \neq f \in k[W]$, we similarly define $\operatorname{ord}_p(f)$ as the order of the image of f in $\mathcal{O}_{W,p}$.

Definition 4. *With the notation above, for any integer $b \in \mathbb{N}$, the locus of order at least b of the vanishing ideal $I_{X,W}$ is*

$$\operatorname{Sing}(I_{X,W}, b) = \{p \in X \mid \operatorname{ord}_p(I_{X,W}) \geq b\}$$

Remark 1 ([4], Chapter III). Note that the loci $\operatorname{Sing}(I_{X,W}, b)$ are closed since the function

$$X \longrightarrow \mathbb{N}, \ p \longmapsto \operatorname{ord}_p(I_{X,W})$$

is Zariski upper semi-continuous.

Lemma 3 ([1]). *With the notation above, X is singular if*

$$\operatorname{Sing}(I_{X,W}, 2) \neq \emptyset$$

Lemma 4. *Let X be a locally of finite type k-scheme. Then X is smooth if and only if X is smooth at every closed point.*

Proposition 1. *Let k be a field and let X be a k-scheme locally of finite type and $x \in X$. Then $\dim_x(X) = \dim(\mathcal{O}_{X,x}) + \operatorname{trdeg}_k(\kappa(x))$.*

Proposition 2. *Assume that k is perfect. Let X be a k-scheme locally of finite type and $x \in X$, then X is smooth at x if and only if $\mathcal{O}_{X,x}$ is regular.*

By Lemma 4, it is enough to test smoothness for closed points of X. For our purposes we want to find the closed points of $\mathrm{Sing}(I_{X,W}, 2)$ in a neighborhood of a closed point $p \in X$. In order to do this we use derivatives with respect to a regular system of parameters of W at p.

Let $p \in W$. By our assumptions on W and 1, the local ring $\mathcal{O}_{W,p}$ has dimension d and since W is smooth, by Proposition 2 it is also regular. Thus we can find a regular system of parameters $X_{p,1}, \ldots, X_{p,d}$ for $\mathcal{O}_{W,p}$. That is, $X_{p,1}, \ldots, X_{p,d}$ form a minimal set of generators for $\mathfrak{m}_{W,p}$. By the Cohen structure theorem, we may think of the completion $\widehat{\mathcal{O}_{W,p}}$ as a formal power series ring in d variables. The map

$$\Phi : k[[y_1, \ldots, y_d]] \longrightarrow \widehat{\mathcal{O}_{W,p}}, \quad y_i \longmapsto X_{p,i}$$

is an isomorphism of local rings. In particular, the order of an element $f \in k[W]$ at p coincides with the order of the formal power series $\Phi^{-1}(f)$. The latter in turn can be computed as follows.

Lemma 5 ([2,3]). *Let $R = k[[y_1, \ldots, y_d]]$, let $\mathfrak{m} = \langle y_1, \ldots, y_d \rangle$ be the maximal ideal of R, and let $F \in R \setminus \{0\}$. Then*

$$\mathrm{ord}(F) = \min\left\{ m \in \mathbb{N} \;\middle|\; \frac{\partial^\alpha F}{\partial y^\alpha} \notin \mathfrak{m} \text{ for some } \alpha \in \mathbb{N}^d \text{ with } |\alpha| = m \right\}$$

where the derivatives denote the usual formal derivatives in characteristic zero and Hasse derivatives in positive characteristic.

As we focus on $\mathrm{Sing}(I_{X,W}, b)$ with $b = 2$, only first order formal derivatives play a role for us. Since these derivatives coincide with the first order Hasse derivatives, we do not need Hasse derivatives here.

Definition 5. *In the situation above, we use the isomorphism Φ of the Cohen structure theorem to define first order derivatives of elements $f \in \widehat{\mathcal{O}_{W,p}}$ with respect to the regular system of parameters $X_{p,1,\ldots,X_{p,d}}$. Set*

$$\frac{\partial f}{\partial X_{p,j}} := \Phi\left(\frac{\partial \Phi^{-1}(f)}{\partial y_j}\right) \in \widehat{\mathcal{O}_{W,p}} \text{ for } j = 1, \ldots, d$$

Remark 2. If $I_{X,W}$ is given by a set of generators $f_{r+1}, \ldots, f_s \in k[W] \setminus \{0\}$, and if $p \in X$, then $p \in \mathrm{Sing}(I_{X,W}, 2)$ iff $\mathrm{ord}_p(f_j) > 1$ for all j.

Furthermore, if $0 \neq f \in k[W]$ is any element, $p \in W$ is any closed point, and $X_{p,1,\ldots,X_{p,d}}$ is a regular system of parameters for $\mathcal{O}_{W,p}$, then $\mathrm{ord}_p(f) > 1$ iff

$$1 \notin \Delta_p(f) := \left\langle f, \frac{\partial f}{\partial X_{p,1}}, \ldots, \frac{\partial f}{\partial X_{p,d}} \right\rangle_{\widehat{\mathcal{O}_{W,p}}} \subset \widehat{\mathcal{O}_{W,p}}$$

Lemma 6. *([1]) Let k be algebraically closed. Let $I_W = \langle f_1, \ldots, f_r \rangle \subset k[x_1, \ldots, x_n]$ and let $\mathcal{J} = \left(\frac{\partial f_i}{\partial x_j}\right)$ be the Jacobian matrix of W. Then there is a finite covering of W by distinguished open subsets $D(h) \subset \mathbb{A}_k^n$ such that:*

(1) Each polynomial h is a maximal minor of $\mathcal{J}$
(2) For each h, the variables x_j not used for differentiation in forming the minor h induce by translation a regular system of parameters for every local ring $\mathcal{O}_{W,p}$, where $p \in W \cap D(h)$ are closed points.

Remark 3. From now on we retain the notation of the lemma above and we assume that $h = \det(M)$ involves the last r columns of $\mathcal{J}$. Furthermore fix one element $0 \neq f \in k[W]$.

The goal now is to find an ideal $\Delta(f) \subset \mathcal{O}_W(W \cap D(h))$ such that

$$\Delta(f)\widehat{\mathcal{O}_{W,p}} = \Delta_p(f) \text{ for each closed point } p \in W \cap D(h)$$

where $\Delta_x(f)$ is defined as in (2).

Construction 1. *([1]) We want to construct a polynomial $\tilde{f} \in k[x_1, \ldots, x_n]$ whose image in $\mathcal{O}_W(W \cap D(h))$ coincides with that of f and whose partial derivatives $\frac{\partial \tilde{f}}{\partial x_i}$, for $i = d+1, \ldots, n$, are mapped to zero in $\mathcal{O}_W(W \cap D(h))/\langle f \rangle$.
For this let A be the matrix of cofactors of M. Then*

$$A \cdot M = h \cdot E_r$$

where E_r is the $r \times r$ identity matrix. Moreover, if $I \subset k[x_1, \ldots, x_n]$ is the ideal generated by the entries of the vector $(\tilde{f}_1, \ldots, \tilde{f}_r)^t = A \cdot (f_1, \ldots, f_r)^t$, then the extended ideal $I\mathcal{O}_{\mathbb{A}_k^n}(D(h))$ and $I_W\mathcal{O}_{\mathbb{A}_k^n}(D(h))$ coincide since h is a unit in $\mathcal{O}_{\mathbb{A}_k^n}(D(h))$.

Let $\tilde{\mathcal{J}} = \left(\frac{\partial \tilde{f}_i}{\partial x_j}\right)$ be the Jacobian matrix of $\tilde{f}_1, \ldots, \tilde{f}_r$. Then the matrix $\tilde{\mathcal{J}}|_{W \cap D(h)}$ obtained by mapping the entries of $\tilde{\mathcal{J}}$ to $\mathcal{O}_W(W \cap D(h))$ is of type

$$\tilde{\mathcal{J}}|_{W \cap D(h)} = (\ * \ | \ h \cdot E_r)$$

In $\mathcal{O}_{\mathbb{A}_k^n}(D(h))$, the polynomial $\hat{f} = h \cdot f$ represents the same class as f. Moreover modulo f, each partial derivative of $\hat{f}$ is divisible by h. Hence after suitable row operations, the partial derivatives in the right hand lower block of the Jacobian matrix of $\tilde{f}_1, \ldots, \tilde{f}_r, \hat{f}$ are mapped to zero in $\mathcal{O}_W(W \cap D(h))/\langle f \rangle$

$$\left(\begin{array}{c|ccc} & h & & 0 \\ * & & \ddots & \\ & 0 & & h \\ \hline \frac{\partial \hat{f}}{\partial x_1} \cdots \frac{\partial \hat{f}}{\partial x_d} & \frac{\partial \hat{f}}{\partial x_{d+1}} & \cdots & \frac{\partial \hat{f}}{\partial x_n} \end{array}\right) \longmapsto \left(\begin{array}{c|ccc} & h & & 0 \\ * & & \ddots & \\ & 0 & & h \\ \hline H_1 \ \cdots \ H_d & 0 & \cdots & 0 \end{array}\right)$$

The row operations correspond of subtracting $k[x_1, \ldots, x_n]$-linear combinations of $\tilde{f}_1, \ldots, \tilde{f}_r$ from $\hat{f}$. In this way, we get a polynomial $\tilde{f}$ as desired. The images of $\tilde{f}$ and f in $\mathcal{O}_W(W \cap D(h))$ coincide, and in $\mathcal{O}_W(W \cap D(h))/\langle f \rangle$ we have :

$$\left(\frac{\partial \tilde{f}}{\partial x_1}, \ldots, \frac{\partial \tilde{f}}{\partial x_n}\right) = (H_1, \ldots, H_d, 0, \ldots, 0)$$

Lemma 7. ([1]) *With notation as above, consider the extended ideal*

$$\Delta(f) = \langle f, H_1, \ldots, H_d \rangle \mathcal{O}_W(W \cap D(h))$$

Then

$$\Delta(f)\widehat{\mathcal{O}_{W,p}} = \Delta_p(f) \text{ for each closed point } p \in W \cap D(h)$$

Remark 4. In the situation of Lemma 7, we write

$$\frac{\partial f}{\partial X_j} := H_j \in k[x_1, \ldots, x_n], \text{ for } j = 1, \ldots, d$$

Lemma 8 ([1]). *Let $f_{r+1}, \ldots, f_s \in k[x_1, \ldots, x_n]$ be a set of generators of the vanishing ideal $I_{X,W}$. Then*

$$\operatorname{Sing}(I_{X,W}, 2) \cap D(h) = \emptyset \iff V\left(I_X + \left\langle \frac{\partial f_i}{\partial X_j} \middle| r \leq i \leq s, \ 1 \leq j \leq d \right\rangle\right) \cap D(h) = \emptyset$$

Lemma 9 ([1]). *Let $f_{r+1}, \ldots, f_s \in k[x_1, \ldots, x_n]$ be generators for the vanishing ideal $I_{X,W}$. With the notation of 3 suppose that $\operatorname{Sing}(I_{X,W}, 2) \cap D(h) = \emptyset$. Then there is a finite covering of $X \cap D(h)$ by distinguished open subsets of the form $D(h \cdot g) \subset \mathbb{A}^n_k$ such that:*

(1) *Each polynomial g is a derivative $\frac{\partial f_i}{\partial X_j}$ of some f_i, for $r+1 \leq i \leq s$*
(2) *If we set $W' := V(f_1, \ldots, f_r, f_i) \subset \mathbb{A}^n_k$, then $W' \cap D(h \cdot g)$ is a smooth complete intersection of codimension $r+1$ in $D(h \cdot g)$*
(3) *We have $X \cap D(h \cdot g) \subset W' \cap D(h \cdot g)$*

Notation 1. *Let k be a field and $X \subset W$ be quasicompact locally of finite type k-schemes. Furthermore let $U_X := \operatorname{Spec} k[x_1, \ldots, x_n]/I_X \subset X$ and $U_W := \operatorname{Spec} k[x_1, \ldots, x_n]/I_W \subset W$ be affine open subsets. We say that U_X and U_W satisfy ($\Diamond$) iff*

(1) $U_X \subset U_W$
(2) U_X *is equidimensional and U_W is a smooth equidimensional complete intersection of $\mathbb{A}^n_k$ of codimension r for an $r \in \mathbb{N}$*
(3) *There are $f_1, \ldots, f_s \in k[x_1, \ldots, x_n]$ with $r \leq s \in \mathbb{N}$ such that*

$$I_W = \langle f_1, \ldots, f_r \rangle \subset I_X = \langle f_1, \ldots, f_s \rangle$$

Now for every equidimensional smooth scheme X, we can find a cover of open affine subschemes that are complete intersections. Indeed for each point $x \in X$, we find open affine $x \in U \subset X$ and complete intersections $U \subset W$ such that $\dim(W) - \dim(U)$ is decreasing at each step until we either find that U is singular or $\dim(W) - \dim(U) = 0$ in which case $U = W$. Since X is smooth we'll always find the latter and thus we find a cover of complete intersections.

Now let k be an perfect field and let $X = \operatorname{Spec} k[x_1,\ldots,x_n]/I_X$ and $W = \operatorname{Spec} k[x_1,\ldots,x_n]/I_W$ be two affine schemes satisfying ($\Diamond$). Then the algorithm to find a cover of affine complete intersections goes as follows:

1. Determine a set M of minors h of the Jacobian matrix of W as in Lemma 6
2. For each $h \in M$
   (a) For each $i \in \{r+1,\ldots,s\}$ and $j \in \{1,\ldots,d\}$, calculate $\frac{\partial f_i}{\partial X_j}$ as in Construction 6
   (b) Find a set $T^{(h)}$ of polynomials $g = \frac{\partial f_i}{\partial X_j}$ as in Lemma 9
   (c) For each $g = \frac{\partial f_i}{\partial X_j} \in T^{(h)}$
      i. Set $W' = \operatorname{Spec} k[x_1,\ldots,x_n]/(I_W + \langle f_i \rangle)$
      ii. Define the isomorphism $\varphi^{(h,g)} : X \cap D(h \cdot g) \to k[x_1,\ldots,x_n,t]/(I_X + \langle g \cdot h \cdot t - 1 \rangle)$
      iii. Start the algorithm again with $X \cap D(h \cdot g)$ and $W' \cap D(h \cdot g)$ to find a complete intersection cover $(W_l^{(h,g)})_{l \in I^{(h,g)}}$ of $X \cap D(h \cdot g)$ and for each $l \in I^{(h,g)}$ let $\lambda_l^{(h,g)} : W_l^{(h,g)} \to k[x_1,\ldots,x_n,t]/(I_X + \langle g \cdot h \cdot t - 1 \rangle)$ be the immersion and let $\mu_l^{(h,g)} : \operatorname{im}(\lambda_l^{(h,g)}) \to W_l^{(h,g)}$ be the left inverse.
3. For each $h,h'M$ and $g \in T^{(h)}$ and $g' \in T^{(h')}$ and $l \in I^{(h,g)}$ and $l' \in I^{(h',g')}$, we get the gluing morphism $W_l^{(h,g)} \to W_{l'}^{(h',g')}$ as $\mu_{l'}^{(h',g')} \circ \varphi^{(h',g')} \circ (\varphi^{(h,g)})^{-1} \circ \lambda_l^{(h,g)}$ wherever that is defined

This is realized in Algorithm 1.

Now if we are given a smooth scheme X with an affine cover $(X_i)_{i \in I}$ and a scheme $W \supset X$ with an affine cover $(W_i)_{i \in I}$ such that $X_i \subset W_i$ and W_i is a smooth complete intersection for each i. Then for each $i \in I$, we get a smooth intersection cover $(U_j^{(i)})_{j \in J^{(i)}}$ of X_i with **AffineCompleteIntersectionCover**(X_i, W_i). We get the gluing map $U_j^{(i)} \to U_{j'}^{(i')}$ as the restriction of the gluing map $X_i \to X_{i'}$ to $U_j^{(i)}$ wherever it maps to $U_{j'}^{(i')}$. Algorithm 2 **CompleteIntersectionCover** computes the complete intersection cover of X.

Algorithm 1: AffineCompleteIntersectionCover

Input: A chart $U_X = \operatorname{Spec} k[x_1,\ldots,x_n]/\langle f_1,\ldots,f_s\rangle$ and a chart $U_W = \operatorname{Spec} k[x_1,\ldots,x_n]/\langle f_1,\ldots,f_r\rangle$ fulfilling $(\Diamond)$

Output: A list L with three entries such that, $L[1]$ is a scheme such that each chart in the cover of it is a complete intersection, $L[2]$ and $L[3]$ are morphisms representing the isomorphism $U_X \longrightarrow L[1]$ and it's inverse

1 find a set
$$S = \left\{(h, I, M) \mid I \subset \{1,\ldots n\},\ |I| = r,\ h = \det(M) \text{ with } M = \left(\frac{\partial f_i}{\partial x_j}\right)_{i\in\{1,\ldots r\}, j\in I}\right\} \text{ such that } 1 \in \langle f_1,\ldots,f_r\rangle + \langle h \mid (h, I, M) \in S\rangle$$

2 list CoverSchemes,PreCoverCharts,PreCoverMaps,PreCoverInvMaps,CoverMaps,CoverInvMaps
3 **foreach** $(h, I, M) \in S$ **do**
4   as in Construction 1 compute $\frac{\partial f_i}{\partial X_j}$ for $r+1 \le i \le s$ and $1 \le j \le d$
5   find a set $T = \left\{(g, i) \mid g = \frac{\partial f_i}{\partial X_j},\ r+1 \le i \le s,\ 1 \le j \le d\right\}$ such that $h \in \langle f_1,\ldots,f_r\rangle + \langle g \mid (g, i) \in T\rangle$
6   **foreach** $(g, i) \in T$ **do**
7   chart $U'_X := \operatorname{Spec} k[x_1,\ldots,x_n, t]/\langle t\cdot h\cdot g - 1, f_1,\ldots,f_s\rangle$
8   chart $U'_W := \operatorname{Spec} k[x_1,\ldots,x_n, t]/\langle t\cdot h\cdot g - 1, f_1,\ldots,f_r, f_i\rangle$
9   chartmap $\varphi = U_X, U'_X, \langle h\cdot g\rangle, \{\{x_1,\ldots,x_n, \frac{1}{h\cdot g}\}\}$
10   chartmap $\psi = U'_X, U_X, \langle 1\rangle, \{\{x_1,\ldots,x_n\}\}$
11   PreCoverCharts=PreCoverCharts+list(U'_X)
12   PreCoverMaps=PreCoverMaps+list(φ)
13   PreCoverInvMaps=PreCoverInvMaps+list(ψ)
14   list L=AffineCompleteIntersectionCover(U'_X, U'_W)
15   CoverSchemes=CoverSchemes+list(L[1])
16   CoverMaps=CoverMaps+list(L[2])
17   CoverInvMaps=CoverInvMaps+list(L[3])
18 scheme ResultScheme
19 morphism ResultMorphism,ResultInvMorphism
20 **for** $i = 1,\ldots,\operatorname{size}(\text{CoverSchemes})$ **do**
21   **for** $j = 1,\ldots,\operatorname{size}(\text{CoverSchemes}[i].\text{cover})$ **do**
22   ResultScheme.cover=ResultScheme.cover+list(CoverSchemes[i].cover[j])
23 int k=0,k'=0
24 **for** $i = 1,\ldots,\operatorname{size}(\text{CoverSchemes})$ **do**
25   **for** $j = 1,\ldots,\operatorname{size}(\text{CoverSchemes}[i].\text{cover})$ **do**
26   k=k+1
27   **for** $i' = 1,\ldots,\operatorname{size}(\text{CoverSchemes})$ **do**
28   **for** $j' = 1,\ldots,\operatorname{size}(\text{CoverSchemes}[i'].\text{cover})$ **do**
29   k'=k'+1
30   ResultScheme.maps[k][k']=CoverMaps[i'].chartmaps[j']∘PreCoverMaps[i']∘PreCoverInvMaps[i]∘CoverInvMaps[i].chartmaps[j]
31 ResultMorphism.preim=U_X
32 ResultInvMorphism.im=U_X
33 ResultMorphism.im=ResultScheme
34 ResultInvMorphism.preim=ResultScheme
35 k=0
36 **for** $i = 1,\ldots,\operatorname{size}(\text{CoverSchemes})$ **do**
37   **for** $j = 1,\ldots,\operatorname{size}(\text{CoverSchemes}[i].\text{cover})$ **do**
38   k=k+1
39   ResultMorphism.chartmaps[1][k]=PreCoverInvMaps[i]∘CoverInvMaps[i].chartmaps[j]
40   ResultInvMorphism.chartmaps[k][1]=CoverMaps[i].chartmaps[j]∘PreCoverMaps[i]
41 **return** list(ResultScheme,ResultMorphism,ResultInvMorphism)

Algorithm 2: CompleteIntersectionCover

Input: A scheme X and a scheme W, such that X and W are covered by the same number of affine charts and for each i, $U_X^{(i)}$ the i-th chart of X and $U_W^{(i)}$ the i-th chart of W fulfill ($\Diamond$). If there is no second input made then we just default $U_W^{(i)}$ to be the chart lying in the same ring as $U_X^{(i)}$ and being given by the zero ideal.

Output: A list L with three entries such that, $L[1]$ is a scheme such that each chart in the cover of it is a complete intersection, $L[2]$ and $L[3]$ are morphisms representing the isomorphism $X \longrightarrow L[1]$ and it's inverse

1 list CoverSchemes,CoverMaps,CoverInvMaps
2 for $i = 1,\ldots,$ size(X.cover) do
3   list L=AffineCompleteIntersectionCover(X.cover[i],W.cover[i])
4   CoverSchemes[i]=L[1]
5   CoverMaps[i]=L[2]
6   CoverInvMaps[i]=L[3]
7 scheme ResultScheme
8 morphism ResultMorphism,ResultInvMorphism
9 for $i = 1,\ldots,$ size(CoverSchemes) do
10 for $j = 1,\ldots,$ size(CoverSchemes[i].cover) do
11   ResultScheme.cover=ResultScheme.cover+list(CoverSchemes[i].cover[j])
12 int k=0,k'=0
13 for $i = 1,\ldots,$ size(CoverSchemes) do
14 for $j = 1,\ldots,$ size(CoverSchemes[i].cover) do
15   k=k+1
16   for $i' = 1,\ldots,$ size(CoverSchemes) do
17   for $j' = 1,\ldots,$ size(CoverSchemes[i'].cover) do
18   k'=k'+1
19   ResultScheme.maps[k][k']=CoverMaps[i'].chartmaps[j']$\circ$ X.maps[i][i']$\circ$CoverInvMaps[i].chartmaps[j]
20 ResultMorphism.preim=X
21 ResultInvMorphism.im=X
22 ResultMorphism.im=ResultScheme
23 ResultInvMorphism.preim=ResultScheme
24 k=0
25 for $i = 1,\ldots,$ size(CoverSchemes) do
26 for $j = 1,\ldots,$ size(CoverSchemes[i].cover) do
27   k=k+1
28   ResultMorphism.chartmaps[i][k]=CoverInvMaps[i].chartmaps[j]
29   ResultInvMorphism.chartmaps[k][i]=CoverMaps[i].chartmaps[j]
30 **return** list(ResultScheme,ResultMorphism,ResultInvMorphism)

Acknowledgement. Gefördert durch die Deutsche Forschungsgemeinschaft (DFG) - Projektnummer 286237555 - TRR 195 [Funded by the Deutsche Forschungsgemeinschaft (DFG, German Research Foundation) - Project- ID 286237555 - TRR 195]. The author acknowledges support of Fraunhofer ITWM.

References

1. Böhm, J., Decker, W., Frühbis-Krüger, A., Pfreundt, F.-J., Rahn, M., Ristau, L.: Towards massively parallel computations in algebraic geometry. Found. Comput. Math. **21**, 767–806 (2021)
2. Bravo, A.-M., Encinas, S., Villamayor, O.: A simplified proof of desingularization and applications. Rev. Mat. Iberoamericana **21**(2), 349–458 (2005)
3. Giraud, J.: Contact maximal en charactéristique positive. Ann. Sci. Éc. Norm. Sup. 4eme série **8**, 201–234 (1975)
4. Hironaka, H.: Resolution of singularities of an algebraic variety over a field of characteristic zero. I, II. Ann. of Math(2) **79**, 109–203, 205–326 (1964)
5. Hironaka, H.: On the characters ν^* and τ^* of singularities. J. Math. Kyoto Univ. **7**(1), 19–43 (1967)

Towards Parallel Algorithms for Gromov-Witten Invariants of Elliptic Curves

Ali Traore[✉]

Ali Traore, Fachbereich Mathematik, RPTU, 67653 Kaiserslautern, Germany
atraore@rptu.de

Abstract. We present a parallel enhanced algorithm for exploring mirror symmetry for elliptic curves through the correspondence of algebraic and tropical geometry, focusing on Gromov-Witten invariants of elliptic curves and, in particular, Hurwitz numbers. We present a new highly efficient algorithm for computing generating series for these numbers. A sequential version of the algorithm has been implemented using SINGULAR and OSCAR. The implementations in [1] outperform by far the previous methods [3] provided in SINGULAR. In this note, we describe work towards the natural next step, which is parallelization. We have integrated our algorithm with GPI-SPACE, a workflow management system for high-performance computing developed at Fraunhofer ITWM. This allows us to run our algorithm simultaneously on a large number of cores. This facilitates computation of quasi-modular representations of the respective generating series.

Keywords: Mirror symmetry · Gromov-Witten invariants · Feynman integrals · massively parallel computations · parallel algorithms for combinatorics · GPI-SPACE

1 Introduction

This paper centers on a use case of the principle of mirror symmetry, which establishes a connection between Gromov-Witten invariants of an elliptic curve and certain integrals over Feynman graphs. Our goal is to provide parallel algorithms for computing Gromov-Witten invariants of elliptic curves. Tropical methods fit into the picture of mirror symmetry through the famous Gross-Siebert program [6].

Here, we present the special case of elliptic curves, for which mirror symmetry is best understood [7]. The tropical side of mirror symmetry for elliptic curves was discovered within the Collaborative Research Center TRR195 which is also responsible for the development of OSCAR [4,8,9].

One consequence of mirror symmetry for elliptic curves is the equality of the generating function of Hurwitz numbers (resp. more generally, of descendant Gromov-Witten invariants) to certain Feynman integrals which are complex analytic path integrals whose construction is governed by combinatorics.

In this paper, we first recall our enhanced algorithm, which has been implemented in [2]. This algorithm computes generating series of Hurwitz numbers via Feynman integrals. Hurwitz numbers enumerate branched covers of non-singular curves with a specified ramification profile over fixed points. One key application is to find the representation of the generating series as quasi-modular forms. This requires to solve a linear system of equations which becomes determined if we know Hurwitz numbers of large enough degree. In the case of descendant Gromov-Witten invariants homogeniety of the quasimodular form is an interesting question. To expedite the computation, which can be very time consuming for covers of high degree and graphs of high genus, we utilize GPI-SPACE, a Petri net-based workflow management system development by the HPC group at Fraunhofer ITWM, to parallelize our algorithm.

2 Gromov-Witten Invariants and Feynman Integrals

2.1 Theoretical Background and Elementary Algorithm

Let $\mathcal{E}$ be an arbitrary complex elliptic curve and $\mathcal{C}$ a non-singular curve of genus g and $\phi : \mathcal{C} \to \mathcal{E}$ a cover.

Definition 1. *(Hurwitz numbers) Fix $2g-2$ points $p_1, \ldots, p_{2g-2}$ in $\mathcal{E}$. We define the Hurwitz number $N_{d,g}$ to be the weighted number of (isomorphism classes of) simply ramified covers $\phi : \mathcal{C} \to \mathcal{E}$ of degree d, where $\mathcal{C}$ is a connected curve of genus g, and the branch points of ϕ are the points p_i, $i = 1, \ldots, 2g-2$. We count each such cover ϕ with weight $|\operatorname{Aut}(\phi)|$.*

Lemma 1. *Fix a Feynman graph Γ and an order Ω, and a tuple $(q_1, \ldots, q_{3g-3})$ as in Definition 14. We express the coefficient of $q^{2 \cdot \underline{a}}$ in $I_{\Gamma,\Omega}(q_1, \ldots, q_{3g-3})$. Assume k is such that the entry $a_k = 0$, and assume the edge q_k connects the two vertices x_{k_1} and x_{k_2}. Choose the notation of the two vertices x_{k_1} and x_{k_2} such that the chosen order Ω implies $\left|\frac{x_{k_1}}{x_{k_2}}\right| < 1$ for the starting points on the integration paths. Then the coefficient of $q^{2 \cdot \underline{a}}$ equals the constant term of the series*

$$\prod_{k|a_k=0} \left(\sum_{w_k=1}^{\infty} w_k \cdot \left(\frac{x_{k_1}}{x_{k_2}}\right)^{w_k}\right) \cdot \prod_{k|a_k \neq 0} \left(\sum_{w_k|a_k} w_k \cdot \left(\left(\frac{x_{k_1}}{x_{k_2}}\right)^{w_k} + \left(\frac{x_{k_2}}{x_{k_1}}\right)^{w_k}\right)\right) \tag{1}$$

For a Taylor series

$$F(x_1, \ldots, x_n) = \sum_{\substack{1 \leq j \leq n \\ a_j \geq 0}} \alpha(a_1, \ldots, a_n) x_1^{a_1}, \ldots, x_n^{a_n}$$

we write

$$\operatorname{coeff}_{[x_1^{a_1}, \ldots, x_n^{a_n}]}(F) = \alpha(a_1, \ldots, a_n).$$

Definition 2 (The propagator and Feynman integral). *We define the propagator as a (formal) series in x and q,*

$$P(x,q) = \sum_{w=1}^{\infty} w \cdot x^w + \sum_{a=1}^{\infty} \left(\sum_{w|a} w(x^w + x^{-w}) \right) q^a \quad (2)$$

Fix a Feynman graph Γ and an order Ω of vertices and write

$$P_{\Gamma,\Omega} = \prod_{k=1}^{3g-3} \left(-P\left(\frac{x_{k_1}}{x_{k_2}}, q \right) \right).$$

Then we define Feynman integral as

$$I_{\Gamma,\Omega}(q) = \text{coeff}_{[x_1^0,\ldots,x_{2g-2}^0]}(P_{\Gamma,\Omega}).$$

2.2 Improvement of the Algorithm

In the standard algorithm, the Feynman integral is computed for a given branch type a, by computing the propagator of a for each permutation of vertices, while in the enhanced algorithm works by assigning to each cover a so-called flip signature by considering the permutation assigning vertices of the source tropical curve to branch points on the target elliptic curve, observing that covers with the same signature lead to the same integral. Further improvements are achieved by direct computation of coefficients of the Laurent series of each propagator, avoiding explicit expansion all propagator together and then compute the coefficient. The enhanced algorithm outperforms by far the standard algorithm and has been implemented in OSCAR .

Consider the example of a source curve in Fig. 1.
$a = (1, 1, 2, 0, 3, 1)$, $\Omega = (1, 3, 4, 2)$ here $x_4 < x_2$, $d = \text{sum}(a)$.

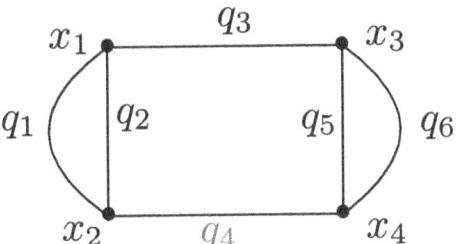

Fig. 1. Caterpillar genus 3

Construction 1. *(Standard Algorithm)* *In the standard algorithm, the propagator is calculated for every possible permutation of the order of vertices.*

```
P = permutations(O) # 4! elements.
for p in P
    Propagator *= propagator(G, a, p)
    # For each p, we check if £x_4 < x_2£.
end
```

The Feynman integral at branch a is then calculated as:

$$I_q(\Gamma, a) = \text{coeff}_{[x_1^0],\ldots,x_4^0}\, \text{propagator}$$

Construction 2 (Enhanced Algorithm). *In the enhanced algorithm, we first compute the flip_signature and group all vertex orders that lead to the same branch type a.*

$$\text{flip\_signature}(a, p) = \begin{cases} 12 \times (1, 1, 2, 0, 3, 1) & \text{if } x_4 < x_2 \\ 12 \times (1, 1, 2, -1, 3, 1) & \text{if } x_4 > x_2 \end{cases}$$

For example, vertex orders lead to either $(1, 1, 2, 0, 3, 1)$ *or* $(1, 1, 2, -1, 3, 1)$. *We then compute the local Feynman integral:*

$$I_{q,1}(\Gamma, a) = \text{coeff}_{[x_1^0],\ldots,x_4^0}\, \text{propagator}(\Gamma, (1, 1, 2, 0, 3, 1))$$

$$I_{q,2}(\Gamma, a) = \text{coeff}_{[x_1^0],\ldots,x_4^0}\, \text{propagator}(\Gamma, (1, 1, 2, -1, 3, 1))$$

Finally, the Feynman integral at branch a is obtained by summing $I_{q,1}$ *and* $I_{q,2}$:

$$I_q(\Gamma, a) = I_{q,1} + I_{q,2}$$

The computation becomes complicated and slow when we attempt to calculate the degree of a Feynman integral for a given graph Γ with a fixed degree d.

To address this issue, we utilize GPI-SPACE, a High-Performance Computing (HPC) system currently under development at ITWM Fraunhofer Kaiserslautern. This enables us to parallelize our algorithm, allowing us to compute the Feynman integral simultaneously for all combinations a of degree d ($d = \sum(a)$).

3 Petri Nets and GPI-SPACE

While sequential algorithms are typically described in a step-by-step manner, Petri nets offer a representation that mirrors both the structure of the algorithms and the current state of computation. Consequently, they provide a means to automatically exploit parallel structures. A Petri net is essentially a directed bipartite graph with two types of vertices: places (depicted as circles) and transitions (illustrated as rectangular boxes). The transitions represent elementary functional units, while the places can contain marked tokens, which are akin to representing data pieces (in the context of colored Petri nets). These transitions serve to connect places, consuming one token from each input place and depositing one token onto each output place. As such, a transition is only capable of firing if all input places hold a token. Figure 2 presents an example of a Petri net.

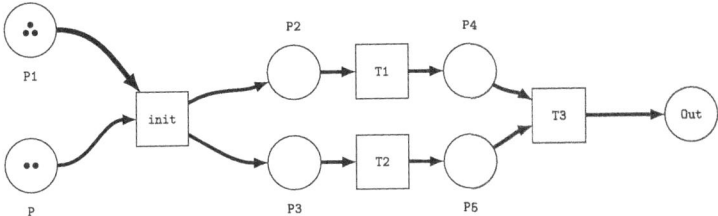

Fig. 2. Task and data parallelism in a Petri net

4 Parallel Enumeration of Combinatorial Objects

To compute the Feynman integral of degree d, the typical approach involves partitioning d into n edge parts, resulting in

$$\binom{d+n-1}{d}$$

elements. However, this partitioning process and the subsequent computation can be time-consuming.

To address this efficiently, we first compute d elements $(b_1, \ldots, b_d)$ of the partition using the gen_block(n, d) function (see Algorithm 1). For each element b_i, we compute its next partition, denoted as m_i, using the iterate(b_i) function (see Algorithm 3). This parallel computation strategy significantly improves the effectiveness of computing the Feynman integral.

Algorithm 1: gen_block

Input: Integer n, Integer d.
Output: Vector of vectors ru.
1 **begin**
2 $\quad ru =$ empty vector of vectors;
3 $\quad$ **for** $e = 0$ **to** $d-1$ **do**
4 $\quad\quad x = [d-e; \text{fill}(0, n-2); e]$;
5 $\quad\quad$ push x to ru;
6 $\quad$ **return** ru;

Example 1. Suppose $d = 4$ and $n = 3$ and we consider as a small example the graph in Fig. 3. Table 4 illustrates how we can parallelize our algorithm.

Utilizing GPI-SPACE [5] alongside the Petri net depicted in Fig. 5 enables us to compute the feynman_integral(G,a) by independently computing each row in the table in Fig. 4 determining a from the preceding one. Subsequently, we aggregate these computations at feynman_degree_sum(G,d) which, in the example, gives 720.

Algorithm 2: next_partition

Input: Vector a representing a partition.
Output: Next partition.

1 **begin**
2     $n = \text{sum}(a)$;
3     $k = \text{length}(a)$;
4   **for** $i = k$ **to** 1 **do**
5   **if** $i = k$ **and** $a[i] = n$ **then**
6   **return** a;
7   **else**
8   **for** $j = i-1$ **to** 1 **do**
9   **if** $a[j] \neq 0$ **then**
10                 $a[j]-=1$;
11                 $a[k] = 0$;
12   t $a[j+1] = ak+1$;
13   **return** a;

Algorithm 3: iterate

Input: vector x.
Output: Vector of vectors ru.

1 **begin**
2     $k = \text{length}(x)$;
3     $d = \text{sum}(x)$;
4     $e = d - x[1]$;
5     $n = \binom{d+k-1}{d}$;
6     $y = [x[1]-1; 0; ...; 0; e+1]$;
7   **for** $i = 1$ **to** n **do**
8   **if** $x \neq y$ **then**
9           $x = \text{next\_partition}(x)$;
10   push x to ru;
11   **else**
12   Break;
13     $vec = [d; 0; ...; 0]$;
14   **if** $x == vec$ **then**
15   push x to the front of ru;
16   **return** ru;

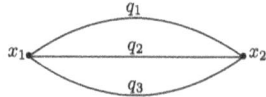

Fig. 3. Caterpillar genus 2

gen_block(d,n)	iterate(vec)	feynman_integral(G,a)
[4, 0, 0]	[4, 0, 0]	84
	[3, 1, 0]	40
[3, 0, 1]	[3, 0, 1]	40
	[2, 2, 0]	68
	[2, 1, 1]	8
[2, 0, 2]	[2, 0, 2]	68
	[1, 3, 0]	40
	[1, 2, 1]	8
	[1, 1, 2]	8
[1, 0, 3]	[1, 0, 3]	40
	[0, 4, 0]	84
	[0, 3, 1]	40
	[0, 2, 2]	68
	[0, 1, 3]	40
	[0, 0, 4]	84

Fig. 4. Parallel enumeration of combinations leading to parallelization of Feynman integral evaluation

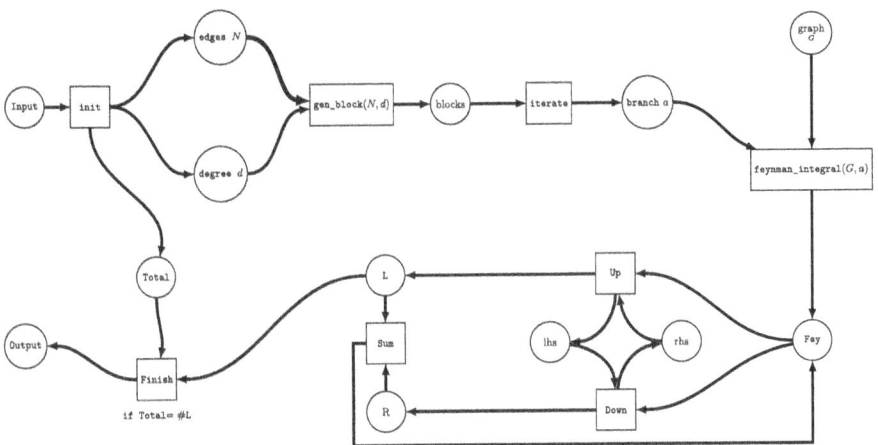

Fig. 5. Petri-net

Acknowledgement. Gefördert durch die Deutsche Forschungsgemeinschaft (DFG) - Projektnummer 286237555 - TRR 195 [Funded by the Deutsche Forschungsgemeinschaft (DFG, German Research Foundation) - Project- ID 286237555 - TRR 195]. We acknowledge support of Project A13 of TRR 195. The work was supported by the German Academic Exchange Service (DAAD) through the Mathematics in Industry and Commerce (MIC) program.

References

1. Böhm, J., Dastur, F., Hoffmann, A., Markwig, H., Traore, A.: GromovWitten.jl. An OSCAR-based package computing Gromov-Witten invariants of elliptic curves (2023). https://github.com/singular-gpispace/tropicalfeynman
2. Böhm, J., Dastur, F., Hoffmann, A., Markwig, H., Traore, A.: Algorithms for Gromov-Witten Invariants of Elliptic Curves (2023). arXiv:2311.11381 [math.AG]
3. Böhm, J., Bringmann, K., Buchholz, A., Markwig, H.: ellipticcovers.lib. A Singular 4 library for Gromov-Witten invariants of elliptic curves (2018)
4. Böhm, J., Bringmann, K., Buchholz, A., Markwig, H.: Tropical mirror symmetry for elliptic curves. J. Reine Angew. Math. **732**, 211–246 (2017)
5. Fraunhofer ITWM, Competence Center High Performance Computing. GPI-Space, September 2024. https://www.gpi-space.de
6. Gross, M., Siebert, B.: Mirror Symmetry via logarithmic degeneration data I. J. Differential Geom. **72**, 169–338 (2006). arXiv:math.AG/0309070
7. Dijkgraaf, R.: The moduli space of curves, in Mirror symmetry and elliptic curves, Birkhäuser Boston. Progr. Math. **129**, 149–163 (1995)
8. Böhm, J., Goldner, C., Markwig, H.: Counts of (tropical) curves in $E \times P^1$ and Feynman integrals. Ann. Inst. Henri Poincaré D. **9**(1), 121–158 (2022). arXiv:1812.04936
9. Böhm, J., Goldner, C., Markwig, H.: Tropical mirror symmetry in dimension one. SIGMA. **18**, 046 (2022). arXiv:1809.10659

Computer Algebra Applications
in the Life Sciences

A SageMath Package for Elementary and Sign Vectors with Applications to Chemical Reaction Networks

Marcus S. Aichmayr[1](✉), Stefan Müller[2], and Georg Regensburger[1]

[1] Institute of Mathematics, University of Kassel, Kassel, Germany
{aichmayr,regensburger}@mathematik.uni-kassel.de
[2] Faculty of Mathematics, University of Vienna, Vienna, Austria
st.mueller@univie.ac.at

Abstract. We present our SAGEMATH package elementary_vectors for computing elementary and sign vectors of real subspaces. In this setting, elementary vectors are support-minimal vectors that can be determined from maximal minors of a real matrix representing a subspace. By applying the sign function, we obtain the cocircuits of the corresponding oriented matroid, which in turn allow the computation of all sign vectors of a real subspace. As an application, we discuss sign vector conditions for existence and uniqueness of complex-balanced equilibria of chemical reaction networks with generalized mass-action kinetics. The conditions are formulated in terms of sign vectors of two subspaces arising from the stoichiometric coefficients and the kinetic orders of the reactions. We discuss how these conditions can be checked algorithmically, and we demonstrate the functionality of our package sign_vector_conditions in several examples.

Keywords: elementary vectors · sign vectors · oriented matroids · generalized mass-action systems · deficiency zero theorem · robustness

1 Elementary Vectors of a Subspace

For real subspaces, elementary vectors are nonzero vectors with minimal support, as introduced in [13]. Since we also deal with parameters, we consider elementary vectors for subspaces Q^n of the quotient field Q of an integral domain R. The *support* of a vector $x \in Q^n$ is the index set of nonzero components,

$$\operatorname{supp} x = \{i \mid x_i \neq 0\}.$$

The SAGEMATH packages are available at: https://github.com/MarcusAichmayr/
The examples of this document are available at: https://marcusaichmayr.github.io/sign_vector_conditions/.

© The Author(s), under exclusive license to Springer Nature Switzerland AG 2024
K. Buzzard et al. (Eds.): ICMS 2024, LNCS 14749, pp. 155–164, 2024.
https://doi.org/10.1007/978-3-031-64529-7_17

Definition 1. *For a subspace $\mathcal{V}$ of Q^n, we call a nonzero vector $v \in \mathcal{V}$ elementary if, for all nonzero vectors $x \in \mathcal{V}$,*

$$\operatorname{supp} x \subseteq \operatorname{supp} v \quad \textit{implies} \quad \operatorname{supp} x = \operatorname{supp} v.$$

Elementary vectors with the same support are easily seen to be multiples. Therefore, a subspace contains only finitely many elementary vectors up to multiples. Further, a subspace is always generated by a finite set of elementary vectors.

We represent a subspace as the kernel of a matrix $M \in R^{d \times n}$ with rank d over Q. If the kernel of a matrix is 1-dimensional, it has exactly one elementary vector. We apply this fact by considering submatrices consisting of $d+1$ columns. To find kernel vectors, we solve systems of linear equations. In particular, we compute maximal minors and apply Cramer's Rule. By inserting additional zeros, we extend the kernel vectors to kernel vectors of the initial matrix. The resulting vectors are elementary if they are nonzero.

For $I \subseteq [n] = \{1, \ldots, n\}$, we denote by M_I the submatrix of M that consists only of the columns corresponding to the indices I. We obtain a formula for computing elementary vectors in R^n.

Proposition 1 (cf. equation (2.1) in [6]). *For a matrix $M \in R^{d \times n}$ with rank d and $I \subseteq [n]$ with $|I| = d+1$, define the vector $v \in R^n$ by*

$$v_i = \begin{cases} (-1)^{|\{k \in I \,:\, k < i\}|} \det M_{I \setminus \{i\}}, & \textit{if } i \in I, \\ 0, & \textit{otherwise.} \end{cases}$$

The vector $v \in \ker M$ is elementary if $\operatorname{rank} M_I = d$.

For computing all elementary vectors in $\ker M$, we need to compute $\binom{n}{d}$ determinants of all $d \times d$ submatrices of M (maximal minors). In contrast, for Gaussian Elimination, we would need to compute the kernel of a $(d+1) \times d$ matrix for each of the $\binom{n}{d+1}$ index sets I. In our implementation, it turns out that the approach using maximal minors is more efficient for computing elementary vectors for medium-sized matrices.

If all maximal minors are nonzero, there is exactly one elementary vector for each index set I. This gives an upper bound of $\binom{n}{d+1}$ elementary vectors with pairwise distinct support of an $(n-d)$-dimensional subspace. If the rank of M_I is not maximal, we obtain the zero vector. Proposition 1 suggests an algorithm for computing elementary vectors, which we have implemented in our SAGEMATH package `elementary_vectors` [1]. We demonstrate it by an example.

```
sage: from elementary_vectors import *
sage: M = matrix([[1, 1, 2, 0], [0, 0, 1, 2]])
sage: M.minors(2)
[0, 1, 2, 1, 2, 4]
sage: elementary_vectors(M)
[(1, -1, 0, 0), (4, 0, -2, 1), (0, 4, -2, 1)]
```

Note that the first maximal minor is zero. This is the reason why we obtain only 3 and not $\binom{4}{3} = 4$ elementary vectors.

Solvability of Linear Inequality Systems. A fundamental theorem for deciding the solvability of linear inequality systems is given in [13,14]. Note that every linear inequality system can be written as an intersection of a subspace and a Cartesian product of intervals. By iterating over elementary vectors, we check whether such an intersection is empty. For more details, we refer to our manuscript [3] (see also [7]).

Theorem 1 ("Minty's Lemma", Theorem 22.6 in [14]). *For a subspace $\mathcal{V}$ of $\mathbb{R}^n$ and nonempty intervals $\mathcal{I} = I_1 \times \cdots \times I_n$,*

either there exists a vector $x \in \mathcal{V} \cap \mathcal{I}$,
or there exists an elementary vector $v \in \mathcal{V}^\perp$ with $v^\top z > 0$ for all $z \in \mathcal{I}$.

Based on Theorem 1, the function exists_vector decides the solvability of such systems.

```
sage: from vectors_in_intervals import *
sage: M = matrix([[1, 0, 1, 0], [0, 1, 1, 1]])
sage: I = intervals_from_bounds([2,5,0,-oo],[5,oo,8,5],
....:      [True,True,False,False],[False,False,False,True])
sage: I
[[2, 5), [5, +oo), (0, 8), (-oo, 5]]
sage: exists_vector(M, I)
True
```

Sign Vectors. We call elements in $\{-, 0, +\}^n$ *sign vectors*. For $x \in \mathbb{R}^n$, we define the sign vector $\text{sign}(x) \in \{-, 0, +\}^n$ by applying the sign function componentwise. For a set $S \subseteq \mathbb{R}^n$, we obtain the set of sign vectors

$$\text{sign}(S) = \{\text{sign}(x) \mid x \in S\} \subseteq \{-, 0, +\}^n$$

by applying sign to each vector of S. As for real vectors, we introduce the *support* $\text{supp}\,\sigma = \{i \mid \sigma_i \neq 0\}$ of a sign vector σ.

Further, we obtain a partial order on $\{-, 0, +\}^n$ by defining $0 < -, +$. As in [9], the *(lower) closure* of a set of sign vectors $T \subseteq \{-, 0, +\}^n$ is the set

$$\overline{T} = \{\sigma \in \{-, 0, +\}^n \mid \sigma \leq \tau \text{ for some } \tau \in T\}.$$

A (realizable) *oriented matroid* is the set of sign vectors that correspond to a real subspace. We call the elements of an oriented matroid *covectors*. To obtain them, we apply the sign function to the elements in a subspace.

The sign vectors corresponding to the elementary vectors are called *cocircuits*. They generate all elements of an oriented matroid, just like the elementary vectors generate all elements of a subspace. Since the cocircuits are determined by the signs of the maximal minors, called *chirotopes*, we can express many sign vector conditions in terms of chirotopes. For further details on oriented matroids, we refer to [4, Chapter 7], [15, Chapters 2 and 6], [12] and the encyclopedic study [5]. Our package also offers several functions for oriented matroids. As an example, we show the computation of cocircuits.

```
sage: from sign_vectors.oriented_matroids import *
sage: M = matrix([[1, 1, 2, 0], [0, 0, 1, 2]])
sage: cocircuits_from_matrix(M)
{(-+00), (-0+-), (+0-+), (+-00), (0-+-), (0+-+)}
sage: covectors_from_matrix(M)
{(0000), (-+00), (--+-), (+-00), (+---+), (-0+-), (+0-+),
    (0-+-), (0+-+), (++-+), (-+-+), (+-+-), (-++-)}
```

2 Applications to Chemical Reaction Networks

For chemical reaction networks (CRNs) with generalized mass-action kinetics, we recall basic notions from [11]. See also [9,10].

A *generalized mass-action system* $(G_k, y, \tilde{y})$ is given by a simple directed graph $G = (V, E)$, positive edge labels $k \in \mathbb{R}_>^E$, and two maps $y \colon V \to \mathbb{R}^n$ and $\tilde{y} \colon V_s \to \mathbb{R}^n$, where $V_s = \{i \mid i \to i' \in E\} \subseteq V$ denotes the set of source vertices. Every vertex $i \in V$ is labeled with a *(stoichiometric) complex* $y(i) \in \mathbb{R}^n$, and every source vertex $i \in V_s$ is labeled with a *kinetic-order complex* $\tilde{y}(i) \in \mathbb{R}^n$. Further, every edge $(i \to i') \in E$ is labeled with a *rate constant* $k_{i \to i'} > 0$ and represents the *chemical reaction* $y(i) \to y(i')$. If every component of G is strongly connected, G and $(G_k, y, \tilde{y})$ are called *weakly reversible*. The associated ODE system for the positive *concentrations* $x \in \mathbb{R}_>^n$ (of n chemical species) is given by

$$\frac{dx}{dt} = \sum_{(i \to i') \in E} k_{i \to i'}\, x^{\tilde{y}(i)} \big(y(i') - y(i)\big). \tag{1}$$

The sum ranges over all reactions, and every summand is a product of the *reaction rate* $k_{i \to i'}\, x^{\tilde{y}(i)}$, involving a monomial $x^{\tilde{y}} = \prod_{j=1}^{n}(x_j)^{\tilde{y}_j}$ determined by the kinetic-order complex of the educt, and the *reaction vector* $y(i') - y(i)$ given by the stoichiometric complexes of product and educt.

Let $I_E, I_{E,s} \in \mathbb{R}^{V \times E}$ be the incidence and source matrices of the digraph G, respectively, and

$$A_k = I_E \operatorname{diag}(k)(I_{E,s})^\top \in \mathbb{R}^{V \times V}$$

be the Laplacian matrix of the labeled digraph G_k. (This definition is used in dynamical systems. In other fields, the Laplacian matrix is defined as $A_k^\top$, $-A_k$, or $-A_k^\top$.) Now, the right-hand-side of (1) can be decomposed into stoichiometric, graphical, and kinetic-order contributions,

$$\frac{dx}{dt} = Y I_E \operatorname{diag}(k)(I_{E,s})^\top x^{\tilde{Y}} = Y A_k\, x^{\tilde{Y}}, \tag{2}$$

where $Y, \tilde{Y} \in \mathbb{R}^{n \times V}$ are the matrices of stoichiometric and kinetic-order complexes, and $x^{\tilde{Y}} \in \mathbb{R}_>^V$ denotes the vector of monomials, that is, $(x^{\tilde{Y}})_i = x^{\tilde{y}(i)}$. Clearly, the change over time lies in the *stoichiometric subspace* $S = \operatorname{im}(YI_E)$, that is, $\frac{dx}{dt} \in S$. Equivalently, trajectories are confined to cosets of S, that is, $x(t) \in x(0) + S$. For positive $x' \in \mathbb{R}_>^n$, the set $(x' + S) \cap \mathbb{R}_>^n$ is called a *stoichiometric class*.

The (stoichiometric) *deficiency* is given by

$$\delta = \dim(\ker Y \cap \operatorname{im} I_E) = |V| - \ell - \dim(S),$$

where $|V|$ is the number of vertices, and ℓ is the number of connected components of the digraph. Analogously, we introduce the *kinetic-order subspace* $\widetilde{S} = \operatorname{im}(\widetilde{Y} I_E)$ and the *kinetic(-order) deficiency* $\widetilde{\delta} = \dim(\ker \widetilde{Y} \cap \operatorname{im} I_E) = |V| - \ell - \dim(\widetilde{S})$.

A steady state $x \in \mathbb{R}^n_>$ of (2) that fulfills $A_k x^{\widetilde{Y}} = 0$ is called a positive *complex-balanced equilibrium* (CBE). On the one hand, if $\delta = 0$, then every equilibrium is complex-balanced. On the other hand, if there exists a CBE, then the underlying graph is weakly reversible, see the comment at the end of Sect. 4 in [11], and cf. [10, Proposition 2.18].

Theorem 2 *(robust $\delta = \widetilde{\delta} = 0$ theorem, Theorem 46 in [9]).* *For a generalized mass-action system, there exists a unique positive CBE in every stoichiometric class, for all rate constants and for all small perturbations of the kinetic orders, if and only if $\delta = \widetilde{\delta} = 0$, the network is weakly reversible, and $\operatorname{sign}(S) \subseteq \operatorname{sign}(\widetilde{S})$.*

We consider a CRN given by a graph with 5 vertices and 6 edges in 2 connected components, and we label the vertices with stoichiometric and kinetic-order complexes, respectively. (The two resulting labeled graphs are shown below.) The kinetic-order complexes involve parameters $a, b, c \in \mathbb{R}$.

$$A + B \rightleftarrows C \quad A \rightleftarrows E \qquad aA + bB \rightleftarrows C \quad A \rightleftarrows E$$
$$\searrow \downarrow \qquad\qquad\qquad \searrow \downarrow$$
$$\quad D \qquad\qquad\qquad\qquad cA + D$$

The resulting stoichiometric and kinetic-order subspaces are given by

$$S = \operatorname{im} \begin{array}{c} A \\ B \\ C \\ D \\ E \end{array} \begin{pmatrix} -1 & 0 & -1 \\ -1 & 0 & 0 \\ 1 & -1 & 0 \\ 0 & 1 & 0 \\ 0 & 0 & 1 \end{pmatrix} \quad \text{and} \quad \widetilde{S} = \operatorname{im} \begin{pmatrix} -a & c & -1 \\ -b & 0 & 0 \\ 1 & -1 & 0 \\ 0 & 1 & 0 \\ 0 & 0 & 1 \end{pmatrix}.$$

Clearly, the network is weakly reversible, and it is easy to verify that $\delta = \widetilde{\delta} = 0$. To study existence and uniqueness of complex-balanced equilibria, we compute the sign vectors of S and $\widetilde{S}$.

```
sage: from sign_vectors.oriented_matroids import *
sage: S = matrix([[-1,-1,1,0,0],[0,0,-1,1,0],[-1,0,0,0,1]])
sage: covectors_from_matrix(S, kernel=False)
{(00000), (00+-0), (0-0+-), (+000-), (+-++-), (0--+-),
 (+0+--), (+++--), (0-+--), (---+-), (--+0-), (--+--),
 (++0-0),  ..., (++-++), (--+0+), (---++), (-+0-+)}
```

Since $\widetilde{S}$ depends on the parameters a, b and c, the sign vectors also depend on these parameters and hence, we cannot compute them directly. Thus, we consider specific values for the parameters and determine the corresponding sign vectors.

```
sage: var('a, b, c');
sage: St = matrix([[-a,-b,1,0,0],[c,0,-1,1,0],[-1,0,0,0,1]])
sage: covectors_from_matrix(St(a=2, b=1, c=1), kernel=False)
{(00000), (+-++-), (0--+-), (+0+--), (+++--), (0-0+-),
    (0-+--), (---+-), (00-++), (+000-), (+0-+0), (--+--),
    (0++-0), ..., (++-+0), (-0+-0), (-+0-+), (-0+--)}
```

For $a = 2$, $b = 1$ and $c = 1$, $\text{sign}(S) \subseteq \overline{\text{sign}(\widetilde{S})}$. Consequently, the conditions of Theorem 2 are satisfied in this specific case. Obviously, we cannot cover all possible cases for the parameters that way. However, by expressing this condition in terms of maximal minors of the kernel matrices, we can compute with the parameters directly.

Proposition 2 (Proposition 32 in [9]) *For subspaces $S, \widetilde{S} \subseteq \mathbb{R}^n$ of dimension $n - d$ and matrices $W, \widetilde{W} \in \mathbb{R}^{d \times n}$ with $S = \ker W$, $\widetilde{S} = \ker \widetilde{W}$, and rank d, the following are equivalent:*

1. $\text{sign}(S) \subseteq \overline{\text{sign}(\widetilde{S})}$.
2. $\det W_I \neq 0$ *implies* $\det W_I \det \widetilde{W}_I > 0$ *for all subsets* $I \subseteq [n]$ *with* $|I| = d$ *(or "< 0" for all I).*

In the example, $S = \ker W$ and $\widetilde{S} = \ker \widetilde{W}$ with

$$W = \begin{pmatrix} 1 & 0 & 1 & 1 & 1 \\ 0 & 1 & 1 & 1 & 0 \end{pmatrix} \quad \text{and} \quad \widetilde{W} = \begin{pmatrix} 1 & 0 & a & a-c & 1 \\ 0 & 1 & b & b & 0 \end{pmatrix}.$$

By applying Proposition 2, we obtain several conditions on a, b and c. We use package [2] to compute this (and several other) sign vector condition(s).

```
sage: W  = matrix([[1, 0, 1, 1, 1], [0, 1, 1, 1, 0]])
sage: var('a, b, c');
sage: Wt = matrix([[1, 0, a, a - c, 1], [0, 1, b, b, 0]])
sage: from sign_vector_conditions import *
sage: condition_closure_minors(W, Wt)
[{a - c > 0, b > 0, a > 0}]
```

Hence, the network has a unique positive CBE if and only if $a, b > 0$ and $a > c$.

Uniqueness of CBE. Also the uniqueness of CBE (in every stoichiometric class and for all rate constants) — and hence its converse: multiple CBE – can be characterized in terms of a sign vector condition. For further details on sign vector conditions for injectivity in the context of CRNs and further references, we refer to [8].

Proposition 3 (cf. Proposition 3.1 in [10]) *For a generalized mass-action system, there exists at most one positive CBE in every stoichiometric class, for all rate constants, if and only if*

$$\text{sign}(S) \cap \text{sign}(\widetilde{S}^\perp) = \{0\}. \tag{3}$$

Because of the parameters, we cannot directly compute the sign vectors in the kernel of $\widetilde{S}$. Again, we use maximal minors to express (3).

Corollary 1 (cf. Corollary 4 in [9]) *For subspaces $S, \widetilde{S} \subseteq \mathbb{R}^n$ of dimension $n - d$ and matrices $W, \widetilde{W} \in \mathbb{R}^{d \times n}$ with $S = \ker W$, $\widetilde{S} = \ker \widetilde{W}$, and rank d, the following are equivalent:*

1. $\operatorname{sign}(S) \cap \operatorname{sign}(\widetilde{S}^\perp) = \{0\}$.
2. *Either* $\det W_I \det \widetilde{W}_I \geq 0$ *for all* $I \subseteq [n]$ *with* $|I| = d$, *or* $\det W_I \det \widetilde{W}_I \leq 0$ *for all* $I \subseteq [n]$ *with* $|I| = d$.

Comparing the maximal minors yields:

```
sage: condition_uniqueness_minors(W, Wt)
[{a - c >= 0, a >= 0, b >= 0}]
```

Hence, positive CBE are unique if and only if $a, b \geq 0$ and $a \geq c$.

Unique Existence of CBE. We discuss a novel algorithm for checking a certain degeneracy condition for subspaces that is part of a characterization of the unique existence of CBE. The result also involves the set of nonnegative sign vectors $T_\oplus = T \cap \{0, +\}^n$ of a set of sign vectors $T \subseteq \{-, 0, +\}^n$.

Theorem 3 ($\delta = \widetilde{\delta} = 0$ *theorem, Theorem 45 in [9]*). *For a generalized mass-action system, there exists a unique positive CBE in every stoichiometric class, for all rate constants, if and only if $\delta = \widetilde{\delta} = 0$, the network is weakly reversible, and*

1. $\operatorname{sign}(S) \cap \operatorname{sign}(\widetilde{S}^\perp) = \{0\}$;
2. *for all nonzero $\widetilde{\tau} \in \operatorname{sign}(\widetilde{S}^\perp)_\oplus$, there is a nonzero $\tau \in \operatorname{sign}(S^\perp)_\oplus$ with $\tau \leq \widetilde{\tau}$;* *and*
3. $(S, \widetilde{S})$ *is nondegenerate.*

We discussed the first condition above. To check the second condition, we use nonnegative cocircuits. For the third condition, we reformulate Definition 13 in [9], regarding the (non-)degeneracy of two subspaces.

Definition 2 *A pair $(S, \widetilde{S})$ of subspaces of $\mathbb{R}^n$ is called* degenerate *if there exists $z \in \widetilde{S}^\perp$ with a positive component such that*

1. *for all $I_\lambda = \{i \in [n] \mid z_i = \lambda\}$ for some $\lambda > 0$, there exists $\pi \in \operatorname{sign}(S)_\oplus$ such that $\pi_i = +$ iff $i \in I_\lambda$; and*
2. *for all nonzero $\tau \in \operatorname{sign}(S^\perp)_\oplus$, we have $\operatorname{supp} \tau \not\subseteq \operatorname{supp} z$.*

That is, the pair $(S, \widetilde{S})$ is degenerate if there exists $z \in \widetilde{S}^\perp$ such that its equal positive components are covered by nonnegative sign vectors of S, and there is no nonzero, nonnegative sign vector in $S^\perp$ such that its support is contained in $\operatorname{supp} z$.

Since we simply cannot iterate over the subspace $\widetilde{S}^\perp$, we have to find a different approach. Instead, we consider sets of nonnegative covectors in $\mathrm{sign}(S)$. Then, we use Theorem 1 (Minty's Lemma) to decide whether a vector z exists that has positive equal components on the support of each of these covectors.

We reformulate Definition 2 by demanding the existence of a set of nonnegative covectors that cover all equal positive components of a vector z.

Definition 3 *A pair* $(S, \widetilde{S})$ *of subspaces in* $\mathbb{R}^n$ *is degenerate if there exists a set of nonzero covectors* $T \subseteq \mathrm{sign}(S)_\oplus$ *with disjoint support and a vector* $z \in \widetilde{S}^\perp$ *such that*

1. *(a) for all $\pi \in T$, $\lambda_\pi = z_i = z_j > 0$ for all $i, j \in \mathrm{supp}\,\pi$,*
   (b) for all $i \notin \bigcup_{\pi \in T} \mathrm{supp}\,\pi$, $z_i \leq 0$, and
   (c) the λ_π's are pairwise distinct;
   and
2. *for all nonzero $\tau \in \mathrm{sign}(S^\perp)_\oplus$, we have $\mathrm{supp}\,\tau \not\subseteq \mathrm{supp}\,z$.*

If two λ_π's are equal, we compose the corresponding covectors to cover equal components of z. Therefore, Condition (1c) is redundant. For the same reason, the supports of the covectors do not need to be disjoint. Since nonnegative covectors can be represented as a composition of nonnegative cocircuits, it suffices to consider cocircuits instead of covectors. Further, note that we can check Condition 3 using cocircuits. Following these observations, we obtain another reformulation of Definition 2.

Definition 4 *A pair* $(S, \widetilde{S})$ *of subspaces in* $\mathbb{R}^n$ *is degenerate if there exists a set of nonnegative cocircuits* $C \subseteq \mathrm{sign}(S)_\oplus$ *and a vector* $z \in \widetilde{S}^\perp$ *such that*

1. *(a) for all $\pi \in C$, $z_i = z_j > 0$ for all $i, j \in \mathrm{supp}\,\pi$, and*
   (b) for all $i \notin \bigcup_{\pi \in C} \mathrm{supp}\,\pi$, $z_i \leq 0$;
   and
2. *for all cocircuit $\tau \in \mathrm{sign}(S^\perp)_\oplus$, we have $\mathrm{supp}\,\tau \not\subseteq \mathrm{supp}\,z$.*

To check degeneracy algorithmically, we iterate over sets of nonnegative cocircuits. In particular, Algorithm 1 below recursively constructs such sets and determines whether a corresponding z exists. One could store the cocircuits and construct a subspace of $\widetilde{S}^\perp$ using the conditions on equal components. Here, we modify the subspace such that its elements are equal on the components corresponding to the cocircuits in each step (line 11) and keep track of the positive entries using an index set. If this subspace contains a vector that is positive on exactly this index set (line 13), condition 1 holds. If this vector also satisfies condition 2 (line 13), it certifies degeneracy. For efficiency, we use cocircuits to check this condition. Note that we apply exists_vector from Sect. 1 to efficiently check for existence in line 13 and 18.

Algorithm 1: subspaces degenerate

1 Function degenerate($S, \widetilde{S}$):
2 $\quad C :=$ set of nonnegative cocircuits of sign(S)
3 $\quad \mathcal{V} := \widetilde{S}^\perp$
4 $\quad$ global $is\_degenerate := False$
5 $\quad$ recursive($C, \mathcal{V}, \emptyset$)
6 $\quad$ return $is\_degenerate$

7 Function recursive($C, \mathcal{V}, I$):
8 $\quad$ while $C \neq \emptyset$ do
9 $\quad\quad$ choose any $\pi \in C$
10 $\quad\quad$ $C := C \setminus \{\pi\}$
11 $\quad\quad$ $\overline{\mathcal{V}} :=$ subspace of $\mathcal{V}$ where vectors are equal on supp π
12 $\quad\quad$ $\overline{I} := I \cup \text{supp}\,\pi$
13 $\quad\quad$ if $z \in \overline{\mathcal{V}}$ exists with $z_i > 0$ iff $i \in \overline{I}$ then
14 $\quad\quad\quad$ for $\sigma \in \text{sign}(\overline{\mathcal{V}})$ with $\sigma_i = +$ iff $i \in \overline{I}$ do
15 $\quad\quad\quad\quad$ if supp $\tau \not\subseteq$ supp σ for all cocircuits $\tau \in \text{sign}(S^\perp)_\oplus$ then
16 $\quad\quad\quad\quad\quad$ $is\_degenerate := True$
17 $\quad\quad\quad\quad\quad$ return
18 $\quad\quad$ else if $z \in \overline{\mathcal{V}}$ exists with $z_i > 0$ if $i \in \overline{I}$ then
19 $\quad\quad\quad$ recursive($C, \overline{\mathcal{V}}, \overline{I}$)
20 $\quad\quad$ if $is\_degenerate$ then return
21 $\quad$ return

We consider Example 20 from [9]. Here, we have matrices

$$W = \begin{pmatrix} 0 & 0 & 1 & 1 & -1 & 0 \\ 1 & -1 & 0 & 0 & 0 & -1 \\ 0 & 0 & 1 & -1 & 0 & 0 \end{pmatrix} \quad \text{and} \quad \widetilde{W} = \begin{pmatrix} 1 & 1 & 0 & 0 & -1 & a \\ 1 & -1 & 0 & 0 & 0 & 0 \\ 0 & 0 & 1 & -1 & 0 & 0 \end{pmatrix}.$$

The existence of a unique positive CBE depends on $a > 0$.

```
sage: var('a'); assume(a > 0);
sage: W=matrix(3,6,[0,0,1,1,-1,0,1,-1,0,0,0,-1,0,0,1,-1,0,0])
sage: Wt=matrix(3,6,[1,1,0,0,-1,a,1,-1,0,0,0,0,0,0,1,-1,0,0])
```

The first two conditions of Theorem 3 are independent of a.

```
sage: condition_uniqueness_sign_vectors(W, Wt)
True
sage: condition_faces(W, Wt)
True
```

Condition 3 holds iff $a \in (0,1) \cup (1,2)$ as we demonstrate for specific values.

```
sage: condition_nondegenerate(W, Wt(a=1/2))
True
sage: condition_nondegenerate(W, Wt(a=2))
False
```

Acknowledgement. This research was funded in part by the Austrian Science Fund (FWF), grant 10.55776/P33218 to SM.

Disclosure of Interests.. The authors have no competing interests to declare.

References

1. Aichmayr, M.S.: elementary_vectors. https://github.com/MarcusAichmayr/elementary_vectors
2. Aichmayr, M.S.: sign_vector_conditions. https://github.com/MarcusAichmayr/sign_vector_conditions
3. Aichmayr, M.S., Regensburger, G.: How to certify solvability of linear inequality systems with elementary vectors (2024, in preparation)
4. Bachem, A., Kern, W.: Linear Programming Duality: An Introduction to Oriented Matroids. Springer-Verlag, Berlin (1992). https://doi.org/10.1007/978-3-642-58152-6
5. Björner, A., Las Vergnas, M., Sturmfels, B., White, N., Ziegler, G.: Oriented Matroids, 2nd edn. Cambridge University Press, Cambridge (1999). https://doi.org/10.1017/CBO9780511586507
6. Brualdi, R.A., Friedland, S., Pothen, A.: The sparse basis problem and multilinear algebra. SIAM J. Matrix Anal. Appl. **16**, 1–20 (1995)
7. Minty, G.J.: A 'from scratch' proof of a theorem of Rockafellar and Fulkerson. Math. Program. **7**, 368–375 (1974)
8. Müller, S., Feliu, E., Regensburger, G., Conradi, C., Shiu, A., Dickenstein, A.: Sign conditions for injectivity of generalized polynomial maps with applications to chemical reaction networks and real algebraic geometry. Found. Comput. Math. **16**, 69–97 (2016). https://doi.org/10.1007/s10208-014-9239-3
9. Müller, S., Hofbauer, J., Regensburger, G.: On the bijectivity of families of exponential/generalized polynomial maps. SIAM J. Appl. Algebra Geom. **3**, 412–438 (2019). https://doi.org/10.1137/18M1178153
10. Müller, S., Regensburger, G.: Generalized mass action systems: complex balancing equilibria and sign vectors of the stoichiometric and kinetic-order subspaces. SIAM J. Appl. Math. **72**, 1926–1947 (2012). https://doi.org/10.1137/110847056
11. Müller, S., Regensburger, G.: Generalized mass-action systems and positive solutions of polynomial equations with real and symbolic exponents (invited talk). In: Gerdt, V.P., Koepf, W., Seiler, W.M., Vorozhtsov, E.V. (eds.) CASC 2014. LNCS, vol. 8660, pp. 302–323. Springer, Cham (2014). https://doi.org/10.1007/978-3-319-10515-4_22
12. Richter-Gebert, J., Ziegler, G.M.: Oriented matroids. In: Handbook of Discrete and Computational Geometry, pp. 111–132. CRC Press, Boca Raton, FL (1997)
13. Rockafellar, R.T.: The elementary vectors of a subspace of $\mathbb{R}^n$. In: Combinatorial Mathematics and Application, Proceedings of Conference on University North Carolina 1967, pp. 104–127 (1969)
14. Rockafellar, R.T.: Convex Analysis. Princeton University Press, Princeton (1970)
15. Ziegler, G.M.: Lectures on Polytopes. GTM, vol. 152. Springer, New York (1995). https://doi.org/10.1007/978-1-4613-8431-1

Machine Learning Within Computer Algebra Systems

Symbolic Integration Algorithm Selection with Machine Learning: LSTMs Vs Tree LSTMs

Rashid Barket[1](✉), Matthew England[1], and Jürgen Gerhard[2]

[1] Coventry University, Coventry, UK
{barketr,matthew.england}@coventry.ac.uk
[2] Maplesoft, Waterloo, ON, Canada
jgerhard@maplesoft.com

Abstract. Computer Algebra Systems (e.g. Maple) are used in research, education, and industrial settings. One of their key functionalities is symbolic integration, where there are many sub-algorithms to choose from that can affect the form of the output integral, and the runtime. Choosing the right sub-algorithm for a given problem is challenging: we hypothesise that Machine Learning (ML) can guide this sub-algorithm choice. A key consideration of our methodology is how to represent the mathematics to the ML model: we hypothesise that a representation which encodes the tree structure of mathematical expressions would be well suited. We trained both an LSTM and a TreeLSTM model for sub-algorithm prediction and compared them to Maple's existing approach. Our TreeLSTM performs much better than the LSTM, highlighting the benefit of using an informed representation of mathematical expressions. It is able to produce better outputs than Maple's current state-of-the-art meta-algorithm, giving a strong basis for further research.

Keywords: Computer Algebra · Symbolic Integration · Machine Learning · LSTM · TreeLSTM · Data Generation

1 Introduction

Machine Learning (ML), and specifically deep learning, has seen a surge of applications in many domains, but only recently have there been applications to computer algebra. One can take two possible approaches when using ML in this field: to directly make predictions to solve a problem, or to aid existing algorithms in their free choices to improve an objective function.

We focus on the Computer Algebra System (CAS) Maple, and its main symbolic integration algorithm, `int`, which is essentially a meta-algorithm to choose from 12 possible sub-algorithms Maple has for indefinite integration[1]. The names of each sub-algorithm are available later in Fig. 4. Each can produce very different, but mathematically equivalent answers, as in the example in Fig. 1. Some can also take much longer to execute than others.

---

[1] https://www.maplesoft.com/support/help/Maple/view.aspx?path=int

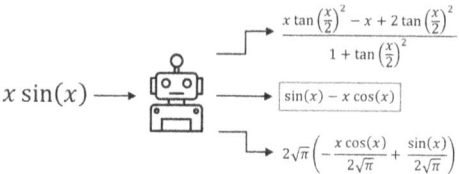

Fig. 1. The output of $\int x\sin(x)dx$ from three successful sub-algorithms. The optimal output is the shortest expression from the second sub-algorithm.

In this paper, we will train Long Short-Term Memory (LSTM) and TreeLSTM models to select the sub-algorithm that produces the optimal length answer for given problems. This is important to a Maple user who would prefer a simpler expression if available. Because the objective function is based on the output length, multiple sub-algorithms may be optimal. Thus, this is a multi-label classification problem.

We proceed with a brief literature review in Sect. 2, before outlining our machine learning methodology and data generation processes in Sects. 3 and 4. We give our experimental results in Sect. 5 and then conclude in Sect. 6.

2 Literature Review

The problem of using ML to improve computer algebra algorithms has gained traction within the last decade. One of the first uses was for the cylindrical algebraic decomposition algorithm where the choice of variable ordering can be key to tractability. Huang et al. made the first attempt in 2014 using a Support Vector Machine [3], with more recent work on this problem involving reinforcement learning and graph neural networks [4] and Explainable AI techniques [6].

The most relevant work for our problem was by Lample & Charton [5] who trained a transformer to calculate integrals directly, learning from a large quantity of (integrand, integral) pairs. They could calculate some integrals that CASs could not, but there are critiques of their data generation and testing methods [7].

Simpson et al. gave an example of ML for computer algebra algorithm selection in [8]: training ML to select from four algorithms to calculate the resultant of two polynomials, reducing CPU time in both Maple and Mathematica.

3 Machine Learning

When integrating with the 12 different sub-algorithms in Maple, each one can produce an answer or output failure. Our objective function, chosen in consultation with Maplesoft, is to minimise the length of the output. With this objective, multiple sub-algorithms may be optimal making this a multi-label classification problem.

The baseline approach of binary relevance is used in this experiment. If L is the set of possible labels, we train $|L|$ binary classifiers $C_1, ..., C_{|L|}$. Each C_j is responsible for predicting the binary classification for each $l_j \in L$. Each binary model C_j is trained independently from the rest. We will consider two ML models to tackle the problem introduced in the following sections.

This problem has not been tackled by ML before. To judge the quality of our models we will compare them to Maple's meta-algorithm implemented for `int`.

3.1 LSTM

Long-Short Term Memory networks were proposed in [2], and were the model of choice for many natural language tasks such as sentiment analysis and machine translation until transformers rose to popularity more recently. Like recurrent neural networks, LSTMs calculate an output y_t based on the input at that time step, x_t, and the hidden state from the previous step, h_{t-1}. The key difference is the inclusion of a long-term memory which is maintained and updated over a long sequence of time. Three new gating mechanisms (the forget, input, and output gates) are introduced to maintain this long-term memory. As sub-algorithm selection for symbolic integration is not a task that has been explored before, there is no baseline to compare to. Thus, we will use LSTM as the baseline to compare against other models in this and future research.

3.2 TreeLSTM

A limitation of the LSTM network is that it can only process sequential information (e.g. text, audio, time-series data). We hypothesise that embedding data with a tree structure (how a CAS like Maple stores mathematical expressions) would be beneficial.

TreeLSTM is a variant of the LSTM network proposed by Tai et al. [9] for tree-structured inputs. Like LSTM, it contains a long-term memory with a gating architecture. However, the hidden state and cell memory of step t is dependent on arbitrarily many children instead of a single child from step $t - 1$. There are forget gates for each child so the network can learn which information is important from which child to add to the hidden state and cell memory. The structure of the TreeLSTM is compared to LSTM in Fig. 2.

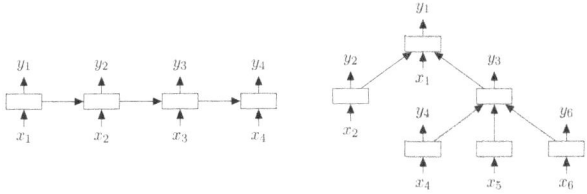

Fig. 2. Visual representation of an LSTM (left) and TreeLSTM (right) [9].

4 Generating Data

As with any ML problem we need a source of data, in our case integrable mathematical expressions. We need a rich variety so that the model can generalise.

4.1 Existing Methods

Lample & Charton introduced three data generation methods in [5]:
- **FWD:** Integrate an expression f through a CAS to get F and add the pair (f, F) to the dataset.
- **BWD:** Differentiate f to get f' and add the pair (f', f) to the dataset.
- **IBP:** Given two expressions f and g, calculate f' and g'. If $\int f'g$ is known then the following holds (integration-by-parts): $\int fg' = fg - \int f'g$. Thus, we add the pair $(fg', fg - \int f'g)$ to the dataset.

These offer a starting point but do not generate the rich variety we need. A thorough discussion of the shortcomings of these methods was given in [1].

4.2 New Methods

We presented a new data generation method, **RISCH**, for elementary integrable expressions in [1]. This is based on the Risch method for symbolic integration and was shown to generate different types of integrals from the previous three.

We have developed another method, **SUB**, which like IBP uses an existing method to generate more data: this time the substitution rule.

Theorem 1 (Substitution Rule). *If $u = g(x)$ is a differentiable function whose range is an interval I and f is continuous on I, then*

$$\int f(g(x))g'(x)dx = \int f(u)du.$$

This means that if $g(x)$ is any differentiable function and $f(x)$ is any integrable function, then we can use Theorem 1 to generate an (integrand, integral) pair. Indeed, if $\int f = F$, then by Theorem 1, we have $\int f(g(x))g'(x)dx = F(g(x))+C$. Like the IBP method, SUB requires a dataset of integrable expressions for $f(x)$. We can use the FWD, BWD, and RISCH methods to generate such a dataset. This method is simple but effective at generating many expressions.

4.3 Datasets of Integrals

Existing datasets of integrals were thoroughly discussed in Sect. 2.2 of [1].

Maplesoft curates a dataset for indefinite integration for testing their integration function during each release: 47,750 examples exist ranging from very simple expressions to complex expressions with many parameters and special functions. Rather than train on this data, we will keep it aside for validation. This is important to show that ML models can generalise outside of the data they are trained on. As our domain for the data generation methods only consists of elementary functions, we only use the 7413 elementary expressions in the Maple test suite for validation at the moment.

5 Experimental Results

5.1 Experiment Setup

Dataset Preparation: Before training any model, we must first prepare a dataset to train on. In Sect. 4 we discussed five different data generation methods: FWD, BWD, IBP, RISCH, and SUB. The ML models will be trained on 20,000 examples from each of the five methods for a total of 100,000 samples.

To label the data, we must first decide on what it means to be optimal. Maple stores all its expressions as a directed acyclic graph (DAG). It differs from a binary tree representation in that whenever a node is repeated, it does not make a new child for that node. Rather, any common sub-expressions only generate one node in memory and are referenced in multiple places where they occur in the expression. This is useful for avoiding redundant storage of identical sub-expressions. An example of a DAG for an expression is shown in Fig. 3.

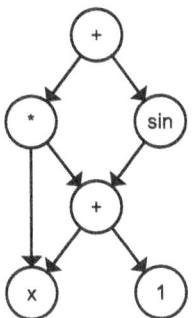

Fig. 3. The DAG structure for the expression $x * (x + 1) + \sin(x + 1)$. Note how there is repeated use of the node x and the sub-tree $x + 1$.

To measure the DAG size, we use a custom Maple function: `expr -> length(sprintf("%lm",expr))`. This is measuring the length of the serialized format of a DAG in Maple. We optimise on this when training ML. Figure 4 shows the distribution of optimal sub-algorithms. The dataset is imbalanced: more data generation ideas and data balancing tools are future work.

Preprocessing: We pre-process the dataset by first removing any expressions where every single algorithm was unsuccessful. We then replace the integers within each expression in the following manner:

1. if the integer is in the range $[-2, 2]$, then nothing changes;
2. if the integer is single-digit not in the range $[-2, 2]$, replace by CONST token;
3. if the integer has two digits, replace by a CONST2 token; and
4. for all other cases, replace by a CONST3 token.

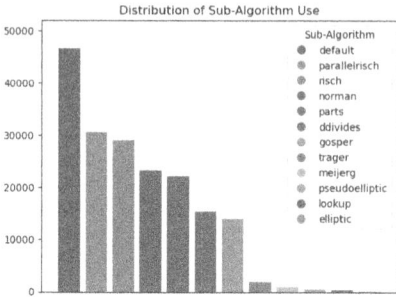

Fig. 4. Optimal sub-algorithms for 100k integrands.

The rationale here is that while normally the coefficients of an expression would not change, this is not the case for small exponents ($(x+1)^1$ and $(x+1)^{-1}$ integrate very differently). The range $[-2, 2]$ captures these special properties, while for the rest, we only differentiate by the number of digits. In [7] the authors critique [5] for having many examples equivalent up to constants, risking data leakage. We only keep unique copies of expressions after replacing the integers with CONST tokens, reducing the training data by roughly 10%.

Model Design: As discussed earlier, we perform binary relevance for each sub-algorithm and train both LSTM and TreeLSTM models. All hyperparameters are kept the same between the two models so we can directly compare the architectures. Note that these hyperparameters have not been fully optimised for these preliminary results. Each model consists of an embedding layer, and two layers (LSTM or TreeLSTM) with the first having 64 cells and the second having 32 cells. We also include 40% dropout on the second layer to help avoid overfitting. We then have a fully connected with ReLU activation, and finally a single cell Sigmoid-activated output of the probability. Both LSTM and TreeLSTM were implemented with Pytorch as the backend, and the TreeLSTM utilised a specific graph learning library called Deep Graph Learning [10].

Evaluation: Each expression will have a probability vector given from the model (the probability of the sub-algorithm being optimal). To evaluate our models we let them select the sub-algorithm with the highest probability; and if that algorithm is not successful we try the next most probable.

We will test against each other and Maple's existing meta-algorithm. We maintain a separate test set of 25,000 expressions with an equal split from all five generators. The test has only integrands that will be successful so there is always an answer. We will also evaluate on the 7413 examples from the Maple test suite to determine if the models can generalise to different data.

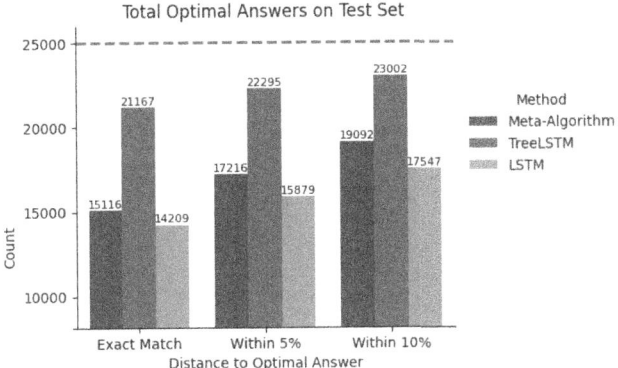

Fig. 5. The number of times each ML model and Maple's meta-algorithm produced the optimal answer, or came close to it, on the testing dataset. The optimal answer is the sub-algorithm that produces the smallest DAG size (an integer) compared to all other sub-algorithms.

5.2 Experiment Results

A code implementation of the experiments used to generate the results is available at https://github.com/rbarket/Int_Algo_Selection. Note that the Maple Test Suite is proprietary so the code implementation is available for the generated test data.

The models took 312 s and 178 s to train for the TreeLSTM and LSTM models respectively, when averaged over training each binary classifier on a single GPU.

Results on the Testing Dataset: Results for the 25,000 test cases are summarised in Fig. 5. TreeLSTM is the dominant model, predicting 84.6% of exact optimal answers compared to 60.5% and 56.8% from Maple's meta-algorithm and LSTM respectively. This is the case even when we allow for the algorithms to only get close to the optimal and still be considered a success.

We analysed how many problems each uniquely predict correctly: 441 for Maple's meta-algorithm, 367 for the LSTM and 2631 for the TreeLSTM. So there are only small quantities of examples where the TreeLSTM is outperformed.

This results are promising, especially considering the low amount of training time and relatively small size of the dataset (compared to other ML applications). Scaling both these factors up should improve performance further. This clearly validates our hypothesis on the benefits of the tree-based representation for mathematical expressions when compared to a sequence of tokens. Before we conclude the benefits of an ML approach, we should judge generalisability of the models by validating performance on independent data.

Results on the Maple Test Suite: Performance on the Maple test suite is summarised in Fig. 6 and paints a somewhat different picture. The TreeLSTM still has the most optimal answers, however, this time allowing a 5% or

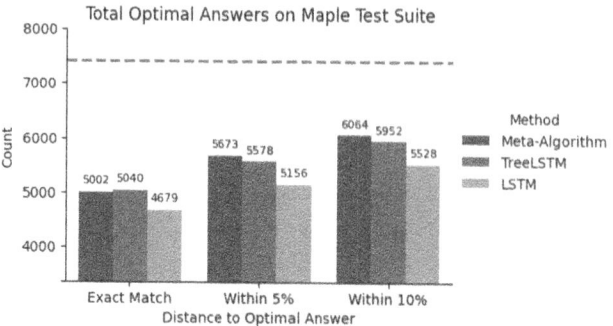

Fig. 6. The number of times each ML model and Maple's meta-algorithm produced the optimal answer, or came close to it, on the Maple Test Suite.

10% margin of error lets Maple's meta-algorithm do better. Both outperform the LSTM. These results demonstrate that the ML models generalise to independently sourced data. Comparing the Maple's meta-algorithm only to the TreeLSTM: the former has 578 uniquely optimal answers and the latter 633.

6 Conclusion

We are able train a ML model to do sub-algorithm selection for symbolic integration to outperform the current state-of-the-art meta-algorithm for the task in Maple. The representation of our data plays a crucial role: the TreeLSTM and LSTM models were the same up to their unique architecture layers, demonstrating the benefit of a tree embedding over a simple sequence of tokens.

Importantly, the TreeLSTM is also competitive with Maple's meta-algorithm on data produced independently from the training set. This is important to show the value of pursuing such an approach for use by Maple in a general-purpose integration routine. Such generalisation was something [5] failed to demonstrate.

We are confident that with an increase in the quantity of training data, and hyperparameter optimisation, the ML models can improve further still.

References

1. Barket, R., England, M., Gerhard, J.: Generating elementary integrable expressions. In: Boulier, F. (ed.) Computer Algebra in Scientific Computing. LNCS, vol. 14139, pp. 21–38. Springer Nature, Switzerland (2023). https://doi.org/10.1007/978-3-031-41724-5_2
2. Hochreiter, S., Schmidhuber, J.: Long short-term memory. In: Neural Computation, vol. 9, pp. 1735-1780. MIT Press, Cambridge, MA, USA (1997). https://doi.org/10.1162/neco.1997.9.8.1735
3. Huang, Z., England, M., Wilson, D., Davenport, J.H., Paulson, L.C., Bridge, J.: Applying machine learning to the problem of choosing a heuristic to select the

variable ordering for cylindrical algebraic decomposition. In: Watt, S.M., Davenport, J.H., Sexton, A.P., Sojka, P., Urban, J. (eds.) CICM 2014. LNCS (LNAI), vol. 8543, pp. 92–107. Springer, Cham (2014). https://doi.org/10.1007/978-3-319-08434-3_8

4. Jia, F., Dong, Y., Liu, M., Huang, P., Ma, F., Zhang, J.: Suggesting variable order for cylindrical algebraic decomposition via reinforcement learning. In: Thirty-seventh Conference on Neural Information Processing Systems (NeurIPS) (2023). https://openreview.net/forum?id=vNsdFwjPtL

5. Lample, G., Charton, F.: Deep learning for symbolic mathematics. In: Proceedings of International Conference on Learning Representations (ICLR) (2020). https://doi.org/10.48550/arxiv.1912.01412

6. Pickering, L., del Río Almajano, T., England, M., Cohen, K.: Explainable AI insights for symbolic computation: a case study on selecting the variable ordering for cylindrical algebraic decomposition. J. Symb. Comput. **123**, 102276 (2024). https://doi.org/10.1016/j.jsc.2023.102276

7. Piotrowski, B., Urban, J., Brown, C.E., Kaliszyk, C.: Can neural networks learn symbolic rewriting? In: Proceedings of Artificial Intelligence and Theorem Proving (AITP) (2019). https://doi.org/10.48550/arXiv.1911.04873

8. Simpson, M.C., Yi, Q., Kalita, J.: Automatic algorithm selection in computational software using machine learning. In: 15th International Conference Machine Learning and Applications (ICMLA), pp. 355–360 (2016). https://doi.org/10.1109/ICMLA.2016.0064

9. Tai, K.S., Socher, R., Manning, C.D.: Improved semantic representations from tree-structured long short-term memory networks. In: Proceedings of 53rd Annual Meeting of the ACL and the 7th International Joint Conference on NLP, vol. 1, pp. 1556–1566. Association for Computational Linguistics (2015). https://doi.org/10.3115/v1/P15-1150

10. Wang, M., et al.: Deep graph library: a graph-centric, highly-performant package for graph neural networks. arXiv:1909.01315 (2019). https://www.dgl.ai/

Exploring Alternative Machine Learning Models for Variable Ordering in Cylindrical Algebraic Decomposition

Rohit John and James Davenport(✉)

University of Bath, Bath BA2 7AY, UK
masjhd@bath.ac.uk

Abstract. Cylindrical Algebraic Decomposition is a computer algebra tool with many applications, from robotics to biochemistry. But it can be very sensitive to the ordering of the variables, which may be partially prescribed by the problem, but generally has at least some freedom. While various algorithmic heuristics to choose the best variable order exist, we are looking at using new machine learning models to pick the variable order directly. Of those machine learning methods we have currently implemented, Feed-forward networks seem the most successful (though some others are nearly as good), and much better than a traditional hand crafted heuristic such as Brown's. We also explore an implementation of Graph Neural Networks as well as possible data pollution in current CAD datasets.

Keywords: Machine Learning · Computer Algebra · Cylindrical Algebraic Decomposition · Variable Ordering

1 Introduction

Cylindrical Algebraic Decomposition (CAD) is a method introduced by George Collins [5] for solving real polynomial systems and performing real quantifier elimination (QE). It segments multi-dimensional spaces into simpler regions, or cells and while introduced as a tool for quantifier elimination, it now has many uses in the resolution of complex algebraic challenges. CAD's utility spans various scientific disciplines, from robotics to epidemic modeling and the validation of economic hypotheses.

CAD comprises two primary phases: projection and lifting. The projection phase reduces the problem's dimensionality by eliminating variables in a specific order, while the lifting phase constructs the decomposition based on this variable ordering. If we are using CAD for QE, the structure of the quantifiers imposes constraints on the choice of variable ordering, but generally there are multiple options for the variable ordering. The choice of variable ordering significantly affects the CAD process's performance and complexity with [3] highlighting how different orderings can lead to vastly different computational complexities, ranging from doubly exponential to constant size. For example, CAD applied to

$\exists x.ax^2 + bx + c = 0$ yields 27 cells using the ordering $a \prec b \prec c$ and 115 using the opposite (note that we have no choice over x: it must be the first variable eliminated).

There is no known method, short of trying all possibilities, for deciding what is the absolute best variable ordering for a given problem. Various algorithmic suggestions, effectively heuristics, exist: commonly used ones are Brown's [2], sotd [6] and ndrr [1].

In such a situation, we can consider using Machine Learning, either to choose which of the algorithmic suggestions to use, or to choose the variable ordering directly. This paper will explore various models in directly selecting optimal variable orderings, as well as a new implementation of Graph Neural Networks (GNN). We will also briefly cover improvements to existing datasets of CAD problems. Fuller details are in the first author's dissertation [12][1]

2 Background

2.1 CAD and Variable Ordering

Cylindrical Algebraic Decomposition (CAD) is a foundational algorithm in computational algebraic geometry, specifically designed for decomposing multi-dimensional space into distinct, non-overlapping regions, or cells. These cells are described through semi-algebraic conditions, which means they are defined by a finite set of polynomial equations and inequalities.

The cylindrical projection property of CAD, which ensures that the projection of any two cells is either identical or disjoint, is defined with respect to a specific ordering, which determines the hierarchical projection of cells onto lower-dimensional spaces.

Let us say there are n cells in $\mathbf{R}^1$ (arranged in increasing order), and above the ith such cell there are n_i cells in $\mathbf{R}^2$ (arranged in increasing order of the new coordinate), and above the jth such cell there are $n_{i,j}$ cells in $\mathbf{R}^3$ (arranged in increasing order of the new coordinate), and so on.

Definition 1 (Implicit in [15]). *The sequence* $[n, n_1, \ldots, n_n, n_{1,1}, \ldots, n_{1,n_1}, n_{2,1}, \ldots, n_{n,n_n}, n_{1,1,1}, \ldots]$ *is called the* tree structure *of the CAD.*

If two problems P_1 and P_2 are such that, for every variable ordering, the CADs for P_1 and P_2 have the same tree structures, then [15] say that P_1 and P_2 are duplicates.

The primary function of CAD is to facilitate Quantifier Elimination (QE) in the real number domain. It does this by producing a decomposition where the truth value of a given logical formula remains consistent within each cell. This consistency allows for the transformation of complex, quantified formulas into simpler, quantifier-free versions by examining a finite number of sample points within the cells.

---

[1] The code for all of these models alongside datasets which will be released shortly are available on https://github.com/rohitpj/New_ML_Models_for_CAD..

The algorithm operates in two main phases: the projection phase, which reduces the dimensionality of the polynomials, and the lifting phase, which incrementally reconstructs the higher-dimensional decompositions based on the outcomes of the projection phase. This process ensures that each cell is appropriately invariant for the initial set of polynomials, maintaining constant signs, truth values or other invariants across it.

The choice of variable ordering in CAD is not merely procedural but can significantly influence the complexity and efficiency of the decomposition. This ordering affects both the theoretical and practical performance of CAD, impacting the number of cells produced and, consequently, the computational resources required for QE tasks. Over the years, various heuristics have been developed to optimize variable ordering.

There is no known algorithm (short of brute-force enumeration) for finding the absolute best ordering for a given problem, hence we need a heuristic method. Traditional heuristics like sotd [6], ndrr [1], and Brown's [2] have been developed to guide variable ordering. Although sotd and ndrr might provide more accurate results by utilizing the projection phase, their real-world efficiency is debatable due to the computational resources required. Brown's heuristic[2], based on simple statistics derived from the polynomials, strikes a balance between accuracy and resource efficiency. Additionally it has been demonstrated that most (if not all) heuristics developed can be misled, and so the need for a computationally efficient and accurate heuristic has developed. As noted by the referees, [13,15] introduce new heuristics which seem to outperform Brown.

2.2 Machine Learning in CAD

Machine learning (ML) encompasses a diverse set of algorithms and methods designed for creating models that learn patterns and rules from data, rather than relying on explicit programming for every decision. This approach is particularly valuable in scenarios where the relationships within the data are complex or not well-understood in advance. When we say "learn from data", what we often mean is "learn from certain developer-selected features of the data", and the choice of features is often critical.

1. Prior research by the second author's team, culminating in [9], considered the input as a set of polynomials, and had features such as "the maximum degree of x_1 among the variables" (Table 1). The dataset was problems in three variables taken from the nlsat dataset[3]. This was looking for a meta-heuristic, i.e. should we use sotd, ndrr or Brown's? Both sotd and ndrr need substantial additional computations to decide their recommended ordering, so we should not use them if Brown's is adequate.
    2. [7] used the same dataset and set of features, but looked at direct choice of the ordering.

---

[2] [11] classifies "triangular" [4] with Brown's, this seems to be a misapprehension: [4] describes a different method from that of projection/lifting.
[3] Originally at https://cs.nyu.edu/~dejan/nonlinear/.

N.B. Both the above were essentially limited to a fixed number of variables, i.e. we could train, and ask questions, about three variables (as these did), but four variables would require a fresh set of features and a fresh training set. Conversely the next approach is suited to sparser problems in any number of variables.
3. A very different approach was taken in [11], which abstracted a problem as a graph whose nodes were the variables, with an edge between x_i and x_j if there is a polynomial in the problem which contains both x_i and x_j. Graph Neural Networks (GNN) have been applied in conjunction with Reinforcement Learning in [11] to choose the variable ordering, however the authors are not aware of an approach that does not use Reinforcement Learning for solving this problem. In this GNN-based method, each variable becomes a node in the graph, and we therefore need per-variable features (see Table 2).

3 Our Methodology

We implemented several Machine Learning based models, among them Linear Regression (LR), K-Nearest Neighbours (KNN), Decision Trees (DT), Extreme Gradient-Boosting (XGB) and Feed-Forward Networks (FFN) which used both regression and classification.

Additionally we developed a new method of using Graph Neural Networks to directly select optimal orderings, and while our results were comparable to implementations that partially used GNNs, they did not hold up to other models. All of these models were used to directly select the most likely ordering.

3.1 Datasets

The issue of insufficient training data has been explored in [8,16]: both using similar methods to balance and augment the MetiTarski and QF_NRA datasets respectively. We will be comparing our data to [8] in this paper. By permuting the variables names in each polynomial set 6 new sets can be formed (assuming 3 variables) each with a different label.

Our main dataset used polynomial sets from [8] who we thank for their work. As stated in [8], the original MetiTarski dataset is heavily imbalanced, with the ordering (x_3, x_2, x_1) being 4 times more likely to be the optimal ordering than any of the others.

One of the minor limitations we encountered was the dataset only labelling one ordering as correct while other orderings share identical computation times and total cells. This may be leading to issues in training where correctly selected orderings may be erroneously discarded and depending on the tie-break method between orderings, can lead to to over fitting. We identify 15% of the dataset with duplicate lowest computation times.

In our experiments in data pollution we trained our models on 4 iterations of the original dataset, Unbalanced, Balanced, Augmented and Shuffled.

Let us suppose there are $5N$ problems $P_1, \ldots, P_{5N}$. We can consider these problems in various ways.

Unbalanced. The original data set, where (as pointed out in [8]) (x_3, x_2, x_1) is nearly four times as likely to be the right choice, compared with (x_1, x_2, x_3). Split 80:20, i.e. $P_1, \ldots, P_{4N}/P_{4N+1}, \ldots, P_{5N}$, for training/testing.

Balanced. The original data set, but each problem P_i is replaced by P'_i: the same problem under a random permutation of variables. Split 80:20 i.e. $P'_1, \ldots, P'_{4N}/P'_{4N+1}, \ldots, P'_{5N}$, for training/testing.

Augmented. The original data set, but each problem P_i is replaced by six problems $P_{i,1}, \ldots, P_{i,6}$, i.e. all six permutations of the three variables. Split 80:20 i.e. $P_{1,1}, P_{1,2} \ldots, P'_{4N,6}/P'_{4N+1,1}, \ldots, P'_{5N,6}$, for training/testing.

Shuffled. As Augmented, except that the data set is shuffled before being split, so each P_i would typically have 4 or 5 representatives in training, and 1 or 2 in testing.

We would echo the statement in [11]: "Besides, the lack of sufficiently large datasets is also a matter of urgency". A lot of their data was generated by use of Maple `randpoly` function, which generates, by default (which appears to be what was used), polynomials with five terms, irrespective of the number of variables. Obviously one can generate large amounts of random data but these may not be representative of actual problems.

3.2 Training and Features

Two feature sets were compared for these models, the first were used in [8,10] with 11 features based on the degrees and proportion of each variable in the equations.

Table 1. General Feature Set, from [7,8,10].

Feature Number	Description
1	Number of polynomials.
2	Maximum total degree of polynomials.
3	Maximum degree of x_0 among all polynomials
4	Maximum degree of x_1 among all polynomials.
5	Maximum degree of x_2 among all polynomials
6	Proportion of x_0 occurring in polynomials.
7	Proportion of x_1 occurring in polynomials.
8	Proportion of x_2 occurring in polynomials.
9	Proportion of x_0 occurring in monomials.
10	Proportion of x_1 occurring in monomials.
11	Proportion of x_2 occurring in monomials.

The second feature set was adapted from the one used in [11] with the 5th and 6th features being removed after performing variable importance analysis. Each variables features were concatenated for 36 total features.

Table 2. Per Variable Feature Set, from [11].

Number	Description
1	Number of other variables occurring in the same polynomials
2	Number of polynomials containing the variable
3	Maximum degree of the variable among all polynomials
4	Sum of degree of the variable among all polynomials
5	Maximum degree of all terms containing the variable
6	Sum of degree of all terms containing the variable
7	Sum of degree of leading coefficient of the variable
8	Sum of number of terms containing the variable
9	Proportion of the variable occurring in polynomials
10	Proportion of the variable occurring in terms
11	Maximum number of other variables occurring in the same term
12	Maximum number of other variables occurring in the same polynomial

Training took place using the Google Colab resource and Hex, the University of Bath's GPU cloud. In line with previous work we additionally performed hyperparameter tuning using grid-search with 5-fold cross validation.

3.3 Implementation of General Models

The implementation of the commonly used Classification models (LR, KNN, DT, GBM, etc.) used the Python sklearn libraries and were relatively simple to implement. We implemented two Feed-Forward Networks for this problem, one aiming to directly classify the most optimal ranking in the same manner as the previously described models, and the other ranking the possible orderings using regression and selecting the highest ranked: a similar model was described in [14]. These models were based on TensorFlow and had identical architectures apart from the final activation functions, with softmax being used for the first model and a linear function for the second. Categorical cross-entropy was used as the loss function for the first model and mean-squared-error for the second.

3.4 Implementation of GNN Models

We began by converting our polynomial sets into graph form, which could then be used to train our GNN. The approach we took has been described in [11], which viewed each variable in the set as a node, and an edge existing between two nodes if the corresponding variables appeared in the same polynomial. We experimented with two forms of GNN, mirroring the style of our FFN.

Our initial implementation aimed to classify the whole graph into one of the 6 possible orderings, however this failed to utilise the most appealing aspect of GNNs, their ability to process graphs of different sizes which would allow our model to function on polynomial sets with larger variable sizes.

We additionally explored node classification as an alternative. This method assigned each node one of the possible positions in the ordering instead of classifying the graph as a whole, and this method performed better than the initial implementation.

3.5 Exploration of Data Pollution

Our initial implementation was trained and tested using the balanced_processed file from [8], however this dataset was shuffled. Simply splitting this dataset into training and testing sets would lead to permutations of polynomial sets being spread across the dataset, possibly leading to data pollution.

To correct for this issue and test the possible effects of data pollution we first sorted the dataset by file id and then split it, resulting in train/test sets with no mixing of permutations. We then compared models trained on the shuffled dataset (with testing instances removed) with models trained on our sorted dataset and a balanced dataset, which was the result of removing all but one of the permutations of every unique polynomial set in the sorted dataset. The original, unbalanced MetiTarski dataset was also trained as a comparison, however this performed very poorly.

Unfortunately this issue was discovered relatively late in the timeline of our work, and so was only tested on a subsection of our models.

4 Results

We compare our models with a random choice approach, and with Brown's heuristic. Brown's heuristic only requires information from the initial polynomials, making it a fair comparison with our heuristics, whereas sotd and ndrr

Table 3. Comparison of Model Accuracy Across Different Feature Sets

Model	Original Features	Extended Features
Random	16.67%	16.67%
Brown	32.15%	32.15%
LR	42.12%	49.09%
KNN	54.12%	56.56%
DT	54.44%	58.48%
XGB	56.55%	59.08%
FFN Classifying	57.60%	**59.95%**
FFN Ranking	**57.67%**	59.91%
Graph GNN	—	23.20%
Node GNN	—	36.78%

perform substantial additional computations which were not published in the dataset we used. We measure the proportion of times the models predict the optimal variable ordering. GNN models were not trained on the original feature sets since they require one feature set for each node

Feed-Forward Networks performed the best on both feature sets, with both classification and regression methods achieving very similar levels of accuracy. While every model saw an increase in accuracy when trained on the extended feature set, Linear Regression improved the most but still falls short of the more sophisticated models.

Table 4. Comparison of Model Accuracy Across Different Training Sets

Model	Augmented	Balanced	Unbalanced	Shuffled
LR	43.22	**44.67**	23.94	47.28
KNN	26.11	25.96	15.01	61.68
DT	36.48	35.35	22.41	59.10
XGB	**47.43**	40.54	**37.18**	61.49
FFN	42.64	32.7	30.09	**62.00**

In Table 4 we can clearly see the effect of training our model on a shuffled dataset since the test set contains permutations of polynomial sets which it already been trained on. When the models are trained on a sorted dataset with no mixing of permutations there is a large drop in accuracy compared to both the Shuffled model and the original models shown in Table 3. For certain models data augmentation may not be useful, and potentially even harmful. In light of the issues demonstrated in Table 4, Table 3 should be viewed as a comparison of the two feature sets.

While our implementation of GNNs were not as accurate as our other models with a test accuracy of 38%, we aim to explore this model further to ensure our implementation is not fundamentally flawed. This may be due to the training data's small graph size, possible making it difficult for the model to distinguish between variables. We will also explore the feature set used to ensure our conversion from polynomial sets to node features were correct. Their ability of GNNs to process polynomials with different numbers of variables makes them more practical for implementations as opposed to models which would have to be trained separately for each and so this an area we hope to explore further.

5 Conclusion and Future Work

In this paper we apply Machine Learning to directly select optimal variable orderings in Cylindrical Algebraic Decomposition which greatly outperform the manual heuristics, as well as develop a new GNN model to directly select orderings. We additionally demonstrate potential data pollution when training models using current datasets. Our results corroborate previous literature along with our own thesis, however more work needs to be done in turning GNNs into a viable model for selecting variable orderings.

We have obtained access to the dataset underlying [11], and intend to experiment with that as well. [16, Sect. 5.2] also points to a further problem with the MetiTarski datatset beyond that in [8]: different problems can be duplicates in the sense that they always have the same tree structure (see Definition 1). [15, Sect. 4.1] made this point and state that it can lead to unequal valuation of selection mechanisms if one does well on a problem with many duplicates.

Disclosure of Interests. The authors have no competing interests to declare that are relevant to the content of this article.

Acknowledgement. Davenport was partially funded by the UK's EPSRC under grant EP/T015748/1. We are grateful to Matthew England and Tereso del Río (Coventry University) for useful comments, as well as Yuhang Dong and Fuqi Jia for access to their dataset. We are grateful to the referees for many comments and drawing our attention to [13].

References

1. Bradford, R., Davenport, J., England, M., Wilson, D.: Optimising problem formulation for cylindrical algebraic decomposition. In: Carette, J., et al. (eds.) Proceedings CICM 2013, pp. 19–34 (2013). https://doi.org/10.1007/978-3-642-39320-4_2
2. Brown, C.: Tutorial handout at ISSAC 2004 (2004). https://www.usna.edu/Users/cs/wcbrown/research/ISSAC04/Tutorial.html
3. Brown, C., Davenport, J.: The complexity of quantifier elimination and cylindrical algebraic decomposition. In: Brown, C. (ed.) Proceedings of ISSAC 2007, pp. 54–60 (2007). https://doi.org/10.1145/1277548.1277557
4. Chen, C., et al.: Computing the real solutions of polynomial systems with the RegularChains library in MAPLE: ISSAC 2011 software demo. Commun. Comput. Algebra 3(45), 166–168 (2011). https://doi.org/10.1145/2110170.2110174
5. Collins, G.E.: Quantifier elimination for real closed fields by cylindrical algebraic decompostion. In: Brakhage, H. (ed.) GI-Fachtagung 1975. LNCS, vol. 33, pp. 134–183. Springer, Heidelberg (1975). https://doi.org/10.1007/3-540-07407-4_17
6. Dolzmann, A., Seidl, A., Sturm, T.: Efficient projection orders for CAD. In: Gutierrez, J. (ed.) Proceedings ISSAC 2004, pp. 111–118 (2004). https://doi.org/10.1145/1005285.1005303
7. England, M., Florescu, D.: Comparing machine learning models to choose the variable ordering for cylindrical algebraic decomposition. In: Kaliszyk, C., Brady, E., Kohlhase, A., Sacerdoti Coen, C. (eds.) Proceedings CICM 2019, pp. 93–108 (2019). https://doi.org/10.1007/978-3-030-23250-4_7
8. Hester, J., Hitaj, B., Passmore, G., Owre, S., Shankar, N., Yeh, E.: An augmented MetiTarski dataset for real quantifier elimination using machine learning. In: Dubois, C., Kerber, M. (eds.) Proceedings CICM 2023. LNCS, vol. 14101, pp. 297–302. Springer, Cham (2023). https://doi.org/10.1007/978-3-031-42753-4_21

9. Huang, Z., England, M., Wilson, D., Davenport, J., Paulson, L.: Using machine learning to improve cylindrical algebraic decomposition. Math. Comput. Sci. **13**, 461–488 (2019). https://doi.org/10.1007/s11786-019-00394-8
10. Huang, Z., England, M., Wilson, D., Davenport, J., Paulson, L., Bridge, J.: Applying machine learning to the problem of choosing a heuristic to select the variable ordering for cylindrical algebraic decomposition. In: Watt, S.M., et al. (eds.) Proceedings of CICM 2014, pp. 92–107 (2014). https://doi.org/10.1007/978-3-319-08434-3_8
11. Jia, F., Dong, Y., Liu, M., Huang, P., Ma, F., Zhang, J.: Suggesting variable order for cylindrical algebraic decomposition via reinforcement learning. NIPS **36**, 76098–76119 (2023). https://proceedings.neurips.cc/paper_files/paper/2023/file/efcb5b06ce8bb672ffa26b9dc5cdd0f9-Paper-Conference.pdf
12. John, R.: Exploring Alternative Machine Learning Models for Variable Ordering in Cylindrical Algebraic Decomposition. BSc. Dissertation, University of Bath (2024)
13. Pickering, L., Del Río Almajano, T., England, M., Cohen, K.: Explainable AI insights for symbolic computation: a case study on selecting the variable ordering for cylindrical algebraic decomposition. J. Symb. Comput. Article **102276**, 123 (2024). https://doi.org/10.1016/j.jsc.2023.102276
14. del Río, T., England, M.: Lessons on Datasets and Paradigms in Machine Learning for Symbolic Computation: A Case Study on CAD. https://arxiv.org/abs/2401.13343 (2024)
15. del Río, T., England, M.: New heuristic to choose a cylindrical algebraic decomposition variable ordering motivated by complexity analysis. In: Boulier, F., England, M., Sadykov, T.M., Vorozhtsov, E.V. (eds.) Computer Algebra in Scientific Computing CASC 2022. LNCS, vol. 13366, pp. 300–317. Springer, Cham (2022). https://doi.org/10.1007/978-3-031-14788-3_17
16. del Río, T., England, M.: Data augmentation for mathematical objects. In: Ábrahám, E., Sturm, T. (eds.) Proceedings of the 8th SC-Square Workshop, CEUR-WS Proceedings, vol. 3455, pp. 29–38 (2023). https://arxiv.org/abs/2307.06984

Constrained Neural Networks for Interpretable Heuristic Creation to Optimise Computer Algebra Systems

Dorian Florescu[1] and Matthew England[2(✉)]

[1] University of Bath, Bath, UK
dmf36@bath.ac.uk
[2] Coventry University, Coventry, UK
Matthew.England@coventry.ac.uk

Abstract. We present a new methodology for utilising machine learning technology in symbolic computation research. We explain how a well known human-designed heuristic to make the choice of variable ordering in cylindrical algebraic decomposition may be represented as a constrained neural network. This allows us to then use machine learning methods to further optimise the heuristic, leading to new networks of similar size, representing new heuristics of similar complexity as the original human-designed one. We present this as a form of ante-hoc explainability for use in computer algebra development.

Keywords: computer algebra · cylindrical algebraic decomposition · machine learning · explainable AI · interpretability · XAI

1 Introduction

1.1 Machine Learning Within Computer Algebra Systems

Machine Learning (ML) refers to tools and techniques that learn rules from data, thus allowing a system to improve its performance on a task without any change to the explicit programming. ML underpins recent AI advances and is applied in an increasing number of domains. Mathematics is no exception: ML has been employed to directly perform mathematical computation, such as [24] who used ML to integrate expressions and solve ODEs, [2] who used ML to find the real discriminant locus, and [19] who surveys ML to predict properties from mathematical structures such as groups and graphs. However, it has been observed that mathematical reasoning is an area ML finds difficult.

Computer Algebra Systems (CASs) are not an obvious domain for ML: their unique selling point is that their answers are exactly correct and so developers are unlikely to replace symbolic computation algorithms with ML[1]. However,

---

[1] Experiments like [24] conflate two very different causes of failure: timeout and giving the wrong answer. For a CAS: the former would be a shame; the latter a disaster.

CAS algorithms often come with choices that have no effect on the mathematical correctness of the end result but can have a big impact on the resources required to obtain it, and on how it is presented. Such choices are often made by human designed heuristics or *"magic constants"* [7] (sometimes not scientifically validated or even documented) but may be better made by ML.

Examples in the literature include [23] which used a Monte-Carlo tree search to find the representation of polynomials most efficient to evaluate, [32] which used ML classifiers to pick from algorithms that compute the resultant, and [21] which used ML to decide whether to precondition input for a CAS.

1.2 Gaining Additional Mathematical Insight from Machine Learning

It seems clear that ML can offer optimisation to CASs, but can it offer any further insight into the underlying mathematics/algorithms?

- [11] suggested that ML can help pure mathematicians with the development of new theorems by uncovering patterns in the data. This led to new results in knot theory and representation theory.
- [30] described a form a genetic programming where algorithms were *evolved* with a large language model performing the crossover step. This resulted in new state-of-the-art heuristics for two NP-hard problems.
- [28] trained ML to select the next S-pair in Buchberger's algorithm to build a Gröbner Basis. An analysis of the agent showed a preference for pairs whose S-polynomials are monomials or low degree: prior human-designed heuristics for the problem considered only the S-pairs and not the polynomials themselves, so this represents a novel strategy.
- [29] described how the SHAP tool [25] may be used to analyse ML models that make a heuristic choice for a CAS and then inform *human-level* heuristics — heuristics that can be expressed in natural language in a similar amount of text as a heuristic designed by a human — that can be operated without any ML architecture.

The tool used in the final paper is from the growing field of Explainable AI.

1.3 Explainable AI

Explainable AI (often abbreviated to XAI) may be defined as those ML techniques whose decisions can be explained (at least partially). Work in the field is usually motivated by the need to error check ML decisions, and to generate greater user trust in ML. However, in the case of mathematics we hypothesise that XAI tools may be used to give guidance or new understanding.

XAI is a new field: there have been several attempts to give a taxonomy such as [1]. One distinction in XAI methods that has firmly emerged [33] is between: **ante-hoc explainability** which refers to ML methods that are themselves by-design transparent in their decisions; and **post-hoc explainability** which use a secondary analysis of an opaque (i.e. black-box) ML model to generate explanations for it.

The need for the latter is driven by the so called *performance-explainability trade-of* whereby those ML techniques which allow ante-hoc explainability are thought to give lower accuracy in general. Although often presented as fact, this trade-off is disputed [31] and is likely application dependent [20]. We should also remember that explainability is inherently-audience dependent (compare explainable to an expert with explainable to the general public) [1].

SHAP is an example of the post-hoc explainability, forming its explanation through experiments involving perturbations in the input. In the present paper, we consider an alternative approach to the same application as [29], but aiming for an ante-hoc explainability approach.

1.4 Plan of the Paper

The paper continues in Sect. 2 with a brief introduction to the CAS choice that we study. Then in Sect. 3 we recap a process presented in [16] to represent instances of our problem as feature vectors: suitable for use in ML. Our new contributions then follow in Sect. 4, where we interpret a well known human-designed heuristic as a small neural network, and in Sect. 5 where we search through a family of similar networks to identify an improved human-level heuristic, in effect defining a new type of ante-hoc explainability technique for such problems.

2 Our Application: Variable Ordering Choice for CAD

2.1 Cylindrical Algebraic Decomposition

Cylindrical Algebraic Decomposition (CAD) was proposed in [9] as a method to perform real quantifier elimination. Given an input in n ordered variables, CAD will decompose the corresponding real space into cells (connected regions of $\mathbb{R}^n$) arranged cylindrically (the projections of a pair of cells with respect to the variable ordering are equal or disjoint) with each semi-algebraic (described by polynomial constraints). The input to CAD is a set of polynomials and the CAD guarantees that each polynomials will have invariant sign upon each cell of the decomposition. CAD has been applied in many fields ranging from robotics [26] through biology [3] to economics [27]. However, CAD has worst-case complexity doubly exponential in the number of variables [10] and thus any work to optimise its implementation can bring swathes of new applications into scope.

2.2 CAD Variable Ordering

The CAD variable ordering controls both the algorithm flow and output format (defining the cylindrical structure). It can have a huge impact, both practically [14] and in terms of theoretical complexity [6]. There exist human-designed heuristics to choose the ordering, e.g. [5], and [13] which use simple statistics of the input. Other heuristics perform increasing amounts of algebraic computation [4,14,34] which we do not consider here in preference for a cheap heuristic.

In the last decade ML models have also been trained to select the variable ordering for CAD. The first attempt was made by [22] with a support vector machine. Later, the present authors experimented with a wider range of models [15], methods for feature engineering [16] and improved metrics for hyperparameter selection [17], culminating in a machine learning pipeline available to use for the task [18]. Separately [8] experimented with deep learning for variable selection.

3 Feature Generation Process

A challenge when using ML to optimise a CAS is the communication between them: the former uses symbolic expressions, and the latter vectors of numerical data called *features*. We summarise next the feature generation process of [16].

3.1 Formalising Brown's Heuristic

Our work was based on an analysis of the Brown heuristic [5] which uses metrics:

1. the overall degree in the input of a variable v;
2. the maximum total degree of monomials in which variable v occurs; and
3. the number of terms which contain the variable v.

It orders on the earlier metrics, breaking ties with the subsequent ones.

In the following we use index p to refer to polynomials in a problem instance, and index m to refer to the monomials in such a polynomial. We consider a CAD problem instance as a set of polynomials: $\boldsymbol{Pr} = \{\mathcal{P}_p \,|\, p = 1, \ldots, P\}$. A generic polynomial P_p is then given by a sum of monomials, $\mathcal{P}_p = \sum_{m=1}^{M_p} c^{m,p} \cdot \prod_{i=1}^{n} x_i^{d_i^{m,p}}$ where $d_i^{m,p}$ is the degree of variable x_i in monomial m of polynomial p.

Thus the polynomials are defined by the series $[c^{m,p}, (d_1^{m,p}, d_2^{m,p}, d_3^{m,p})]$, for $m = 1, \ldots, M_p$. This allows formalising the problem set of all polynomials as

$$\mathbb{S}_{\boldsymbol{Pr}} = \{\{\, [c^{m,p}, (d_1^{m,p}, d_2^{m,p}, d_3^{m,p})] \,|\, m = 1, \ldots, M_p\} \,|\, p = 1, \ldots, P\}.$$

Then the three metrics of the Brown heuristic above are formalised as

1. $F^1(d_v) := \max_{m,p} d_v^{m,p}$,
2. $F^2(d_v) := \max_{m,p} \operatorname{sgn}(d_v^{m,p}) \cdot (d_1^{m,p} + d_2^{m,p} + d_3^{m,p})$,
3. $F^3(d_v) := \sum_{m,p} \operatorname{sgn}(d_v^{m,p})$.

Here max and $\sum$ are the maximum and sum functions with the subscript m, p indicating that they are applied over all monomials in all polynomials; while $\operatorname{sgn}(x)$ is the function which takes values in $\{-1, 0, 1\}$ according to the sign of its input (used to identify which terms contain the input – in our situation the sign is only ever positive or zero).

3.2 Generating Similar Features

We notice that the features used by the Brown heuristic are simple to compute using only max, $\sum$, and sgn applied to degrees over monomials and polynomials. We define variants of the functions $\max_x$ and $\sum_x$ where x indicates whether we sum over monomials, polynomials or both; and we define similarly the averaging functions:

$$\mathrm{av}_m = \frac{1}{M_p}\sum_m, \quad \mathrm{av}_p = \frac{1}{P}\sum_p, \quad \mathrm{av}_{m,p} = \frac{1}{P}\sum_p \frac{1}{M_p}\sum_m;$$

then we can express all the features used in [22] in this formalisation [16].

In [16] we generalised this to create a larger set of features: all those of the form $f(\boldsymbol{Pr}) = (g_4 \circ g_3 \circ g_2 \circ g_1 \circ h^{m,p})(\boldsymbol{Pr})$, where $h^{m,p}(\boldsymbol{Pr}) \in \{d_v^{m,p}, \mathrm{sgn}(d_v^{m,p}), (\sum_{v'} d_{v'}^{m,p}) \mid v = 1, 2, 3\}$ and g_1, g_2, g_3, g_4 are taken from $\{\max_p, \max_m, \max_{m,p}, \sum_p, \sum_m, \sum_{m,p}, \mathrm{av}_p, \mathrm{av}_m, \mathrm{av}_{m,p}, \mathrm{sgn}, \mathrm{Id}\}$, with Id as the identity function.

Many of the features generated will be equivalent: either at a mathematical level or for the dataset in question and so before using these for ML we should identify a unique subset. In [16] it was demonstrated that we may use these in a ML pipeline to improve the performance compared to using just a human crafted set. These features have also been the basis for further work in [12, 29].

4 Interpreting Brown's Heuristic as a Neural Network

Recall the Brown heuristic from Sect. 3.1 which used 3 rules with different priority levels. We claim this can be equivalently represented as a dense 2-layer neural network with summation activation functions, as visualised in Fig. 1 for the three variable case (in the general case there will be n inputs into each of the nodes of the first layer). The summations are weighted as in Fig. 1 where the weights are defined in terms $w > 0$ which we select such that

$$F^i(d_v) < w - 1, i, v \in \{1, 2, 3\} \tag{1}$$

for all problems in the dataset (or large enough to cover all problems of interest). The outputs of the first network layer are

$$y_v = F^1(d_v)w^2 + F^2(d_v)w + F^3(d_v).$$

We will show that the magnitude of y_v orders the variables as Brown's heuristic.

Assume first that $F^1(d_v) > F^1(d_{v'})$, for $v, v' \in \{1, 2, 3\}$. We aim to show $y_v > y_{v'}$ irrespective of the values of the other features. All features are positive integers, meaning our assumption becomes $F^1(d_v) \geq F^1(d_{v'}) + 1$. We start with

$$\begin{aligned}y_{v'} &= F^1(d_{v'})w^2 + F^2(d_{v'})w + F^3(d_{v'}) \\ &\leq (F^1(d_v) - 1)w^2 + F^2(d_{v'})w + F^3(d_{v'})\end{aligned}$$

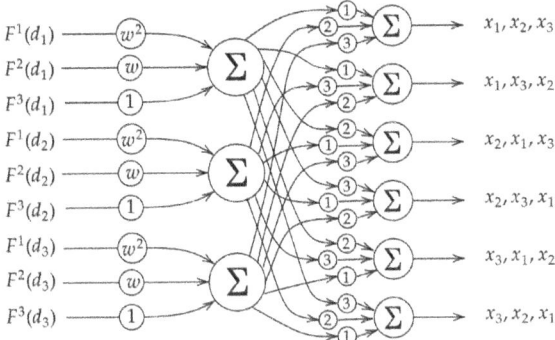

Fig. 1. Neural network inspired by Brown's heuristic

where the inequality follows by our assumption. Then by the repeated use of (1) we have the further strict inequality

$$y_{v'} < F^1(d_v)w^2 - w^2 + (w-1)w + (w-1)$$
$$= F^1(d_v)w^2 - 1 < F^1(d_v)w^2.$$

Then since all features are positive we have that

$$y_{v'} < F^1(d_v)w^2 + F^2(d_v)w + F^3(d_v) = y_v.$$

Now consider the case where $F^1(d_v) = F^1(d_{v'})$ for all $v, v' \in \{1, 2, 3\}$ and $F^2(d_v) > F^2(d_{v'})$. We want to show that under these assumptions $y_{v'} < y_v$ for any realisations of F^3. We proceed similarly to above:

$$y_{v'} = F^1(d_{v'})w^2 + F^2(d_{v'})w + F^3(d_{v'})$$
$$= F^1(d_v)w^2 + F^2(d_{v'})w + F^3(d_{v'})$$
$$\leq F^1(d_v)w^2 + (F^2(d_v) - 1)w + F^3(d_{v'})$$
$$< F^1(d_v)w^2 + (F^2(d_v) - 1)w + (w-1)$$
$$= F^1(d_v)w^2 + F^2(d_v)w - 1$$
$$< F^1(d_v)w^2 + F^2(d_v)w$$
$$< F^1(d_v)w^2 + F^2(d_v)w + F^3(d_v) = y_v.$$

Similarly in the final case where $F^1(d_v) = F^1(d_{v'})$, and $F^2(d_v) = F^2(d_{v'})$ for all $v, v' \in \{1, 2, 3\}$ and feature F^3 selects the ordering. With $F^3(d_v) > F^3(d_{v'})$

$$y_{v'} = F^1(d_{v'})w^2 + F^2(d_{v'})w + F^3(d_{v'})$$
$$= F^1(d_v)w^2 + F^2(d_v)w + F^3(d_{v'})$$
$$\leq F^1(d_v)w^2 + F^2(d_v)w + F^3(d_v) - 1$$
$$< F^1(d_v)w^2 + F^2(d_v)w + F^3(d_v) = y_v.$$

Thus the internal layer will be ordered correctly. The last layer of the network then performs weighted summations of the outputs of the first network layer, $y_v + 2y_{v'} + 3y_{v''}$, with each output neuron labelled corresponding to the weights applied to the variables (see Fig. 1). If $y_v < y_{v'} < y_{v''}$ then the weighted sum $y_v + 2y_{v'} + 3y_{v''}$ is the highest among all output neurons, meaning this neural network may be used to produce the same orderings as the Brown heuristic.

5 Searching Through Similar Constrained Neural Networks

Now we have represented the Brown heuristic as a (severely constrained) neural network we may consider editing this network to see if a superior heuristic of the same complexity as Brown can be obtained.

5.1 Feature Selection

We perform feature selection using a dataset of random 3-variable polynomial problems (as described in [18]). We generated 84 features algorithmically following Sect. 3 and taking all permutations this led to 19,656 possible feature triplets. For each triplet, we computed the variable ordering predicted by the neural network. The triplet that led to the shortest computing time is given by

1. $\sum_p \max_m d_v^{m,p}$, sum of the highest degree of a variable in each polynomial;
2. $\sum_p \max_m \text{sgn}(d_v^{m,p}) \cdot (\sum_{v'} d_{v'}^{m,p})$, the maximum sum of degrees of all variables for the terms in which a given variable exists; and
3. $\sum_p \max_m \text{sgn}(d_v^{m,p})$, the number of polynomials containing the variable.

We note that none of these features are in the Brown heuristic.

This new feature triplet was selected based on performance on random polynomials. But to judge its performance we will evaluate it on the NLSAT dataset of 3-variable polynomials from the real world (non-linear arithmetic satisfiability problems) as described in [15]. The Brown heuristic requires a total computing time of 10,580 s on this dataset, while the network defined by this new triple resulted in a smaller computing time of 10,181 s: which is 399 s shorter.

5.2 Weight Tuning

We next consider changing the weights in the neural network. This will have the effect of a more complicated combination of the three features to be considered: but still weighted sums of the same three pieces of information.

We use the neural network inspired by Brown's heuristic (with $w = 30$) as a starting point. The weights are then trained on the 3-variable random training dataset using the *adam* stochastic gradient-based optimizer with learning rate $2 \cdot 10^{-5}$. From the input data we used the new three features identified in Subsect. 5.1. To avoid overfitting, the performance is evaluated for each step on the independent 3-variable NLSAT dataset.

After only three epochs of training the CAD times decreased to 9908 (after which there was only minimal improvement). The weight matrix was only changed slightly, but this was sufficient to avoid the case of ties on the three metrics (followed by a random choice) which was common in the dataset.

6 Final Thoughts

This paper contributes to the ongoing conversation on how ML can contribute to computer algebra: not only algorithm optimisation but also mathematical discovery. We present an approach, constrained neural networks, that may be viewed as ante-hoc explainability. It allows for heuristics to be uncovered which are human-level in complexity. The methodology could be directly applied to other variable ordering choices in symbolic computation, and we expect it could be adapted for use on other choices also. It remains to be shown whether these more interpretable ML outputs can lead to new mathematical understanding.

Acknowledgements. DF and ME were both supported by EPSRC grant EP/R019622/1: Embedding Machine Learning within Quantifier Elimination Procedures. ME was also supported by EPSRC grant EP/T015748/1: Pushing Back the Doubly-Exponential Wall of Cylindrical Algebraic Decomposition (DEWCAD).

References

1. Barredo Arrieta, A., et al.: Explainable artificial intelligence (XAI): concepts, taxonomies, opportunities and challenges toward responsible AI. Inf. Fusion **58**, 82–115 (2020). https://doi.org/10.1016/j.inffus.2019.12.012
2. Bernal, E.A., Hauenstein, J.D., Mehta, D., Regan, M.H., Tang, T.: Machine learning the real discriminant locus. J. Symb. Comput. **115**, 409–426 (2023). https://doi.org/10.1016/j.jsc.2022.08.001
3. Bradford, R., et al.: Identifying the parametric occurrence of multiple steady states for some biological networks. J. Symb. Comput. **98**, 84–119 (2020). https://doi.org/10.1016/j.jsc.2019.07.008
4. Bradford, R., Davenport, J.H., England, M., Wilson, D.: Optimising problem formulation for cylindrical algebraic decomposition. In: Carette, J., Aspinall, D., Lange, C., Sojka, P., Windsteiger, W. (eds.) CICM 2013. LNCS (LNAI), vol. 7961, pp. 19–34. Springer, Heidelberg (2013). https://doi.org/10.1007/978-3-642-39320-4_2
5. Brown, C.W.: Companion to the tutorial: Cylindrical algebraic decomposition, ISSAC 2004 (2004). http://www.usna.edu/Users/cs/wcbrown/research/ISSAC04/handout.pdf
6. Brown, C., Davenport, J.: The complexity of quantifier elimination and cylindrical algebraic decomposition. In: Proceedings ISSAC 2007, pp. 54–60. ACM (2007). https://doi.org/10.1145/1277548.1277557
7. Carette, J.: Understanding expression simplification. In: Proceedings of ISSAC 2004, pp. 72–79. ACM (2004). https://doi.org/10.1145/1005285.1005298

8. Chen, C., Zhu, Z., Chi, H.: Variable ordering selection for cylindrical algebraic decomposition with artificial neural networks. In: Bigatti, A.M., Carette, J., Davenport, J.H., Joswig, M., de Wolff, T. (eds.) ICMS 2020. LNCS, vol. 12097, pp. 281–291. Springer, Cham (2020). https://doi.org/10.1007/978-3-030-52200-1_28
9. Collins, G.E.: Quantifier elimination for real closed fields by cylindrical algebraic decompostion. In: Brakhage, H. (ed.) GI-Fachtagung 1975. LNCS, vol. 33, pp. 134–183. Springer, Heidelberg (1975). https://doi.org/10.1007/3-540-07407-4_17
10. Davenport, J.H., Heintz, J.: Real quantifier elimination is doubly exponential. J. Symb. Comput. **5**(1-2), 29–35 (1988). https://doi.org/10.1016/S0747-7171(88)80004-X
11. Davies, A., et al.: Advancing mathematics by guiding human intuition with AI. Nature **600**, 70–74 (2021). https://doi.org/10.1038/s41586-021-04086-x
12. del Río, T., England, M.: Lessons on datasets and paradigms in machine learning for symbolic computation: a case study on CAD. Preprint (2024). https://doi.org/10.48550/arXiv.2401.13343
13. del Río, T., England, M.: New heuristic to choose a cylindrical algebraic decomposition variable ordering motivated by complexity analysis. In: Boulier, F. et al. (eds.), Proceedings of CASC 2022. LNCS, vol. 13366, pp. 300–317. Springer International, Cham (2022). https://doi.org/10.1007/978-3-031-14788-3_17
14. Dolzmann, A., Seidl, A., Sturm, T.: Efficient projection orders for CAD. In: Proceedings of ISSAC 2004, pp. 111–118. ACM (2004). https://doi.org/10.1145/1005285.1005303
15. England, M., Florescu, D.: Comparing machine learning models to choose the variable ordering for cylindrical algebraic decomposition. In: Kaliszyk, C., et al. (eds.) Proceedings of CICM 2019. LNCS, vol. 11617, pp. 93–108. Springer International, Cham (2019). https://doi.org/10.1007/978-3-030-23250-4_7
16. Florescu, D., England, M.: Algorithmically generating new algebraic features of polynomial systems for machine learning. In: Abbott, J., Griggio, A. (eds.) Proceedings of SC^2 2019. CEUR-WS 2460 (2019). http://ceur-ws.org/Vol-2460/
17. Florescu, D., England, M.: Improved cross-validation for classifiers that make algorithmic choices to minimise runtime without compromising output correctness. In: Slamanig, D., et al. (eds.), Proceedings of MACIS 2019. LNCS, vol. 11989, pp. 341–356. Springer International, Cham (2020). https://doi.org/10.1007/978-3-030-43120-4_27
18. Florescu, D., England, M.: A machine learning based software pipeline to pick the variable ordering for algorithms with polynomial inputs. In: Bigatti, A., et al. (eds.) Proceedings of ICMS 2020. LNCS, vol. 12097, pp. 302–322. Springer International, Cham (2020). https://doi.org/10.1007/978-3-030-52200-1_30
19. He, Y.H.: Machine-learning mathematical structures. Int. J. Data Sci. Math. Sci. **1**(1), 1–25 (2022). https://doi.org/10.1142/S2810939222500010
20. Herm, L.V., Heinrich, K., Wanner, J., Janiesch, C.: Stop ordering machine learning algorithms by their explainability! A user-centered investigation of performance and explainability. Int. J. Inf. Manage. 102538 (2022). https://doi.org/10.1016/j.ijinfomgt.2022.102538
21. Huang, Z., England, M., Davenport, J.H., Paulson, L.: Using machine learning to decide when to precondition cylindrical algebraic decomposition with Groebner bases. In: Proceeedings of SYNASC 2016, pp. 45–52. IEEE (2016). https://doi.org/10.1109/SYNASC.2016.020

22. Huang, Z., England, M., Wilson, D., Davenport, J.H., Paulson, L., Bridge, J.: Applying machine learning to the problem of choosing a heuristic to select the variable ordering for cylindrical algebraic decomposition. In: Watt, S.M., et al. (eds.) Proceedings of CICM 2014. LNCS, vol. 8543, pp. 92–107. Springer International, Cham (2014). http://dx.doi.org/10.1007/978-3-319-08434-3_8
23. Kuipers, J., Ueda, T., Vermaseren, J.A.M.: Code optimization in FORM. Comput. Phys. Commun. **189**, 1–19 (2015). https://doi.org/10.1016/j.cpc.2014.08.008
24. Lample, G., Charton, D.: Deep learning for symbolic mathematics. In: Mohamed, S., et al. (eds.) Proceedings ICLR 2020 (2020). https://iclr.cc/virtual_2020/poster_S1eZYeHFDS.html
25. Lundberg, S.M., Lee, S.I.: A unified approach to interpreting model predictions. In: Proceedings of NIPS 2017, pp. 4768–4777. Curran Associates Inc. (2017). https://dl.acm.org/doi/10.5555/3295222.3295230
26. Manubens, M., Moroz, G., Chablat, D., Rouillier, F., Wenger, P.: Cusp points in the parameter space of degenerate 3-RPR planar parallel manipulators. J. Mech. Robot. **4**, 041003 (2012). https://doi.org/10.1115/1.4006921
27. Mulligan, C.B., Davenport, J.H., England, M.: TheoryGuru: a mathematica package to apply quantifier elimination technology to economics. In: Davenport, J.H. et al. (eds.) Proceedings of ICMS 2018, LNCS, vol. 10931, pp. 369–378. Springer International, Cham (2018). https://doi.org/10.1007/978-3-319-96418-8_44
28. Peifer, D., Stillman, M., Halpern-Leistner, D.: Learning selection strategies in Buchberger's algorithm. In: Daumé III, H., Singh, A. (eds.) Proceedings of ICML 2020. PMLR 119, pp. 7575–7585 (2020). https://proceedings.mlr.press/v119/peifer20a.html
29. Pickering, L., Del Rio Almajano, T., England, M., Cohen, K.: Explainable AI insights for symbolic computation: a case study on selecting the variable ordering for cylindrical algebraic decomposition. J. Symb. Comput. **123**, 102276 (2024). https://doi.org/10.1016/j.jsc.2023.102276
30. Romera-Paredes, B., et al.: Mathematical discoveries from program search with large language models. Nature **625**, 468–475 (2023). https://doi.org/10.1038/s41586-023-06924-6
31. Rudin, C.: Stop explaining black box machine learning models for high stakes decisions and use interpretable models instead. Nat. Mach. Intell. **1**(5), 206–215 (2019). https://doi.org/10.1038/s42256-019-0048-x
32. Simpson, M.C., Yi, Q., Kalita, J.: Automatic algorithm selection in computational software using machine learning. In: Proceedings of ICMLA 2016, pp. 355–360 (2016). https://doi.org/10.1109/ICMLA.2016.0064
33. Speith, T.: A review of taxonomies of explainable artificial intelligence (XAI) methods. In: Proceedings of FAccT 2022, pp. 2239–2250. ACM (2022). https://doi.org/10.1145/3531146.3534639
34. Wilson, D., England, M., Davenport, J.H., Bradford, R.: Using the distribution of cells by dimension in a cylindrical algebraic decomposition. In: Proceedings of SYNASC 2014, pp. 53–60. IEEE (2014). http://dx.doi.org/10.1109/SYNASC.2014.15

Machine Learning for Number Theory: Unsupervised Learning with L-Functions

Thomas Oliver[✉]

University of Westminster, London, UK
T.Oliver@westminster.ac.uk

Abstract. There is a strong tradition of computation in number theory, with notable data-driven insights including the prime number theorem and the conjecture of Birch and Swinnerton-Dyer. A huge arithmetic online database known as the LMFDB went live in the mid-2010s, to which we began applying machine learning methodologies in 2020. This led to a data scientific perspective on old problems, and the discovery of surprising new structures in arithmetic statistics known as *murmurations*. In this extended abstract, we will apply unsupervised learning techniques to a small dataset taken from the LMFDB, chosen so as to demonstrate one approach to generalising the original experiments.

Keywords: Clustering · PCA · L-functions

1 Introduction

Number theory is an ancient subject that is central in modern mathematics. Notable open questions include the Riemann Hypothesis and the Conjecture of Birch and Swinnerton–Dyer (BSD), both of which are Millennium Prize Problems, both of which are traditionally formulated in terms of L-functions, and both of which are relevant to the present text. Computation has always played a role in number theoretic research. Indeed, the prime number theorem emerged from analysis of logarithm tables, and the BSD emerged from experimentation with number theoretic data on some of the first computers at Cambridge Mathematical Laboratory. More recently, the L-functions and modular forms database (LMFDB) has facilitated data scientific exploration of number theory [11]. In this extended abstract, we will outline one approach to using machine learning within the context of number theory.

There are three standard paradigms of machine learning: supervised, unsupervised, and reinforcement. The methodology of supervised learning was applied to number theory in [1,4–6]. In particular, these papers considered how supervised learning might be used to calculate the rank of an elliptic curve, for which there is no rigorous algorithm. A compilation of this work is available in [12]. In this extended abstract, we offer something similar for the unsupervised learning experiments undertaken in [1,4,7]. We will connect this to the so-called *murmuration* phenomenon, which emerged from data science, but has since been studied by entirely traditional methods [2,3,8,9,14,16]. It is notable that this discussion neglects reinforcement learning, and other forms of data-driven artificial intelligence that are increasingly prominent in societal discourse. Though the discovery of murmurations involved a good level of human intervention, the right blend of large language models with automated theorem proving may accelerate the process in future. Some evidence in this direction, with a connection to number theory, was recently published [13].

We conclude this introduction with a review of the subsequent sections. In Sect. 2, we define the terminology to be used later. In Sect. 3, we describe some illustrative unsupervised learning experiments on a small custom dataset. This dataset is somewhat novel, in that we do not restrict ourselves to specific L-function families, in contrast to the original experimentation in [4–6]. In Sect. 4, we speculate on the utility of this approach in the context of arithmetic discovery, inspired by the role of similar methodology in the emergence of murmurations.

2 A Crash Course in L-Functions

In this section, we review some terminology that is used later in the text. The reader not interested in such details need only know that we will use something called the *root analytic conductor* to order our data in Sect. 3. This quantity is a normalized measure of complexity that is natural from the perspective of number theory. The simpler *conductor* was used to organise the data in prior work, in which all L-functions were sampled from specified families.

For the purposes of this text, an L-function is a Dirichlet series of the form

$$L(s) = \sum_{n=1}^{\infty} a_n n^{-s},$$

where $(a_n)_{n=1}^{\infty}$ is a sequence of complex numbers, which exhibits certain salient behaviour as in [15]. There are two well-established origins for L-functions, namely, motives and automorphic forms. Even the constant sequence $a_n = 1$ is already interesting. In this case $L(s) = \zeta(s)$ is known as the Riemann zeta function, which can be viewed as a prototype for all that follows. We will give further examples below.

We will assume throughout that $L(s)$ converges in some right half-plane of the form $\mathrm{Re}(s) > k$. It is generally conjectured - and sometimes proved - that $L(s)$ admits analytic continuation to $\mathbb{C}$ (with the possible exception of a simple pole at $s = 1$) and satisfies a functional equation of the form:

$$\Lambda(s) = \epsilon \overline{\Lambda}(k-s),$$

where ϵ is a complex number with absolute value 1, and

$$\Lambda(s) = N^{s/2} \prod_{r=1}^{R} \Gamma_{\mathbb{R}}(s + \mu_r) \prod_{c=1}^{C} \Gamma_{\mathbb{C}}(s + \nu_c) L(s),$$

for some integer N, and some complex numbers μ_r, ν_c, in which $\Gamma_\mathbb{R}(s) = \pi^{-s/2}\Gamma(s)$ and $\Gamma_\mathbb{C}(s) = 2(2\pi)^{-s}\Gamma(s)$. The subscripts in the gamma functions are conventions that allude to the archimedean components of the underlying Galois/automorphic representations. The *root number* ϵ appearing in the functional equation above will be mentioned at various points in what follows.

Definition 1. *The conductor (resp. degree) of $L(s)$ is the integer N (resp. $R + 2C$) as per its functional equation. An L-function is described as primitive if it cannot be written as a product of L-functions with lower degree.*

Example 1. In the case that $L(s) = \zeta(s)$ we write $\xi(s) = \Gamma_\mathbb{R}(s)\zeta(s)$ so that $\xi(s) = \xi(1-s)$. In particular, the Riemann zeta function is primitive, with degree 1 and conductor 1.

Example 2. If E is an elliptic curve over $\mathbb{Q}$, then its L-function is more easily described as an Euler product:

$$L(E,s) = \prod_{p \nmid N} \frac{1}{1 - a_p(E)p^{-s} + p^{1-2s}} \prod_{p | N} \frac{1}{1 - a_p(E)p^{-s}},$$

where, in the left-hand product, p ranges over the infinite set of prime numbers not dividing N, in the right-hand product, p ranges over the finite set of prime numbers dividing N, and $a_p(E) = p + 1 - \#E(\mathbb{F}_p)$. The conductor N encodes the primes p for which the reduction of E mod p is singular, and may be calculated with Tate's algorithm. Elliptic L-functions are primitive, and have degree 2. More generally, one may attach an L-function of degree $2g$ to any arithmetic curve over $\mathbb{Q}$ with genus g.

Example 3. If F is a number field, then the Dedekind zeta function is given by

$$\zeta_F(s) = \sum_{n=1}^{\infty} a_n(F) n^{-s},$$

where $a_n(F)$ is the number of ideals in $\mathcal{O}_F$ with ideal norm n. The conductor of $\zeta_F(s)$ is related to the discriminant of F. A number field F is a finite-dimensional vector space over $\mathbb{Q}$, and the degree of $\zeta_F(s)$ is equal to $\dim_\mathbb{Q}(F)$. If $F \neq \mathbb{Q}$, then $\zeta_F(s)$ is imprimitive. Dedekind zeta functions are examples of the Artin L-functions attached to complex Galois representations.

Definition 2. *An L-function is said to be arithmetic if their exists an integer $w \geq 0$ so that, for all $n \geq 1$, $a_n n^{w/2}$ is an algebraic integer. An arithmetic L-function is said to be rational if the number field generated by $(a_n n^{w/2})_{n=1}^{\infty}$ is equal to $\mathbb{Q}$.*

Roughly speaking, the Langlands philosophy is that all arithmetic L-functions are automorphic. Every L-function mentioned so far is rational. On the other hand, Maass forms, for example, are a source of non-arithmetic automorphic L-functions.

Definition 3. *The analytic conductor of $L(s)$ is given by*

$$A = N \exp\left(2 \operatorname{Re}\left(\frac{L'_\infty(1/2)}{L_\infty(1/2)}\right)\right),$$

where $L_\infty(s)$ is the product of gamma factors appearing in $\Lambda(s)$. The root analytic conductor is $\sqrt[d]{A}$, where d is the degree of $L(s)$.

The idea of the analytic conductor is that N describes the non-archimedean part of the underlying representation, and that μ_r, ν_c capture its archimedean part. The root analytic conductor is a measure of complexity that is normalized for degree, which accounts for the fact that higher degree L-functions may naturally have larger conductors. In [4–7], we considered "single-origin" L-functions, for example, elliptic L-functions, or Dirichlet L-functions, and ordered them by the conductor N. In the following section, we will use the root analytic conductor $\sqrt[d]{A}$ to order L-functions with multiple origins.

3 Unsupervised Learning with RStudio

We begin by describing our dataset. After that, we implement two well-known strategies for unsupervised learning, namely, clustering and dimensionality-reduction.

3.1 Data Summary

In this section, we employ a custom dataset built from rational L-functions with similar root analytic conductor. More precisely, we consider the 12 rational L-functions with root analytic conductor in the interval $[0.438, 0.450]$. Each datapoint is the 10-dimensional vector given by $(a_5, a_7, \ldots, a_{37})$. The indices are prime numbers in the range $5 \leq p \leq 37$. This dataset is similar to those utilised in [4–7], however it does not restrict to a specific type of L-function. The interval was chosen so as to generate a small dataset with some interesting variations. In particular, we note that $L_1, L_3, L_5, L_8, L_9, L_{10}, L_{11}, L_{12}$ correspond to 2-dimensional complex Galois representations (L-functions of degree 2 and motivic weight 0) and L_2, L_4, L_6, L_7 correspond to genus 2 curves defined over $\mathbb{Q}$ (L-functions of degree 4 and motivic weight 1). Amongst the genus 2 curves, L_2 is different in the sense that it has root number $\epsilon = -1$. We represent our dataset in Table 1.

3.2 Hierarchical Clustering with RStudio

We implemented hierarchical clustering on Table 1 using Euclidean distance and complete linkage. This groups our L-functions by similarity, as visualised by the dendrogram in Fig. 1. Looking from top to bottom, we may interpret this figure as two big clusters. Looking at the x-axis, we observe that L_4, L_6, L_7 form the right-hand cluster. Recall that these L-functions are all attached to

Table 1. Dataset to be used in experiments below. Each attribute is the specified Dirichlet coefficient, and each tuple is an L-function with root analytic conductor in the interval $[0.438, 0.450]$. The data was taken from [11].

	a_5	a_7	a_{11}	a_{13}	a_{17}	a_{19}	a_{23}	a_{29}	a_{31}	a_{37}
L_1	0	0	0	-2	0	0	0	0	-2	0
L_2	-2	0	-1	-2	0	-7	-5	5	-1	4
L_3	0	0	0	0	0	0	0	0	0	0
L_4	1	-3	-1	1	4	0	6	2	-8	2
L_5	0	0	0	-2	0	0	0	0	0	0
L_6	-2	0	2	-2	5	3	-3	-3	-4	5
L_7	-1	-3	2	2	-3	2	0	-3	-1	3
L_8	1	0	-2	0	0	-2	0	0	2	0
L_9	1	0	-1	0	0	0	1	0	-1	-1
L_{10}	0	1	0	2	0	-1	0	0	2	0
L_{11}	0	1	0	-2	0	1	0	0	-2	0
L_{12}	0	0	0	0	0	0	0	-2	0	0

genus 2 curves. On the other hand, the y-axis indicates that these points are less similar to each other than the Artin L-functions in the left-hand cluster. The remaining genus 2 curve L_2 is included in the left-hand cluster, though it is the last to be linked. The height indicates that it is seemingly as similar to the Artin representations as the curves represented by L_6 and L_7 are to each other. The separation of L_2 from the other genus 2 curves may be due to the fact that the root number ϵ in its functional equation is different to that of L_4, L_6, L_7. Indeed, in similar experiments for low genus curves, we have seen that unsupervised learning is able to detect such differences [6,7], and this will be relevant in subsequent sections.

3.3 PCA with RStudio

Principal component analysis (PCA) is an unsupervised learning technique for dimensionality reduction. More precisely, PCA projects high-dimensional data onto the axes of greatest variation. This is achieved by approximating the covariance matrix by one with lower rank. The covariance matrix is a real symmetric positive-definite matrix, and hence its eigenvalues are real and positive. In particular, these eigenvalues can be ordered, and the low rank approximation is most effective when there is a significant drop in the eigenvalue. In Figs. 2 and 3, we depict the outcome of two-dimensional PCA for the data in Table 1. In Fig. 3, the image has been clustered using 2-nearest neighbours. In both Figures, we note the genus 2 curves are all close to the boundary, and that the Artin representations are in the middle. In Fig. 3, we note that three of the genus 2 curves are grouped together. Whilst the fourth is grouped with the Artin repre-

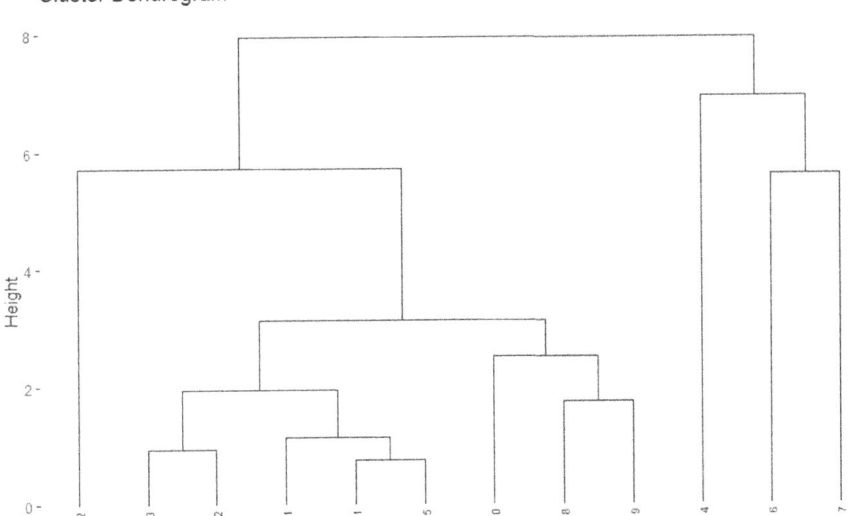

Fig. 1. The output of hierarchical clustering for the data in Table 1. Image generated using RStudio.

sentations, it is somewhat distanced from them. In fact, in [6,7], it is observed that PCA distinguishes between curves of different rank, which, according to the parity conjecture, is a quantity connected to the root number ϵ appearing in the functional equation.

4 Outlook

In the context of our small custom dataset, we have seen that unsupervised learning may detect arithmetic structures in the coefficients of L-functions. This is consistent with prior work which, on one hand, considered much larger datasets, with much larger conductors, and in much higher dimensions, but, on the other, placed much greater restrictions the L-functions involved by specifying their source of origin. By utilising the root analytic conductor, one may speculate that such restrictions can be eliminated.

It is natural to consider implementing the experiments presented here on larger datasets of rational L-functions, with a view towards arithmetic discovery. By way of justification, we briefly review the connection between unsupervised machine learning and the recently-discovered murmuration phenomenon. In [6, 7], large datasets of elliptic L-functions with similar conductors were represented by 100-dimensional vectors. One observation made in [6,7] was that the root number ϵ for an elliptic curve may be determined by machine learning such datasets, in particular, by PCA. We have mentioned above that PCA yields a method to project high-dimensional data to a vector space with lower dimension whilst still maximising variance. The coefficients (or *weightings*) appearing in the

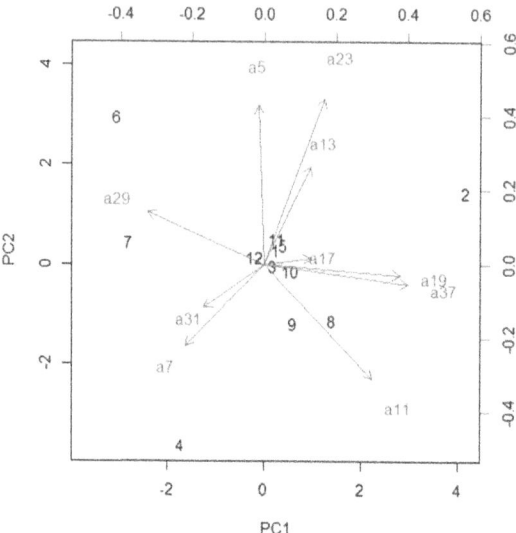

Fig. 2. Two-dimensional PCA for the data in Table 1. In black, we have the image of each data point. In red, we have the image of the attribute vectors. The label PC1 (resp. PC2) refers to the first (resp. second) principal component (axis of greatest variation). Image generated using RStudio.

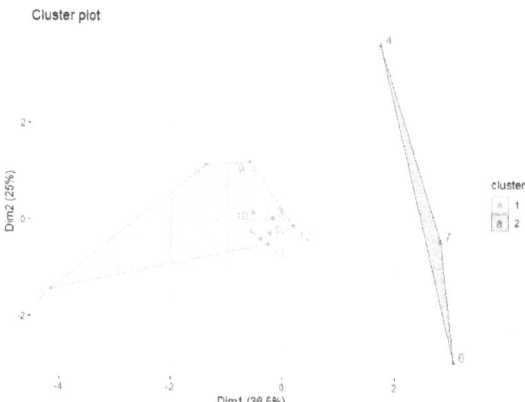

Fig. 3. Clustering the image of two-dimensional PCA for the data in Table 1. We caution the reader that the axes have been rotated 180° from Fig. 2. The percentages on the axes quantify the contribution of these eigenvectors to the variation in the data (which is the ratio of the corresponding eigenvector to the total of the eigenvectors). Image generated using RStudio.

components of this projection are a measure of feature importance. Plotting each prime against its corresponding weight in the first principal component yielded an early hint of the murmuration of elliptic curves [7, Figure 5]. Since their first appearance, murmurations have been observed for much larger conductors and in other arithmetic contexts [8], including the genus 2 curves featured here. Perhaps more importantly, number theorists have proved murmuration theorems in some key cases [2,9,16]. The proven cases do not include the original context of elliptic curves.

Acknowledgement. I am grateful to all organisers of the International Congress on Mathematical Software and specifically to Matthew England and Alexander Kasprzyk for giving me the opportunity to speak in the session titled *Machine Learning within Computer Algebra Systems*. In Sect. 3, I used data downloaded from the LMFDB [11] and implemented the algorithms using RStudio.

References

1. Amir, M., He, Y.-H., Lee, K.-H., Oliver, T., Sultanow, E.: Machine learning class numbers of real quadratic fields. Int. J. Data Sci. Math. Sci. **1**(2), 103–134 (2023)
2. Bober, J., Booker, A., Lee, M., Lowry-Duda, D.: Murmurations of modular forms in the weight aspect, arXiv:2310.07746 (2023)
3. Cowan, A.: Murmurations and explicit formulas. arXiv:2306.10425 (2023)
4. He, Y.-H., Lee, K.-H., Oliver, T.: Machine learning the Sato-Tate conjecture. J. Symbolic Comput. **111**, 61–72 (2022)
5. He, Y.-H., Lee, K.-H., Oliver, T.: Machine learning number fields. Math., Comput. Geometry Data **2**, 49–66 (2022)
6. He, Y.-H., Lee, K.-H., Oliver, T.: Machine learning invariants of arithmetic curves. J. Symbolic Comput. **115**, 478–491 (2023)
7. He, Y.-H., Lee, K.-H., Oliver, T., Pozdnyakov, A.: Murmurations of elliptic curves. arXiv:2204.10140 (2022)
8. He, Y.-H., Lee, K.-H., Oliver, T., Pozdnyakov, A., Sutherland, A.: Murmurations of L-functions, in preparation
9. Lee, K.-H., Oliver, T., Pozdnyakov, A.: Murmurations of Dirichlet characters. arXiv:2307.00256 (2023)
10. Kazalicki, M., Vlah, D.: Ranks of elliptic curves and deep neural networks. Res. Number Theor. **9**, 53 (2023)
11. The L-functions and Modular Forms Database. http://www.lmfdb.org. Accessed 21 Feb 2024
12. Oliver, T.: Supervised learning arithmetic invariants. In He., Y.-H. (ed.) Machine Learning in Pure Mathematics and Theoretical Physics (World Scientific), pp. 331–363 (2023). https://doi.org/10.1142/q0404
13. Romera-Paredes, B., et al.: Mathematical discoveries from program search with large language models. Nature **625**, 468–475 (2024)
14. Sarnak, P.: https://publications.ias.edu/node/2726. Accessed 21 Feb 2024
15. Selberg, A.: Old and new conjectures and results about a class of Dirichlet series. In: Bombieri, E., et al. (eds.) Proceedings of the Amalfi Conference on Analytic Number Theory, Maoiori (1989), pp. 367–385 (1992)
16. Zubrilina, N.: Murmurations. arXiv:2310.07681 (2023)

Numerical Software for Special Functions

Approximation of an Inverse of the Incomplete Beta Function

Michael B. Giles$^{(\boxtimes)}$ and Casper Beentjes

University of Oxford, Oxford, U.K.
mike.giles@maths.ox.ac.uk
https://people.maths.ox.ac.uk/gilesm/

Abstract. Following the methodology previously developed for the inverse Poisson cumulative distribution function (Giles, 2016)., new approximations for the inverse of the incomplete beta function are derived in order to develop efficient evaluations of the inverse binomial cumulative distribution function.

Keywords: approximation · incomplete beta function · binomial distribution · inverse transform

1 Introduction

The motivation for this paper is the development of software for efficiently generating binomial random variables which are used in a wide variety of applications. The cumulative distribution function (CDF) for a binomial random variable K with parameters n, p is defined as

$$\overline{C}(k) \equiv \mathbb{P}(K \leq k) = \sum_{m=0}^{k} \frac{n!}{m!\,(n{-}m)!}\, p^m(1{-}p)^{n-m}, \qquad (1)$$

and we are particularly interested in achieving good efficiency in cases in which n is large. The inverse CDF is defined for $0 < u < 1$ as

$$\overline{C}^{\,-1}(u) = k, \qquad (2)$$

where $0 \leq k \leq n$ is the smallest integer such that

$$u \;\leq\; \sum_{m=0}^{k} \frac{n!}{m!\,(n{-}m)!}\, p^m(1{-}p)^{n-m}. \qquad (3)$$

Given this inverse CDF, a standard approach to random number generation is to generate a random variable U which is uniformly distributed on the open unit interval $(0,1)$ and then define the binomial random variate by $k = \overline{C}^{\,-1}(U)$.

© The Author(s), under exclusive license to Springer Nature Switzerland AG 2024
K. Buzzard et al. (Eds.): ICMS 2024, LNCS 14749, pp. 207–214, 2024.
https://doi.org/10.1007/978-3-031-64529-7_22

Previous work on the inverse Poisson CDF exploited a connection to the incomplete gamma function [4]. In a similar way, we exploit a connection between the binomial distribution and the incomplete beta function defined as

$$C(x) = \frac{n!}{(x-1)!\,(n-x)!} \int_0^{1-p} t^{n-x}(1-t)^{x-1}\,dt,$$

for $0 < x < n+1$. It can be verified that $\overline{C}(k) = C(k+1)$ for integers $0 \le k \le n$. Hence, as illustrated in Fig. 1, $\overline{C}^{-1}(u) = \lfloor C^{-1}(u) \rfloor$, where the floor function $\lfloor x \rfloor$ returns the largest integer less than, or equal to, x.

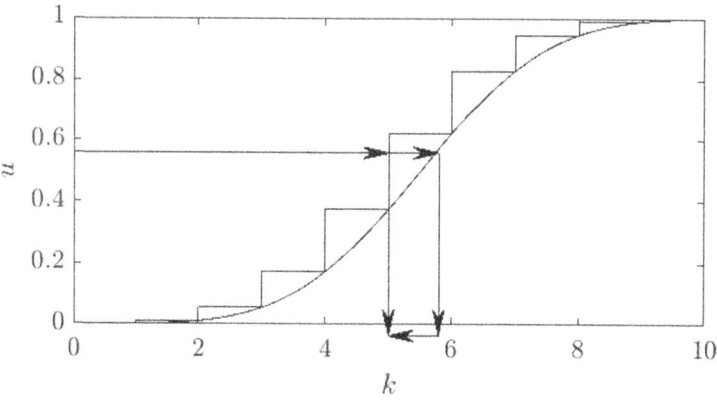

Fig. 1. Plot of $\overline{C}(k)$ and $C(k)$ for $n=10$, $p=0.5$, and an illustration of the rounding down of $C^{-1}(u)$ to give $\overline{C}^{-1}(u)$.

Note that $C(k) \equiv I_{1-p}(n+1-k, k)$, where $I_p(a,b)$ is the regularised incomplete beta function, so $C^{-1}(u)$ corresponds to an inverse of the incomplete beta function, but it is different to others considered in the literature [2,6,9] and implemented in software [1,5] which find the inverse of $I_p(a,b)$ holding fixed two of p, a, b, whereas we hold fixed p and $a+b$.

This paper derives two approximations $\widetilde{Q}(u)$ to the quantile function $Q(u) \equiv C^{-1}(u)$ which in the future will be used as components within an algorithm to efficiently compute the inverse binomal CDF function $\overline{C}^{-1}(u)$ for use in generating binomial random variables.

2 Normal Asymptotic Approximation

Asymptotically, as $n \to \infty$, the binomial CDF approaches a Normal distribution with mean np and variance npq, where $q = 1-p$. This motivates the change of variables

$$x = np + \sqrt{npq}\, y, \quad t = q + \sqrt{pq/n}\,(z-y),$$

with y being the deviation from the mean, normalised by the standard deviation, and z being defined so that $y = z$ at the upper limit of the integral. This leads to

$$C(x) = \frac{1}{\sqrt{2\pi}} \int_{y-\sqrt{nq/p}}^{y} J \, dz,$$

where

$$\log J = \frac{1}{2}\log(2\pi) + \log \Gamma(n+1) - \log \Gamma(x) - \log \Gamma(n-x+1)$$
$$+ (n-x)\log t + (x-1)\log(1-t) + \frac{1}{2}\log(p(1-p)/n).$$

An asymptotic expansion in powers of $n^{-1/2}$, followed by exponentiation and a second expansion in powers of $n^{-1/2}$, yields

$$I(y,z) = \exp(-\frac{1}{2}z^2) \left(1 + \sum_{m=1}^{\infty} n^{-m/2} e_m(p,y,z)\right)$$

where each of the $e_m(p,y,z)$ is polynomial in p, y and z. Integrating by parts then gives

$$C(x) = \Phi(y) + \phi(y)\left(n^{-1/2}\tilde{f}_1(p,y) + n^{-1}\varepsilon^2 \tilde{f}_2(p,y) + n^{-3/2}\varepsilon^3 \tilde{f}_3(p,y) + O(n^{-2})\right)$$

where $\Phi(y)$ is the Normal CDF function, $\phi(y) = \Phi'(y)$ is the Normal probability density function, and $\tilde{f}_1, \tilde{f}_2, \tilde{f}_3$ are polynomial in both p and y.

Inverting this expansion, following the methodology in [4, Appendix A] gives the final asymptotic expansion for $Q(u) \equiv C^{-1}(u)$ in which $w = \Phi^{-1}(u)$,

$$Q(u) = np + \sqrt{npq}\, w + \left(2 + 2p + (q-p)w^2\right)/6$$
$$+ \left((-2 + 14pq)w + (-1 - 2pq)w^3\right)/(72\sqrt{npq})$$
$$+ (p-q)(2+pq)(16 - 7w^2 - 3w^4)/(1620\, npq) + O(n^{-3/2}),$$

providing the following three approximations with increasing cost and accuracy:

$$\tilde{Q}_{N0}(u) = np + \sqrt{npq}\, w + \left(2 + 2p + (q-p)w^2\right)/6$$
$$\tilde{Q}_{N1}(u) = \tilde{Q}_{N0}(u) + \left((-2+14pq)w + (-1-2pq)w^3\right)/(72\sqrt{npq})$$
$$\tilde{Q}_{N2}(u) = \tilde{Q}_{N1}(u) + (p-q)(2+pq)(16 - 7w^2 - 3w^4)/(1620\, npq).$$

The first of these corresponds to the Cornish-Fisher expansion with skewness correction (based on the mean, variance and skew of the binomial distribution) used by [7] as the starting point for a search to determine the inverse of the binomial distribution.

Figures 2 and 3 show the errors given by these approximations when $|w| < 3$, and x and $n+1-x$ are both larger than 10. In particular, Fig. 3 clearly indicates the different asymptotic orders of accuracy of the three approximations.

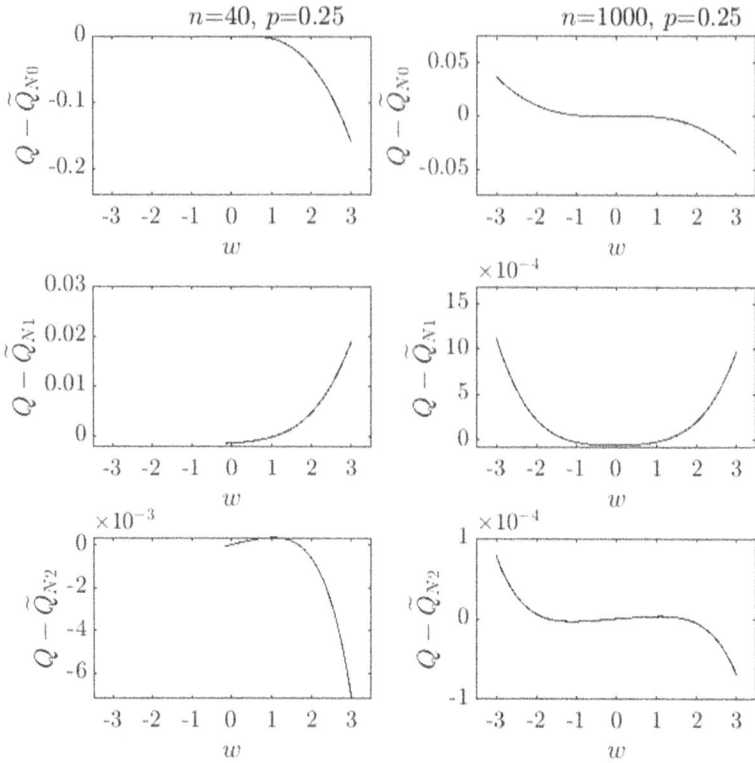

Fig. 2. Errors in Normal approximations $\widetilde{Q}_{N0}$, $\widetilde{Q}_{N1}$, $\widetilde{Q}_{N2}$, for $p = 0.25$, $n = 40, 1000$ and $|w| \leq 3$, $10 < x < n - 9$.

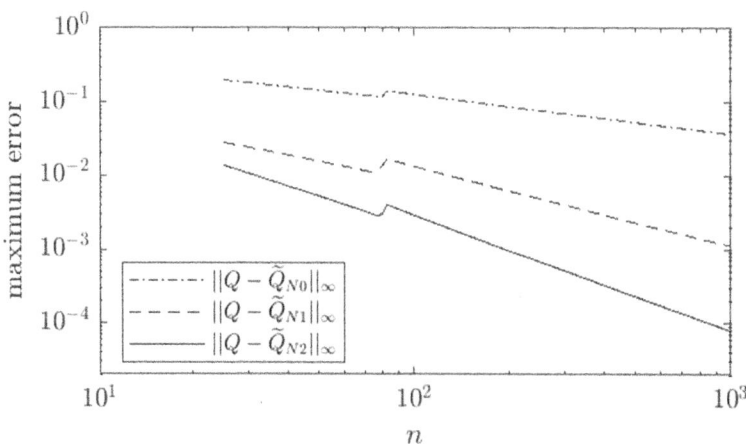

Fig. 3. Maximum errors for approximations $\widetilde{Q}_{N0}$, $\widetilde{Q}_{N1}$, $\widetilde{Q}_{N2}$, for $p = 0.25$ and $|w| \leq 3$, $10 < x < n - 9$.

3 GST Asymptotic Approximation

The Normal asymptotic approximations converge very slowly in powers of $n^{-1/2}$, and can be very inaccurate when $|w| > 3$. An alternative asymptotic expansion which is more accurate, but also more expensive, is based on previous work by Gil, Segura and Temme [3,8].

The expansion starts from the integral representation for $C(x)$ expressed as

$$C(x) = \frac{n!}{(x-1)!\,(n-x)!} \int_0^{1-p} \frac{\exp(-\nu(\xi \log t - (1-\xi)\log(1-t)))}{t(1-t)}\, dt,$$

where $\nu = n+1$, $\xi = x/\nu$. This leads to a representation in terms of the complementary error function [3, Equation (2.1)], which is particularly useful for large ν. In terms of the normal CDF, Φ, this is given by

$$C(x) = \Phi(-\eta\sqrt{\nu}) + R_\nu(\eta),$$

where the remainder $R_\nu(\eta)$ has an expansion given by [3, Equation (A9)] and η is given by

$$\eta = \sqrt{-2\left(\xi \log \frac{p}{\xi} + (1-\xi)\log \frac{1-p}{1-\xi}\right)} \equiv h_p(\xi),$$

with the square root being chosen to have the same sign as $p-\xi$, i.e. $\eta > 0$ if $p > \xi$, and $\eta < 0$ if $p < \xi$. Hence, if $u = C(k)$ and $w = \Phi^{-1}(u)$ then to leading order $-\eta\sqrt{\nu} \approx w$. Following [3] we define $\eta_0 \equiv -w/\sqrt{\nu}$ and thus we have the approximation

$$x = \widetilde{Q}_{T0}(U) \equiv \nu \xi_0 \equiv \nu\, h_p^{-1}(\eta_0).$$

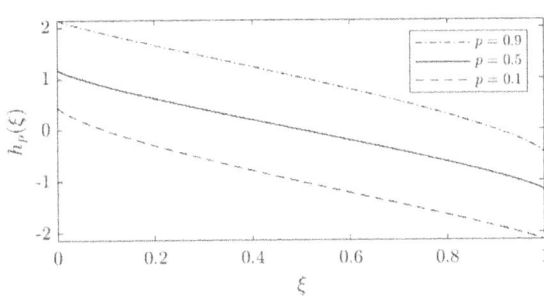

Fig. 4. The function $h_p(\xi)$.

Given η_0, the value $\xi_0 = h_p^{-1}(\eta_0)$ can be obtained by a Newton iteration, requiring no more than 5 iterations when starting from $\xi = p$, for which $h_p(p)=0$, $h_p'(p) = -1/\sqrt{pq}$, and using the log1p function to maintain full precision.

Alternatively, for small values of s it can be obtained from a Taylor expansion of $h_p^{-1}(s)$,

$$h_p^{-1}(s) = p - p(1-p) \sum_{k=0}^{\infty} a_k(p) \left(\frac{s}{\sqrt{p(1-p)}} \right)^k,$$

with coefficients $a_k(p)$ that are polynomials in p [3, Section 2.1].

If $u = C(x) = \Phi(w)$, and $\xi_0 = h_p^{-1}(\eta_0) = h_p^{-1}(-w/\sqrt{\nu})$, then to leading order $x \approx \nu \, \xi_0$. We now seek a more accurate asymptotic expansion of the form

$$x \approx \nu \, \xi_0 + g_p(\eta_0).$$

Following [3] we return to the relation between η and w (or u) and consider the expansion

$$\eta = \eta_0 + \nu^{-1} \eta_1 + O(\nu^{-2}),$$

where the first order correction η_1 is given, using [3, Equations (2.2) and (3.6)], as

$$\eta_1 = \eta_0^{-1} \log f_p(\eta_0), \quad f_p(\eta) = \sqrt{\xi(1-\xi)}\,\eta/(p-\xi).$$

Using this expansion for η we can improve the accuracy of the first order approximation ξ_0 via

$$\xi(\eta) = \xi_0 + \nu^{-1} \eta_1 \frac{d\xi}{d\eta}(\eta_0) + O(\nu^{-2}),$$

where implicit differentiation of the relation $\eta = h_p(\xi)$ provides the expression for the derivative

$$\frac{d\xi}{d\eta} = -\eta \Big/ \left(\log \frac{(1-\xi)p}{(1-p)\xi} \right).$$

This results in a GST approximation $\tilde{Q}_{T1}(u)$ with $O(\nu^{-1})$ error,

$$\tilde{Q}_{T1}(u) = \nu \, \xi_0 + g_p(\eta_0),$$

where

$$g_p(\eta_0) = \left\{ \eta_0^{-1} \log \left(\sqrt{\xi_0(1-\xi_0)}\,\eta_0/(p-\xi_0) \right) \right\} \times \left\{ -\eta_0 \Big/ \left(\log \frac{(1-\xi_0)p}{(1-p)\xi_0} \right) \right\}$$

with $\xi_0 = h_p^{-1}(\eta_0)$ and $\eta_0 = -w/\sqrt{\nu} = -\Phi^{-1}(u)/\sqrt{\nu}$.

The advantage of this GST approximation, compared to the Normal asymptotic approximations, is its accuracy over the full range of possible values for w, not just $|w| < 3$. Figures 5 and 6 show the errors for $k = 10, 100$ and $|w| < 10$ corresponding to u taking values in the range $[10^{-23}, 1-10^{-23}]$.

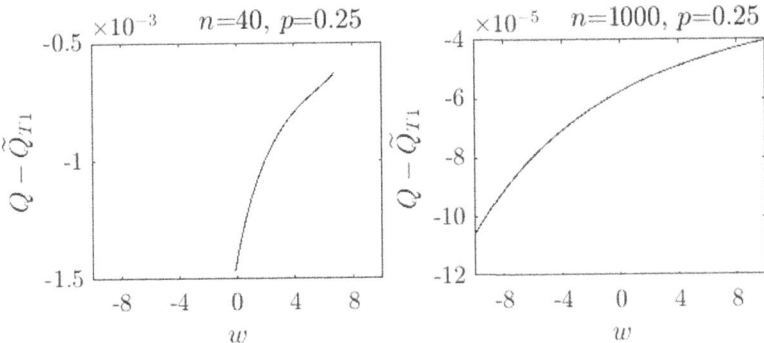

Fig. 5. Error in $\widetilde{Q}_{T1}$ approximation for $p = 0.25$ and $n = 40, 1000$.

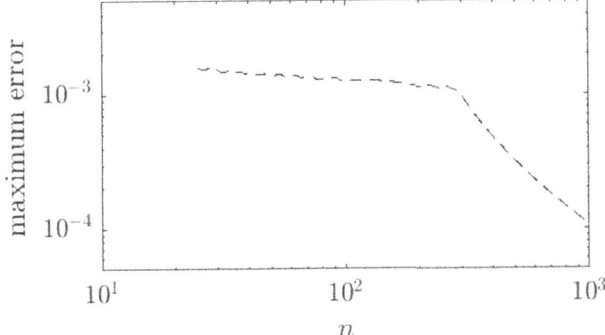

Fig. 6. Maximum $\widetilde{Q}_{T1}$ errors for $p = 0.25$ and $|w| \leq 10$, $10 < x < \nu - 9$.

4 Conclusions

In this paper we have derived two approximations of a particular inverse of the incomplete beta function, different to the inverses previously considered in the literature and in software implementations. This will be used in future work as part of an algorithm to compute the inverse binomial cumulative distribution function. For large values of n, the Normal asymptotic approximation labelled $\widetilde{Q}_{N2}$ is excellent when $|\Phi^{-1}(u)| < 3$, whereas when $|\Phi^{-1}(u)| > 3$, the slightly more expensive GST approximation based on the work of Gil, Segura and Temme [3] provides comparable accuracy. For most values of the uniform random variate u, this will enable the correct binomial random variate to be calculated; in a very small proportion of cases a more expensive secondary step will be required. Full details will be given in a future paper.

Acknowledgements. This research was funded by the UK EPSRC programme grant EP/P020720/1 and by the Hong Kong Innovation and Technology Commission (InnoHK Project CIMDA) and their support is gratefully acknowledged. Much of the paper was written while the first author was visiting the Oden Institute for Computational Engineering & Sciences at the University of Texas at Austin, and their warm hospitality and support is much appreciated too.

References

1. Boost++ library.: Incomplete beta function inverses. Retrieved March 11, 2024 from https://live.boost.org/doc/libs/1_48_0/libs/math/doc/sf_and_dist/html/math_toolkit/special/sf_beta/ibeta_inv_function.html (2024)
2. Gil, A., Segura, J., Temme, N.: Efficient algorithms for the inversion of the cumulative central beta distribution. Num. Algorithms **74**(1), 77–91 (2017). https://doi.org/10.1007/s11075-016-0139-2
3. Gil, A., Segura, J., Temme, N.: Asymptotic inversion of the binomial and negative binomial cumulative distribution functions. Electron. Trans. Numer. Anal. **52**, 270–280 (2020). https://doi.org/10.1553/etna_vol52s270
4. Giles, M.: Algorithm 955: Approximation of the inverse Poisson cumulative distribution function. ACM Trans. Math. Softw. **42**(1) (2016). https://doi.org/10.1145/2699466
5. GNU Scientific Library: Beta pinv/qinv functions. Retrieved March 11, 2024 from https://www.gnu.org/software/gsl/doc/html/randist.html#the-beta-distribution (2024)
6. Majumder, K., Bhattacharjee, G.: Algorithm as 64: Inverse of the incomplete beta function ratio. J. Royal Statistical Society. Series C (Applied Statistics) **22**(3), 411–414 (1973). https://doi.org/10.2307/2346798
7. Mathlib library: Binomial quantile function. Retrieved March 11, 2024 from https://github.com/SurajGupta/r-source/blob/master/src/nmath/qbinom.c (2024)
8. Temme, N.: Uniform asymptotic expansions of the incomplete gamma functions and the incomplete beta function. Math. Comput. **29**(132), 1109–1114 (1975). https://doi.org/10.2307/2005750
9. Temme, N.: Asymptotic inversion of incomplete gamma functions. Math. Comput. **58**(198), 755–764 (1992). https://doi.org/10.1090/S0025-5718-1992-1122079-8

DLMF Standard Reference Tables on Demand

Bonita V. Saunders[1]([✉])[iD], Sean Brooks[2], Ron Buckmire[3],
Rachel E. Vincent-Finley[4], Franky Backeljauw[5], Stefan Becuwe[5],
Bruce Miller[1], Marjorie McClain[1], and Annie Cuyt[5]

[1] National Institute of Standards and Technology, Gaithersburg, MD 20899, USA
bonita.saunders@nist.gov
[2] Coppin State University, Baltimore, MD, USA
[3] Occidental College, Los Angeles, CA, USA
[4] Southern University and A&M College, Baton Rouge, LA, USA
[5] University of Antwerp, Middelheimlaan 1, B-2020, Antwerp, Belgium

Abstract. Mathematical software designers, numerical analysts, and other researchers often need high quality tables of function values, but most current libraries and systems that produce such tables offer limited information about accuracy. To address this void, the National Institute of Standards and Technology (NIST) Applied and Computational Mathematics Division (ACMD) and the University of Antwerp Computational Mathematics (CMA) Research Group are collaborating to build the DLMF Standard Reference Tables on Demand (DLMF Tables) web service. DLMF Tables will provide a standard of comparison for testing numerical software by computing, on demand, special functions to user-defined accuracy with guaranteed error bounds.

Keywords: Numerical software verification · Validated computation · Validated numerics · Reliable computation · Special functions · Multiple precision arithmetic · Floating-point arithmetic · Correct rounding

1 Introduction

In 2010, the National Institute of Standards and Technology (NIST) launched the NIST Digital Library of Mathematical Functions (DLMF) [6] to update and replace the widely cited 1964 Abramowitz and Stegun resource, Handbook of Mathematical Functions with Formulas, Graphs, and Mathematical Tables (A&S) [1]. A&S addressed a critical need for tables of reference values and other information to support the computation and understanding of special functions, but today's reliable computing machines, computer algebra systems, and multiple precision computational packages have diminished the need for reference tables. However, mathematical and physical scientists, numerical analysts, and software developers still need a way to confirm the accuracy of software used to compute mathematical function values. DLMF Standard Reference Tables on Demand (DLMF Tables) is a collaborative project between members of the NIST

Applied and Computational Mathematics Division (ACMD) and the University of Antwerp Computational Mathematics Research Group (CMA) to address this problem. The goal is an online system where users can generate tables of special function values at user-specified precision with an error certification to validate their own algorithms or confirm the accuracy of results from commercial or publicly available packages.

In 2020 the DlMF Tables Project team was joined by three mathematicians from the inaugural African Diaspora Joint Mathematics (ADJOINT) Workshop sponsored by the Mathematical Sciences Research Institute (MSRI) in Berkeley, California. ADJOINT Workshops are designed to encourage collaborations between experienced and early career researchers, especially those of African descent. DLMF Tables Project member B. Saunders led an ADJOINT team that focused on an introduction to validated computation of mathematical functions and explored the application to DLMF Tables. The ADJOINT team members were supported by MSRI (now known as the Simons Laufer Mathematical Sciences Institute) through grants from the National Science Foundation (NSF), National Security Agency (NSA), and Sloan for more than a year. The ADJOINT members have continued a post-workshop collaboration with the DLMF Tables Team.

In this paper we discuss our beta site located at the University of Antwerp, which uses a computational engine based on CMA's MpIeee, a multiple precision IEEE 754/854 compliant C++ floating point arithmetic library. Ultimately, we want a permanent standalone system that is also accessible directly through the NIST DLMF. In Sect. 2 we describe features of CMA's MpIeee library that support its use for DLMF Tables. Section 3 discusses the design of DLMF Tables and its connection to the NIST DLMF. In Sect. 4 we discuss future work and directions for the project.

2 MpIeee

In CMA's MpIeee library, IEEE 754-1985 [11] and IEEE 854-1987 [12] compliance is extended to multiple precision floating point arithmetic in base 2^m, $1 \leq m \leq 24$, and 10^n, $1 \leq n \leq 7$. In line with these standards, MpIeee arithmetic has strong semantics with basic arithmetic operations, remainder and square root operations, and format conversions all rounded exactly. This means that the operation or conversion is performed as if it were first done in infinite precision and the result rounded to obtain the final floating point (or integer, if conversion to integer) representation. Of course, the correctly rounded value will depend on the rounding mode used. Currently, MpIeee offers round to nearest (default), and the directed modes: round up (toward $+\infty$), round down (toward $-\infty$), round toward 0 (truncation), and round away from 0. Exact rounding is a necessary requirement for a reliable and portable mathematical software library [7,9,10].

Furthermore, MpIeee allows mixed precision arithmetic, that is, the precision of each operand and result can all be different. It also handles exceptions such as division by 0 or ∞, underflow, overflow, invalid operations, includes all elementary functions as suggested in IEEE 754-2008 [13] and features a growing list of special functions [2–5].

A variety of numerical methods can be used to compute special function values, including convergent power series, asymptotic expansions, recurrence relations, differential equations, and continued fractions [5,8]. Whether a particular method should be used in a given subdomain of a function will depend on the convergence properties of the method there.

Currently, MpIeee special function evaluation routines use either convergent power series or continued fraction implementations and cover a wide range of arguments on the real line. Tables comparing the effectiveness of both representations for several special functions can be found in Chap. 13 of [5].

3 Design of DLMF Tables

DLMF Tables Version 1.0β is located at http://dlmftables.uantwerpen.be/. In Fig. 1 we show the top level which, similarly to the NIST DLMF, consists of a table of contents listing the chapters available. Currently, the chapter numbers correspond to chapter identifiers in the DLMF.

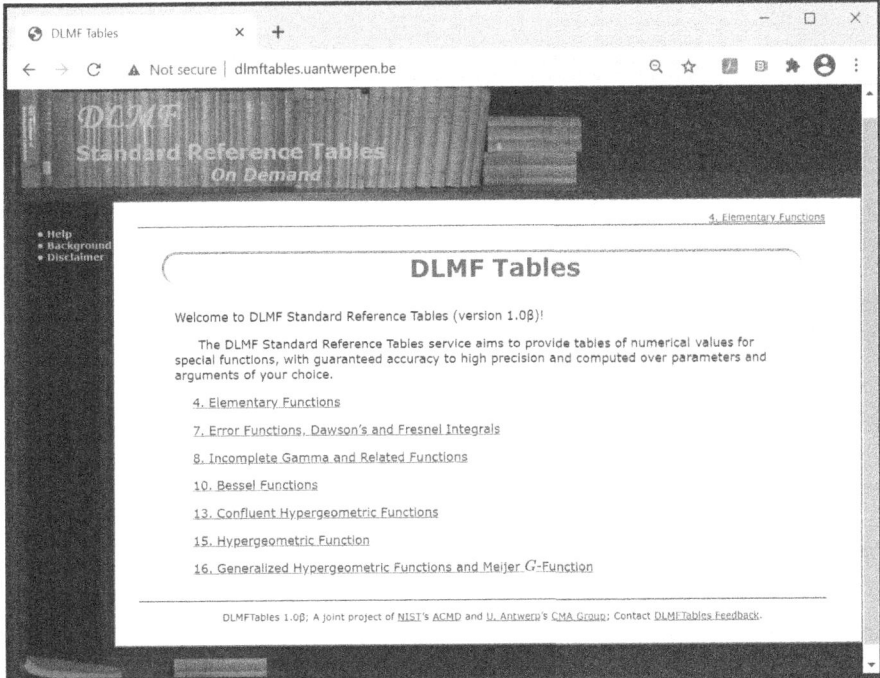

Fig. 1. DLMF Standard Reference Tables on Demand: Table of Contents. Numbers correspond to chapter identifiers in the NIST Digital Library of Mathematical Functions (DLMF).

After a chapter is selected, the input screen for the chapter appears, as shown in Fig. 2; and the user chooses a function and requests either a tabulation or comparison. Each function description panel contains a link to the function's definition in the DLMF and any known restrictions on the evaluation domain.

3.1 Tabulation

The tabulation option lets the user create a table of function values based on parameter and argument data entered directly into the form or uploaded from a

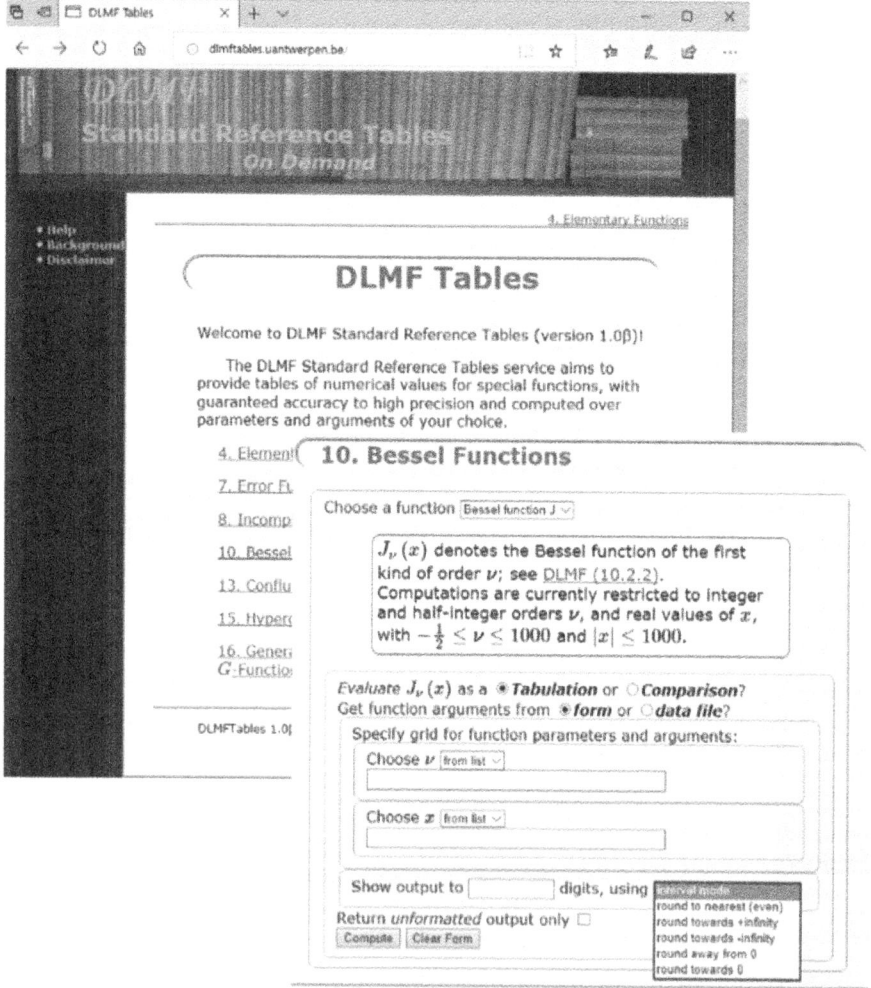

Fig. 2. DLMF Tables Input Screen. Description panel contains link to function definition in NIST DLMF. User may choose a Tabulation: output table of reference values or Comparison: output table of reference values, user input values, and relative errors.

file meeting the site's format specifications. The user then inputs the number of output digits desired, that is, the precision, and chooses the output mode for the computation. The default mode is interval mode, where a high-precision enclosure, upper and lower bound, is displayed for each function value requested. Currently the precision is supplemented with an extra five digits, displayed in a smaller font.

Alternatively, the user may choose that output values be displayed exactly in one of five rounding modes:

- round to nearest (even)
- round up (toward $+\infty$)
- round down (toward $-\infty$)
- round toward 0
- round away from 0.

3.2 Comparison

When the user selects the comparison option a user-supplied data file must be uploaded. The user must also decide which output mode will be used for the

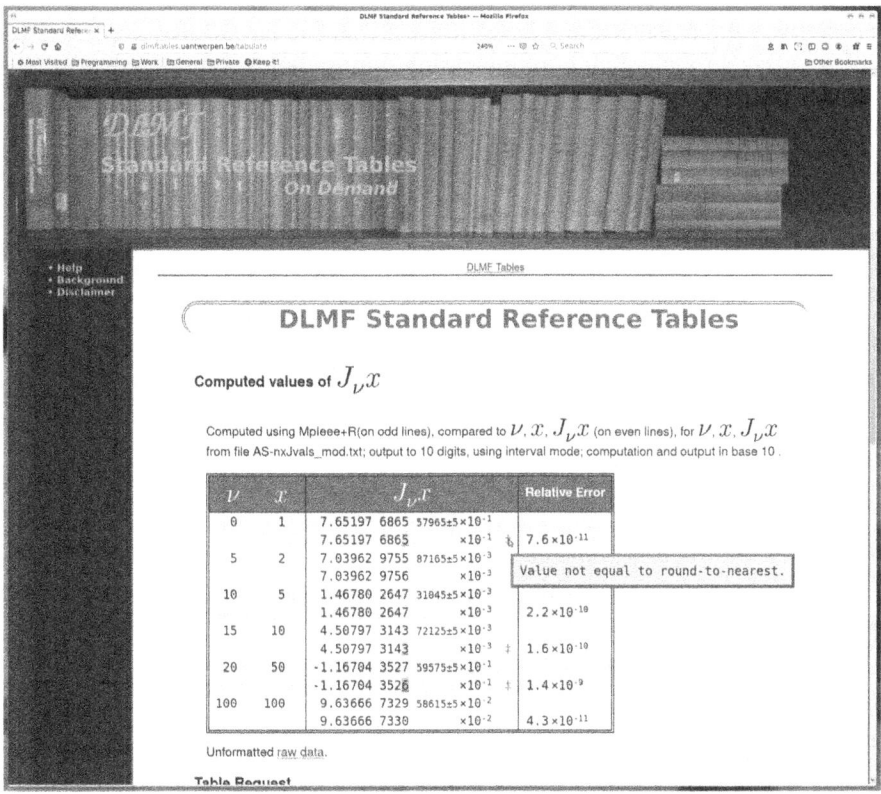

Fig. 3. Bessel function $J_\nu x$ comparison of DLMF Tables interval mode output (odd lines) versus user-supplied values (even lines).

comparison. Figure 3 shows a table of Bessel function J values outputted by DLMF Tables in interval mode compared to user-supplied values from a file. If a user's value falls within the interval, but is not the round to nearest value, one or more digits will be coded yellow as a warning. If the user's value falls completely outside the interval, then one or more digits will be coded red to indicate that the value is incorrect. If a rounding mode is the selected output mode used for comparison, the user's value is considered incorrect if all digits do not match the rounded digits. Incorrect digits will be coded red. Also, note that the relative error between the user's value and the DLMF Tables value is provided in the last column. For interval mode the midpoint of the interval is used as the exact value when computing the relative error.

4 Future Directions

The release of the beta version marked the completion of the initial phase of the project. Computation codes for more functions are in the pipeline and additional site enhancements are being planned. Only real valued functions are included at present, but hopefully we will be able to add complex valued functions in the future. Testing of the site has also started, but the sparsity of systems and software of this type make comprehensive testing challenging. Additional dedicated staff are needed to maintain our momentum, but finding a way to engage strong high school students and undergraduates, along with a postdoc may also be a possibility.

Disclaimer. All references to commercial products are provided only for clarification of the results presented. Their identification does not imply recommendation or endorsement by NIST.

Disclosure of Interests. The authors have no competing interests to declare that are relevant to the content of this article.

Acknowledgement. S. Brooks, R. Buckmire, and R. Vincent-Finley were supported in this work by National Science Foundation, Grant Nos. DMS-1915954 and DMS-2016406; National Security Agency, Grant No. H98230-20-1-0015; and Sloan Foundation, Grant No. G-2020-12602 as participants in ADJOINT 2020 hosted by the Mathematical Sciences Research Institute in Berkeley, California.

References

1. Abramowitz, M., Stegun, I.A. (eds.): Handbook of Mathematical Functions with Formulas, Graphs, and Mathematical Tables. No. 55 in National Bureau of Standards Applied Mathematics Series, U.S. Government Printing Office, Washington, D.C. (1964), corrections appeared in later printings up to the 10th Printing, December, 1972. Reproductions by other publishers, in whole or in part, have been available since 1965

2. Backeljauw, F.: A library for radix-independent multiprecision IEEE-compliant floating-point arithmetic. Tech. Rep. 2009-01 Universiteit Antwerpen (2009)
3. Backeljauw, F., Becuwe, S., Cuyt, A., Van Deun, J., Lozier, D.W.: Validated evaluation of special mathematical functions. Sci. Comput. Programm. **90**, 2 – 20 (09 2014). https://doi.org/10.1016/j.scico.2013.05.006, http://www.sciencedirect.com/science/article/pii/S0167642313001263, special issue on Numerical Software: Design, Analysis and Verification
4. Colman, M., Cuyt, A., Deun, J.V.: Validated computation of certain hypergeometric functions. ACM Trans. Math. Softw. **38**(2) (Jan 2012). https://doi.org/10.1145/2049673.2049675, https://doi.org/10.1145/2049673.2049675
5. Cuyt, A.A., Petersen, V., Verdonk, B., Waadeland, H., Jones, W.B.: Handbook of Continued Fractions for Special Functions. SpringerLink: Springer e-Books, Springer Netherlands (2008). https://doi.org/10.1007/978-1-4020-6949-9, https://books.google.com/books?id=DQtpJaEs4NIC
6. NIST Digital Library of Mathematical Functions. https://dlmf.nist.gov/, Release 1.2.0 of 2024-03-15, F. W. J. Olver, et al. eds.
7. Fousse, L., Hanrot, G., Lefèvre, V., Pélissier, P., Zimmermann, P.: MPFR: a multiple-precision binary floating-point library with correct rounding. ACM Trans. Math. Softw. **33**(2), 13 (2007). https://doi.org/10.1145/1236463.1236468, mPFR: http://mpfr.org
8. Gil, A., Segura, J., Temme, N.M.: Numerical Methods for Special Functions. Society for Industrial and Applied Mathematics, Philadelphia, PA (2007). https://doi.org/10.1137/1.9780898717822
9. Goldberg, D.: What every computer scientist should know about floating-point arithmetic. ACM Comput. Surv. **23**(1), 5–48 (1991). https://doi.org/10.1145/103162.103163
10. Higham, N.J.: Accuracy and Stability of Numerical Algorithms. Society for Industrial and Applied Mathematics, Philadelphia, PA, second edn. (2002). https://doi.org/10.1137/1.9780898718027, https://epubs.siam.org/doi/abs/10.1137/1.9780898718027
11. IEEE: IEEE standard for binary floating-point arithmetic. ANSI/IEEE Std 754-1985, pp. 1–20 (1985). https://doi.org/10.1109/IEEESTD.1985.82928, https://ieeexplore.ieee.org/stamp/stamp.jsp?tp=&arnumber=30711&isnumber=1316
12. IEEE: IEEE standard for radix-independent floating-point arithmetic. ANSI/IEEE Std 854-1987, pp. 1–19 (Oct 1987). https://doi.org/10.1109/IEEESTD.1987.81037
13. IEEE: IEEE standard for floating-point arithmetic. IEEE Std 754-2008, pp. 1–70 (Aug 2008). https://doi.org/10.1109/IEEESTD.2008.4610935

Mathematical Research Data

Integrating Mathematical Data and Resources: Advancements in zbMATH Open for Enhanced Mathematical Research Accessibility and Reproducibility

Maxence Azzouz-Thuderoz[(✉)], Madhurima Deb, Matteo Petrera, Moritz Schubotz, and Olaf Teschke

Department of Mathematics, FIZ Karlsruhe, Berlin, Germany
maxence.azzouz-thuderoz@fiz-karlsruhe.de

Abstract. We report the ongoing efforts of swMATH, an integral part of zbMATH Open, to collect precise referencing software metadata. zbMATH Open is emerging as a unified platform offering a spectrum of mathematical resources, including mathematical software, formulas, reviews, and serial and mathematical item classification. zbMATH Open offers connection to external partners, DLMF and OEIS, via its Links API by indexing approximately 6,330 documents containing 65,069 references to OEIS sequences and 15,858 references to 2,053 DLMF functions. Significantly, the collection of 44,594 software entries from swMATH is entirely accessible through zbMATH Open. Here, we emphasize the accurate referencing of mathematical software in swMATH for maintaining integrity, advancing mathematical research, and enhancing reproducibility. We describe how swMATH is embedded into zbMATH open and elaborate on the relationship of software and other mathematical research data like OEIS and DLMF, ensuring a complete and FAIR resource for the mathematical research community.

Keywords: zbMATH Open · software · metadata

1 The swMATH Journey

1.1 On the Importance of Metadata for Software

Software is taking a growing specific place in research [6]. They are actionable knowledge and often part of the scientific knowledge process. For this, they are unavoidable when one needs to ensure that scientific work is reproducible. Years ago, the FAIR (Findable, Accessible, Interoperable, Reusable) [21] movement hit the rock of science. The FAIR principles [14] stated for research data have been extended to research software to improve the sustainability of research software. Especially, the authors state that *metadata, data, and software should* be well-described such that they can be reused, combined, and extended in different settings.

As an illustration of the importance of FAIR research data, [5] provides a list of mathematical object registries like the *Database of Ring Theory*, *MathRepo*, or *The House of Graphs* and sets their degree of compliance with the FAIR principles. However, like any STEM discipline, mathematics has also been concerned with the importance of mathematics software sustainability. Specifically, [7] suggested a methodology to build metadata to improve the FAIRness of mathematics software. Furthermore, other mathematicians have developed software, such as [15], that implemented models in topology, with the first goal being reusability.

1.2 swMATH as an Extension of zbMATH Open

zbMATH Open is today the world's largest portal for mathematics research information services. Originating in Germany in 1868 as a mathematical review service, the entity later rebranded as Zentralblatt MATH, has evolved into the premier destination for accessing scientific articles in mathematics since its inception in 2021.

Complementing zbMATH Open, the swMATH database service was introduced in 2013 to bolster the discoverability and prominence of research software within the mathematics community [13]. By late 2023, this resource had been seamlessly integrated into and become accessible through the zbMATH Open interface [1].

Scholars now benefit from a consolidated interface on the zbMATH Open website, where they can comprehensively access information about mathematical software. This integration streamlines the research process, enhancing accessibility and efficiency for the global mathematical community.

The zbMATH Open Interface. integrates a collection of interlinked metadata entities as authors, articles, and software, all in-house indexed, forming a collection of persistent identifiers (PIDs) distinct and autonomous from any third-party dependant identifiers such as DOIs or ORCIDs. These mathematics domain-based PIDs are connected to formulas and numbers from registries like OEIS and DLMF and are categorized according to the Mathematics Subject Classification [10]. All these items form a world-class, rich knowledge graph of academic mathematical outputs.

These metadata often support hyperlinks to publications like DOIs and arXiv preprint papers. Lastly, citing mathematics articles in a BibTeX format is facilitated by a dedicated button on the zbMATH Open platform.

Moreover, zbMATH launched in 2021 an Open Archives Initiative Protocol for Metadata Harvesting (OAI-PMH) service to make zbMATH an open and interoperable resource [19]. OAI-PMH is a computer protocol developed by the Open Archives Initiative to exchange metadata. It enables the creation and automatic updating of centralized repositories where metadata from various sources can be queried simultaneously. It leaves the possibility of read-only access to resources.

Additionally, the zbMATH Open Links API is widely usable in tracing bibliographic references related to a specific topic under the present partners of zbMATH Open, DLMF and OEIS. This API enables researchers to access DLMF [3] or OEIS [11] objects cited at zbMATH Open. Furthermore, one can retrieve the links associated with a mathematical subject classification (MSC) code or author ID. 6,312 links from zbMATH Open to DLMF and 67,436 links from zbMATH Open to OEIS are displayed on the zbMATH Open website. This tool is essential for statistically analyzing the bibliographic content in DLMF or OEIS.

Continuously to this work, a REST API service has also been released in 2023 [12], making for the first time swMATH metadata interoperable in a standard way.

In parallel, the German Mathematical Research Data Initiative (MaRDI) portal has also initiated an OAI-PMH service to display rich domain contents. We look forward to reporting further on our work conncecting swMATH to the MaRDI portal, which caters to swMATH as a pioneer research software catalog.

2 swMATH in the EOSC

2.1 Protecting Source Code and Indexing It, a Fruitful Collaboration with Software Heritage

The SIRS [4] report, unveiled in 2020, introduced the ARDC principles, shedding light on the inherent fragility of software and emphasizing the need to establish a robust scholarly infrastructure to ensure the longevity of research software. Aggregators like swMATH are urged to adhere to these principles, with the FAIRCORE4EOSC project striving to enforce FAIR compliance within software catalogs such as swMATH [21]. The authors delineate fundamental principles for software development in academic research, encompassing Archiving, Referencing, Describing, and Citing software. As swMATH serves as a catalog, it references software and its metadata but does not store the source code. This limitation raises concerns about the accessibility and longevity of software hosted on various servers or platforms. Consequently, zbMATH Open may encounter the common issue of link rot, where URL links to software projects become inaccessible over time. To address these challenges, swMATH has partnered with Software Heritage to enhance its ARDC compliance [2]. In 2018, Software Heritage was established as a long-term software preservation library to archive all open-source software projects worldwide. With over 200 million archived projects from platforms like GitHub and BitBucket, Software Heritage plays a crucial role in the FAIRCORE4EOSC project as a universal archive for scientific software repositories. Furthermore, Software Heritage has developed the SWHID [9], an intrinsic persistent identifier for software, facilitating the identification of software projects and their versions. As shown in Fig. 1, more than 13,000 of the 39,749 swMATH software [20] used standard repository services like GitHub or GitLab for development. The source code of these repositories has been archived and indexed thanks to the set of APIs supported by Software Heritage [17].

Table 1. Overview of the identified repository services in swMATH

	Number of swMATH software
No identified repository service	27,942
GitHub	16,373
GitLab	226
Bitbucket	91

These archive source codes are accessible from the zbMATH Open interface with user-friendly URLs [20].

By implementing these principles, swMATH aims to solidify its position as a dependable software catalog within the EOSC ecosystem.

2.2 Disseminating Software Metadata for the Advancement of Open Science

In recent years, software metadata has garnered significant attention within the scholarly community [16]. Recognized as indispensable components for comprehending the functionality and purpose of software, they are essential to guarantee the reproducibility of scientific results. Scholars create and utilize software metadata, but also aggregators like swMATH which are able to carry out information regarding the impact of the software, like software citations in articles or related software projects. While the source code forms the software's core, it alone often proves insufficient for fully grasping the intricacies of its functionality. Emphasizing the importance of robust metadata retrieval is paramount to understanding the impact of software in a specific scientific domain. To facilitate the discovery and utilization of software metadata, platforms such as swMATH require broader dissemination to reach a wider audience. By enhancing the visibility and accessibility of this highest-order scientific output, swMATH can significantly improve the discoverability and utilization of software resources within the academic community.

2.3 swMATH: Towards a Connected Resource into the EOSC Ecosystem

Since June 2022, the Department of Mathematics at FIZ Karlsruhe has actively participated in the European Open Science Cloud Program through its involvement in the FAIRCORE4EOSC project. This initiative is dedicated to advancing the adherence to FAIR principles within digital research outputs, aiming to develop nine core components. Leveraging the zbMATH Open portal and the swMATH catalog, we strive to demonstrate how the implementation of these components enhances the FAIRness of the zbMATH Open portal while also fostering synergies with the MaRDI portal, the German national initiative for mathematics:

- EOSC Research Discovery Graph (RDGraph)
- PID Graph
- EOSC Metadata Schema and Crosswalk Registry (MSCR)
- EOSC Data Type Registry (DTR)
- EOSC PID Meta Resolver
- EOSC Research Activity Identifier Service (RAiD)
- EOSC Research Software APIs and Connectors (RSAC)

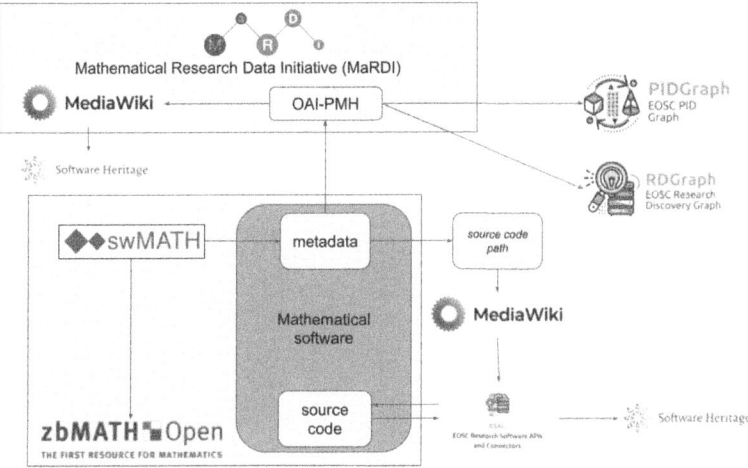

Fig. 1. The workflow to connect swMATH into the EOSC

Through its involvement in the FAIRCORE4EOSC project, zbMATH Open is actively making its metadata accessible to the scientific community, as shown in Fig. 2. The REST API deployed by FIZ Karlsruhe to showcase swMATH metadata serves as the foundation for our ongoing efforts to expose metadata in a standardized manner. This API is an invaluable tool for efficiently populating the OAI-PMH server, which is supported and instantiated by the MaRDI infrastructure. The adoption of the OAI-PMH protocol has emerged as a gold standard among Open Science institutions for managing scholarly archives. Notably, the MaRDI infrastructure is intended to integrate seamlessly into the EOSC ecosystem, contributing directly to the infrastructure of the PIDGraph. As a compliant OAI-PMH service endorsed by DataCite , the PIDGraph facilitates the enhanced interconnection of swMATH metadata with other scholarly entities. Preliminary findings [18] suggest that numerous mappings between entities within the PID-Graph and zbMATH can be further anticipated, resulting in enriched entity linking within the domain-specific knowledge graph of mathematics. Moreover, the RDGraph, a scientific knowledge graph supported by OpenAIRE, aggregates metadata from the PIDGraph, further enhancing the accessibility and discoverability of scholarly resources. In parallel to swMATH, a similar workflow is developed to process the zbMATH article's metadata.

2.4 swMATH Adopted CodeMeta as the Standard Format to Display Software Metadata

Beyond its core initiatives, zbMATH Open enhances its software infrastructure by implementing appropriate metadata standards such as CodeMeta or DataCite. CodeMeta, a recognized standard vocabulary for software metadata, facilitates the comprehensive description of our software through rich metadata. This encompasses critical details such as authorship, versioning, citations from scholarly articles referencing the software, and publications announcing new releases. By embracing these standards, we aim to enhance the accessibility and comprehensibility of our software offerings, fostering greater transparency and collaboration within the academic community.

@context	https://doi.org/10.5063/schema/codemeta-2.0
@type	SoftwareSourceCode
license	https://spdx.org/licenses/GPL-2.0-or-later
codeRepository	https://github.com/sagemath/sage
dateCreated	2005-02-24
datePublished	2005-02-24
dateModified	2023-09-16
issueTracker	https://github.com/sagemath/sage/issues
name	SageMath
version	10.2.beta3
description	Sage (SageMath) is free, open-source math software that supports research and teaching in algebra, geometry, number theory, cryptography, numerical computation, and related areas. Both the Sage development model and the technology in Sage itself are distinguished by an extremely strong emphasis on openness, community, cooperation, and collaboration: we are building the car, not reinventing the wheel. The overall goal of Sage is to create a viable, free, open-source alternative to Maple, Mathematica, Magma, and MATLAB. Computer algebra system (CAS).
applicationCategory	Mathematics
developmentStatus	active
relatedLink	https://epubs.siam.org/doi/book/10.1137/1.9781611975468
programmingLanguage	Python, Cython, C, C++, Lisp, Fortran
operatingSystem	macOS, Windows(WSL), Linux,
softwareRequirements	For detailed informations visit the download link on the SageMath Website, https://doc.sagemath.org/html/en/installation/index.html#

Fig. 2. The metadata of the SageMath software exposed into the CodeMeta vocabulary

The metadata will be accessible and downloadable through a MediaWiki instance supported by the MaRDI infrastructure, providing essential information for convenient software citation. To facilitate this process, swMATH will present citable versions of this metadata in BibLaTeX format. BibLaTeX, widely utilized for citation management in LaTeX documents, has recently seen enhancements with dedicated software entries [8]. We anticipate integrating BibLaTeX utilities into the MediaWiki interface and the zbMATH Open portal, enabling one-click correction of research software citations in LaTeX documents.

3 Conclusion

We showcased the crucial role played by software metadata as a critical hidden pillar of scientific research. swMATH, by becoming an even more connected resource in the open science landscape, is intended to be a reliable resource for mathematicians, discretely advancing mathematics research. More than ever, the interconnection of software source code, metadata, articles, authors, formulae, and other digital mathematics entities is necessary to encompass a scientific knowledge production natively compatible with the FAIR criteria. From that, we look forward to improving research software citations by giving mathematicians the credit they deserve. Future work will also consider the possibility of better associating source code and related metadata.

Glossary

ARDC	Archive, Reference, Describe, Cite
DLMF	Digital Library of Mathematical Functions
DOI	Digital Object Identifier
SIRS	Scholarly Infrastructure for Research Software
EOSC	European Open Science Cloud
FAIR	Findable, Accessible, Interoperable, Reusable
FAIRCORE4EOSC	Core Components Supporting a FAIR EOSC
MaRDI	Mathematical Research Data Initiative
OAI-PMH	Open Archives Initiative Protocol for Metadata Harvesting
OEIS	On-Line Encyclopedia of Integer Sequences
ORCID	Open Researcher and Contributor ID
PID	Persistent identifier
PIDGraph	Persistent Identifier Graph
REST API	Representational State Transfer Application Programming Interface
swMATH	swMATH is a catalog of mathematical software
zbMATH Open	zbMATH Open (formerly known as Zentralblatt MATH) is a portal for the mathematics

Acknowledgements. This research is supported by the FAIRCORE4EOSC (Core Components Supporting a FAIR EOSC) project, funded by the EU's Horizon Europe Research and Innovation Programme under Grant Agreement No. 101057264 and Deutsche Forschungsgemeinschaft (DFG) under Grant Agreement No. 460135501, NFDI, 29/1 "MaRDI - Mathematische Forschungsdateninitiative".

References

1. Azzouz-Thuderoz, M., Schubotz, M., Teschke, O.: Sustaining the swMATH project: Integration into zbMATH open interface and open data perspectives. **126**, 62–64. ISSN 2747-7894, 2747-7908. https://doi.org/10.4171/mag/118, https://euromathsoc.org/magazine/articles/118
2. Barker, M., et al.: Introducing the FAIR principles for research software. **9**(1), 622. ISSN 2052-4463. https://doi.org/10.1038/s41597-022-01710-x. https://www.nature.com/articles/s41597-022-01710-x

3. Cohl, H.S., Teschke, O., Schubotz, M.: Connecting islands: Bridging zbMATH and DLMF with scholix, a blueprint for connecting expert knowledge systems. **120**, 66–67. ISSN 2747-7894, 2747-7908. https://doi.org/10.4171/mag/35, https://euromathsoc.org/magazine/articles/35
4. European Commission, Directorate-General for Research, and Innovation. *Scholarly infrastructures for research software - Report from the EOSC Executive Board Working Group (WG) Architecture Task Force (TF) SIRS*. Publications Office (2020). https://doi.org/10.2777/28598
5. Conrad, T., Ferrer, E., Mietchen, D., Pusch, L., Stegmuller, J., Schubotz, M.: Making mathematical research data FAIR: A technology overview. http://arxiv.org/abs/2309.11829
6. De Roure, D.: UK research software survey (2014). https://eprints.soton.ac.uk/475473/
7. Della Vecchia, A., Joswig, M., Lorenz, B.: A FAIR file format for mathematical software. http://arxiv.org/abs/2309.00465
8. Di Cosmo, R.: Announcing biblatex-software: software citation made easy. **45**(4), 22–23, b. ISSN 0163-5948. https://doi.org/10.1145/3417564.3417570, https://doi.org/10.1145/3417564.3417570. Place: New York, NY, USA Publisher: Association for Computing Machinery
9. Di Cosmo, R.: Archiving and referencing source code with software heritage. In: Bigatti, A.M., Carette, J., Davenport, J.H., Joswig, M., de Wolff, T. (eds.) ICMS 2020. LNCS, vol. 12097, pp. 362–373. Springer, Cham (2020). https://doi.org/10.1007/978-3-030-52200-1_36
10. Dunne, E., Hulek, K.: Mathematics subject classification 2020. 2020-3 **115**, 5–6 (2020). ISSN 1027-488X. https://doi.org/10.4171/NEWS/115/2, https://ems.press/doi/10.4171/news/115/2
11. Ehsani, D., Petrera, M. and Teschke, O.: The integration of OEIS links in zbMATH open. **128**, 60–64. ISSN 2747-7894, 2747-7908. https://doi.org/10.4171/mag/132. https://euromathsoc.org/magazine/articles/132
12. Fuhrmann, M., Müller, F.: A REST API for zbMATH open access, pp. 63–65. https://doi.org/10.4171/mag/174
13. Greuel, G.-M., Sperber, W.: swMATH – an information service for mathematical software. In: Hong, H., Yap, C. (eds.) ICMS 2014. LNCS, vol. 8592, pp. 691–701. Springer, Heidelberg (2014). https://doi.org/10.1007/978-3-662-44199-2_103
14. Hasselbring, W., Carr, L., Hettrick, S., Packer, H., Tiropanis, T.: From FAIR research data toward FAIR and open research software. **62**(1), 39–47. ISSN 2196-7032, 1611-2776. https://doi.org/10.1515/itit-2019-0040. https://www.degruyter.com/document/doi/10.1515/itit-2019-0040/html
15. Jauregui, C.M., Hyun, J., Neofytou, A., Gray, J.S., Kim, H.A.: Avoiding reinventing the wheel: reusable open-source topology optimization software. **66**(6), 145. ISSN 1615-147X, 1615-1488. https://doi.org/10.1007/s00158-023-03589-7, https://link.springer.com/10.1007/s00158-023-03589-7
16. Jiménez, R.C., et al.: Four simple recommendations to encourage best practices in research software. **6** 876. ISSN 2046-1402. https://doi.org/10.12688/f1000research.11407.1, https://f1000research.com/articles/6-876/v1
17. Schubotz, M., Azzouz-Thuderoz, M.: Mapping of swMATH ids and software heritage ids. https://doi.org/10.5281/zenodo.10854038
18. Schubotz, M., Teschke, O.: Mapping between zbMATH open identifiers, DOIs and ORCIDs. https://zenodo.org/record/7378860

19. Schubotz, M., Teschke, O.: zbMATH open: Towards standardized machine interfaces to expose bibliographic metadata. **119**, 50–53, b. ISSN 2747-7894, 2747-7908. https://doi.org/10.4171/mag/12, https://ems.press/doi/10.4171/mag/12
20. The NFDI Consortium of Mathematics. swmath and faircore4eosc
21. Wilkinson, M.D., et al.: The FAIR guiding principles for scientific data management and stewardship. **3**(1), 160018. ISSN 2052-4463. https://doi.org/10.1038/sdata.2016.18, https://www.nature.com/articles/sdata201618

Open Access This chapter is licensed under the terms of the Creative Commons Attribution 4.0 International License (http://creativecommons.org/licenses/by/4.0/), which permits use, sharing, adaptation, distribution and reproduction in any medium or format, as long as you give appropriate credit to the original author(s) and the source, provide a link to the Creative Commons license and indicate if changes were made.

The images or other third party material in this chapter are included in the chapter's Creative Commons license, unless indicated otherwise in a credit line to the material. If material is not included in the chapter's Creative Commons license and your intended use is not permitted by statutory regulation or exceeds the permitted use, you will need to obtain permission directly from the copyright holder.

A FAIR File Format for Mathematical Software

Antony Della Vecchia[1]([✉])[iD], Michael Joswig[1,2][iD], and Benjamin Lorenz[1][iD]

[1] Discrete Mathematics/Geometry, Technische Universität Berlin, Berlin, Germany
vecchia@math.tu-berlin.de
[2] Max Planck Institute for Mathematics in the Sciences, Leipzig, Germany

Abstract. We describe a JSON based file format for storing and sharing results in computer algebra without losing accuracy. Guided by practical usability, some key features are the flexibility to handle data structures unknown at the time of design, a clear method for transitioning to the latest format and a way of separating data of distinct or even contradicting semantics. This is implemented in the computer algebra system OSCAR [5,20], but we also indicate how it can be used in a different context.

We discuss general considerations, with a focus on comprehensibility and long-term storage. General concepts for data serialization, like Protocol Buffers[1] or Julia's [2] Serialization.jl, do not suffice for the rich semantics of computer algebra. Specialized software systems do allow for storing and writing files with mathematical data of a limited number of types, for example the mps file format used in optimization to store linear and integer programs. Hence this allows for sharing data, e.g., in databases such as MIPLIB [16]. However, formats like mps do not lend themselves to more general data. The current standard for computer algebra systems is to use notebooks to store entire computations, Jupyter[2] being the current standard. While these notebooks are very handy they do not provide a proper serialization of the intermediate results which can make certain recomputations undesirable or impossible.

In the late 1990s the OpenMath project [8] developed a general framework for mathematical data. Their effort was confronted with fundamental criticism, e.g., by Fateman [10,11]. In light of the point held against OpenMath in [10], we pick a particular system, namely the new computer algebra system OSCAR written in Julia, and rely on its semantics with no attempt to formalize the semantics in general. We store data as annotated trees which is a common idea amongst comprehensive serialization formats, e.g., Protocol Buffers and OpenMath. Our format extends in a way the current JSON file format of polymake [1], which is a

---

The project was supported by MaRDI, funded by the Deutsche Forschungsgemeinschaft (DFG), project number 460135501, NFDI 29/1 MaRDI - Mathematische Forschungsdateninitiative.
[1] https://protobuf.dev/.
[2] https://jupyter.org.

translation of the original XML version [14]. The syntax is fixed by an extensible JSON schema, which is explained in Sect. 3. In Sect. 4 we discuss how users of other computer algebra systems can make potential use of our format.

Our file format, developed as part of the Mathematics Research Data Initiative (MaRDI) [19], aims to eventually extend beyond OSCAR, for this reason we use the file extension `mrdi`.

1 The File Format by Example

Our running example is a bivariate polynomial over a finite field, namely

$$2y^3z^4 + (a+3)z^2 + 5ay + 1 \in \mathrm{GF}(49)[y,z], \qquad (1)$$

where GF(49), i.e., the finite field with 49 elements, is constructed as a degree two algebraic extension over the prime field $\mathrm{GF}(7) \cong \mathbb{Z}/7\mathbb{Z}$. More precisely, as a $(\mathbb{Z}/7\mathbb{Z})$-algebra, GF(49) is isomorphic to the quotient $(\mathbb{Z}/7\mathbb{Z})[x]/\langle x^2+1 \rangle$, and x^2+1 is a minimal polynomial. In the latter quotient algebra we pick a generator and call it a. As the degree of the field extension equals two, the coefficients have a maximal a-degree of one.

```
{ "_ns": { "Oscar": ["https://github.com/oscar-system/Oscar.jl",
                     "1.0.0" ] },
  "_type": { "name": "MPolyRingElem",
             "params": "a7029443-b1d3-4708-a66f-f68eb6616cf" },
  "data": [[["3", "4"], [["0", "2"]]],
           [["0", "2"], [["0", "3"], ["1", "1"]]],
           [["1", "0"], [["1", "5"]]],
           [["0", "0"], [["0", "1"]]]],
  "_refs": {
    "a7029443-b1d3-4708-a66f-f68eb6616cf": { ... },
    "f2b7cb6b-535a-4a52-a0cc-75f8e93a6719": { ... },
    "23f25330-83f7-43a0-ac74-da6f2caa7eb8": {
      "_type": "FqField",
      "data": { "def_pol": {
        "_type": { "name": "PolyRingElem",
                   "params": "f2b7cb6b-535a-4a52-a0cc-75f8e93a6719" },
        "data": [["0", "1"], ["2", "1"]]
} } } } }
```

Fig. 1. JSON description of the bivariate polynomial $2y^3z^4 + (a+3)z^2 + 5ay + 1$ in the polynomial ring $\mathrm{GF}(49)[y,z]$ from (1). We hide all but one reference describing a quotient field with defining polynomial x^2+1.

Our encoding keeps track of the entire history of the construction. As a consequence, when we store that one polynomial, we also store the univariate polynomial ring $(\mathbb{Z}/7\mathbb{Z})[x]$, the minimal polynomial x^2+1 and the quotient algebra $(\mathbb{Z}/7\mathbb{Z})[x]/\langle x^2+1\rangle$. In this way, we can associate with the algebraic expression (1) an annotated tree which reflects its construction. This has a direct translation into JSON code, shown in Fig. 1.

We distinguish between *basic types*, which do have a fixed normal form. These include standard Julia types but also includes algebraic types such as the integers (ZZRingElem) or the rationals (QQFieldElem). The more interesting types are *parametric*, and the type MPolyRingElem of the polynomial (1) serves as our running example. That type identifies (1) as an element in some multivariate polynomial ring. Its ring of coefficients is referenced as a parameter to the type MPolyRingElem, where it listed under the params property. The base ring of a multivariate polynomial ring can be any ring; in our example this is the quotient of a univariate polynomial ring by some ideal (spanned by one irreducible polynomial). This gives rise to a recursive description because the parameter can have any type. Consequently, parameters may have their own parameters. All the parameter types, their parameters, and so on are stored in the global dictionary _refs, and the params property refers to them via universally unique identifiers (UUIDs). We generate version four UUIDs specified by RFC 4122 on a first save and these persist throughout an active OSCAR session. For instance, this allows us to distinguish between isomorphic copies of a ring, which may play different roles in a specific computation. This is useful, e.g., when we start with two polynomials $p \in \mathbb{Q}[a,b]$ and $q \in \mathbb{Q}[x,y]$, and much later we want to take their product in the ring $\mathbb{Q}[a,b,x,y]$. It occurs in daily computer algebra routine that the "universe" $\mathbb{Q}[a,b,x,y]$ is not known in advance but rather the result of a sequence of several computational steps.

The example of finite fields illustrates that normal forms may not always provide an adequate description due to the incremental nature of certain computer algebra constructions, such as Hensel lifts in number theory; see [3] and [13, §15.4]. UUIDs facilitate tracking these evolving computations.

So, the type and its recursive parameters set the context which specifies the syntax of the serialized data. The actual data is stored in the property with the same name. In our running example the root of the data subtree has four children, one for each term of the polynomial (1). In the JSON code each data subtree is written as a nested list of nested lists, marked by square brackets.

The discussion so far deals with the syntactic aspects of serialization. It is a key design choice that the semantics is fully implicit. In our example the semantics is determined by OSCAR, version 1.0.0, which is specified in the namespace property _ns. Namespaces form the point of entry for the possibility to store data which are foreign to OSCAR. This is the topic of Sect. 4 below.

2 More Examples

To display the range of possibilities arising from our concept, we pick two examples of very different nature.

Non-general Type Surfaces in $\mathbb{P}^4$. It is known that each smooth projective algebraic surface can be embedded in projective 5-space, which we denote as $\mathbb{P}^5$. By a result of Ellingsrud and Peskine [9] there are only finitely many families of surfaces of *non-general type*, i.e., they admit an embedding already in $\mathbb{P}^4$. Decker and Schreyer obtained the number 52 as an explicit degree bound for such surfaces [6]. In loc. cit. the authors also construct 49 non-general type surfaces in $\mathbb{P}^4$ of degree up to 15. That list is available in OSCAR via files stored in our file format.

Here is one such surface (of degree three, with sectional genus zero), which is described as the vanishing locus of an ideal with three homogeneous generators in the polynomial ring $\mathrm{GF}(31991)[x, y, z, u, v]$. The code below shows an interactive Julia session using OSCAR 1.0.0.

```
julia> S = cubic_scroll()
Projective scheme
  over finite field of characteristic 31991
defined by ideal with 3 generators

julia> defining_ideal(S)
Ideal generated by
  31990*x*y + 19122*x*z + 4788*x*u + ... + 20742*u*v + 25408*v^2
  7471*x*y + 23772*x*z + 27471*x*u + ... + 30545*u*v + 9903*v^2
  x^2 + 3601*x*y + 7253*x*z + 7206*x*u + ... + 6535*u*v + 26586*v^2
```

Converting the descriptions of these 49 surfaces from the literature to objects suitable for computation takes some time and is prone to error. So it is desirable to store such data explicitly, without the need of any computation or conversion.

Toric Varieties. The following code constructs two divisors on a toric variety; see [4]. A *toric variety* is an algebraic variety which is implicitly described by the normal fan of a convex lattice polytope, and a *toric divisor* is a formal integer linear combination of facets of that polytope, i.e., rays of its normal fan. The polytope in our example is a triangle and so divisors are given by integer vectors of length three, one entry for each facet.

```
@testset "ToricDivisor" begin
  pp = projective_space(NormalToricVariety, 2)
  td0 = toric_divisor(pp, [1,1,2])
  td1 = toric_divisor(pp, [1,1,3])
  vtd = [td0, td1]
  test_save_load_roundtrip(path, vtd) do loaded
    @test coefficients(td0) == coefficients(loaded[1])
    @test coefficients(td1) == coefficients(loaded[2])
    @test toric_variety(loaded[1]) == toric_variety(loaded[2])
  end
end
```

The test saves and loads `vtd`, which is a vector formed of those two divisors, it then checks if loading yields the same objects. Additionally, the code checks if the underlying toric variety for the two divisors is the same, using UUIDs.

```
{ "$id": "https://oscar-system.org/schemas/mrdi.json",
  "$schema": "https://json-schema.org/draft/2020-12/schema",
  "type": "object",
  "required": ["_type"],
  "properties": { "_ns": { "type": "object" },
                  "_type": {
                    "oneOf": [
                      { "type": "string"},
                      { "type": "object", "properties": {
                        "name": {"type": "string"},
                        "params": {"$ref": "#/$defs/data"}
                      } } ] },
  "data": {"$ref": "#/$defs/data"},
  "_refs": {"type": "object", "patternProperties": {
    "^[0-9a-fA-F]{8}-([0-9a-fA-F]{4}-){3}[0-9a-fA-F]{12}$": {
      "$ref": "#"
    } } } },
  "$defs": { "data": {
    "oneOf": [
      { "type": "string"},
      { "type": "array", "items": { "$ref": "#/$defs/data"} },
      { "type": "object", "not": { "required": [ "_ns" ] },
        "patternProperties": {
          "^[a-zA-Z0-9_]*": {"$ref": "#/$defs/data"} } },
      { "$ref": "https://polymake.org/schemas/data.json"}
    ] } } }
```

Fig. 2. File format specification following the JSON Schema specification [22].

3 Format Specification

JSON Schema [22] is a declarative language for describing JSON file specifications, similar to RELAX NG for XML. Our file specification is shown in Fig. 2. JSON has four types, namely string, array, number, and object (dictionary or hash map). The first occurrence of the type property describes the file itself, where it expects the file to be of type object. The properties and patternProperties keywords are used to describe the specifications for the keys and values of the object. Only the values with keys being matched in the object specification (either exactly or by regular expression) will be checked. The required keyword enforces that the objects have all properties listed in the array, here we enforce that the _type property is present. Some validators can handle common string formats, so we enforce that the keys of _refs should have the format of a UUID. The oneOf keyword is used to specify that one of the specifications in the list is expected. The $ref keyword uses a path or URL to refer to specifications defined elsewhere, the # symbol denotes the root. Other definitions can be described using the $defs section. For example, our definition for data accepts several options including recursive `object` and `array` structures as well as data formatted in accordance with `polymake`'s schema.

We use UUIDs instead of simpler indices so that references are valid throughout an entire session. Consider the following scenario, Alice, computes with several, e.g., multivariate polynomials with coefficients in some fixed finite field, like in Fig. 1. Then she stores a vector of three such polynomials in one file. Further computations then yield a 3×3-matrix, whose coefficients lie in the same polynomial ring. Alice stores that matrix in another file. She sends both files to Bob, who wants to continue that computation, e.g., by multiplying the matrix with the vector. In general, the finite field is constructed as a sequence of field extensions over the prime field. While there is only one finite field of any given order, there are many field towers leading to the same. Since the encoding depends on the details of the construction, the entire context must be present. UUIDs allow for recognizing the same base ring across several files. This is particularly useful for databases and large scale parallel computations.

4 Beyond OSCAR

The `mrdi` file format is meant to have a wide scope, and OSCAR mainly serves as a proof of concept. Here are some aspects not covered yet.

Namespaces. OSCAR is based on and extends Nemo/Hecke [12], GAP [18], polymake, and Singular [7]. So it is concerned with exact computations in number theory, group and representation theory, polyhedral geometry and optimization, as well as commutative algebra and algebraic geometry. Since our file format defers its semantics to a specific version of a specific software system, currently we are considering "algebraic" data only. In [10] Fateman pointed out that "Sin[x]" in Mathematica[3] and "sin(x)" in Maple[4] mean very different things. This makes it difficult to define and make use of any formal semantics covering such data beyond one software. In our file format this problem could be resolved by defining separate namespaces for Mathematica and Maple.

It may even make sense to have data from distinct namespaces in the same file. Any software system is free to interpret what it can understand and ignores the rest. Via the underlying tree structure "the rest" may refer to arbitrary subtrees. In this way, the `mrdi` file format is a flexible container format, which is similar in spirit to the Portable Document Format (PDF).[5] For instance, a PDF file may contain audio data, but not every PDF viewer is capable of playing back sound.

To show how this can work in practice, in Fig. 3 we display a short code fragment which reads a multivariate polynomial with rational coefficients from a `mrdi` file into SageMath [21]. That code is fully functional and complete, without shortcuts or hidden parts.

---

[3] https://www.wolfram.com/mathematica.
[4] https://maplesoft.com/products/maple.
[5] https://pdfa.org/resource/iso-32000-pdf/.

```python
import json
from sage.all import PolynomialRing, QQ, prod

def load_oscar_polynomial(path):
  with open(path) as json_file:
    file_data = json.load(json_file)
    if ("Oscar" in file_data["_ns"] and
        file_data["_ns"]["Oscar"][1].startswith("1.0.")):
      t, d, refs = (file_data[k] for k in ["_type", "data", "_refs"])
      parent_ring_data = refs[t["params"]]["data"]
      base_ring = parent_ring_data["base_ring"]
      if base_ring["_type"] != "QQField":
        raise NotImplementedError("only rational coefficients supported")
      symbols = ",".join(parent_ring_data["symbols"])
      R, gens = QQ[symbols].objgens()
      p = R(0)
      for e, c in d:
        exps = [int(exponent) for exponent in e]
        coeff = QQ(c.replace("//", "/"))
        p += coeff * prod([g**i for g, i in zip(gens, exps)])
      return p
    else:
      raise RuntimeError("can only load OSCAR version 1.0 polynomials")
```

Fig. 3. Python code for SageMath 10.2 to load a rational polynomial from a mrdi file written with OSCAR 1.0

Going through the Python code also allows us to explain how we avoid Fateman's criticism of OpenMath [10,11]. Code like the one in Fig. 3 explicitly translates from OSCAR to SageMath. This requires the programmer to know about both systems. The necessity of direct communication between pairs of computer algebra systems was seen as a drawback in the 1990s, and it was a major motivation behind OpenMath to overcome this obstacle. The price to pay for a centralized communication concept like OpenMath, however, is the need to formally specify the semantics. Each computer algebra system has a rich implicit semantics which is often very difficult to spell out explicitly. Consequently, developing a formalized semantics to govern several computer algebra systems simultaneously seems to be at least as involved as writing an entirely new computer algebra system from scratch. It is therefore our conclusion to stick to the implicit semantics of one system, e.g., OSCAR, and to translate explicitly whenever necessary. In this way Fateman's criticism does not apply.

For rational polynomials the effort to translate from OSCAR to another computer algebra system is quite moderate, as illustrated in Fig. 3. Depending on the data types, other translations might require more effort. Namespaces create the flexibility for every user to pick their own point of departure, i.e., picking a computer algebra system other than OSCAR.

Databases. Any serialization lends itself to storing similar files in some systematic folder hierarchy, mimicking a simple database. Our file format is no exception, and the algebraic surfaces from Sect. 1 form an example in OSCAR. The NoSQL database MongoDB uses a record structure which essentially agrees with JSON objects. In this way, our serialized data can directly be used for storage and retrieval in highly efficient large scale databases. The same concept was already exploited in polymake's database project polyDB [17]. Note that MongoDB requires UTF-8 encoded strings in the JSON objects, whence we restrict our file format to UTF-8, too.

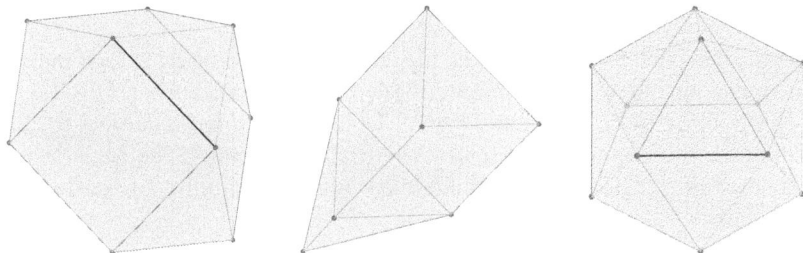

Fig. 4. (a) Triangular cupola, (b) elongated triangular pyramid, (c) gyroelongated pentagonal pyramid

Another interesting collection of mrdi files are the Johnson solids from [15]. The latter are 3-dimensional convex polytopes whose facets are regular polygons. The Johnson solids generalize the Archimedean solids, and there are precisely 92 of them (up to rigid motions and scaling) which are not Archimedean. The dataset [15] comprises exact algebraic numbers as well as approximate floating point numbers as coordinates for those 92 polytopes. Furthermore, the data comes with a Julia script allowing users to read certain properties of a Johnson solid using only standard Julia and JSON parsing, independent from OSCAR. Since the mrdi file format is based on JSON, there are even simpler methods to access basic information about that dataset. For instance, via jq,[6] which is a command-line JSON processor. The following three commands print out the respective number of vertices for the triangular cupola, the elongated triangular pyramid and the gyroelongated pentagonal pyramid. These three Johnson solids are displayed in Fig. 4.

```
> jq '.data.float.VERTICES | length' j3
9
> jq '.data.float.VERTICES | length' j7
7
> jq '.data.float.VERTICES | length' j11
11
```

---

[6] https://jqlang.github.io/jq/.

Version Upgrades. The OSCAR project itself should not be considered monolithic or complete. On the contrary, OSCAR has been designed to keep evolving, with new data types, encodings and use cases. Some of these changes suggest modifications to how the data is serialized, and the file format needs to be able to go along with such changes. To this end our data comes with version numbers. Upgrade scripts provide arbitrary transformations from old data into the current standard, in contrast to Protocol Buffers whose format is forward compatible. Such an upgrade scheme is in place within the polymake project for more than a decade. In 2020 version 4.0 of polymake replaced the previous XML based serialization by JSON, through the same mechanism. This shows that such upgrades are feasible.

Metadata. The current version of the file format can optionally attach metadata that includes an entry for the name of the data and an author ORCID.[7] We think this is sufficient for a first version of the file format and is subject to change depending on the requirements set by the MaRDI portal.[8] The MaRDI portal aims to provide services for the findability and accessibility of mathematical research data.

5 Concluding Remarks

The main point of our design is the lack of dependence on any particular programming language. This by itself sets it apart from Julia's Serialization.jl or Python's pickle module. Converting mathematical data into JSON objects, which are mere strings, always means an overhead. For long term storage space efficiency is often less relevant, while other features are more important. However, once data becomes so large that it hits physical bounds, e.g., as the capacity of a hard drive, it becomes mandatory to think about data compression. Via namespaces our approach allows to compress subtrees of the JSON object and use Base64 binary to text encoding to obtain a new valid JSON object. A more thorough discussion is beyond the scope of the present article.

Acknowledgement. We are grateful to the entire OSCAR developer team for implementing and discussing code; special thanks to Claus Fieker, Tommy Hofmann, and Max Horn. Further we are indebted to Lars Kastner for discussing FAIR principles, to Wolfram Decker for explaining algebraic surfaces in $\mathbb{P}^4$, and to John Abbott, Ewgenij Gawrilow, and Aaruni Kaushik for helpful feedback.

---

[7] https://orcid.org.
[8] https://portal.mardi4nfdi.de/wiki/Portal.

References

1. Assarf, B., et al.: Computing convex hulls and counting integer points with polymake. Math. Program. Comput. **9**(1), 1–38 (2017)
2. Bezanson, J., Edelman, A., Karpinski, S., Shah, V.B.: Julia: a fresh approach to numerical computing. SIAM Rev. **59**(1), 65–98 (2017)
3. Buchberger, B., Loos, R.: Algebraic simplification. In: Buchberger, B., Collins, G.E., Loos, R. (eds.) Computer Algebra, vol. 4, pp. 11–43. Springer, Vienna (1983). https://doi.org/10.1007/978-3-7091-3406-1_2
4. Cox, D.A., Little, J.B., Schenck, H.K.: Toric Varieties. Graduate Studies in Mathematics, vol. 124. American Mathematical Society, Providence (2011)
5. Decker, W., Eder, C., Fieker, C., Horn, M., Joswig, M. (eds.): The Computer Algebra System OSCAR: Algorithms and Examples. Algorithms and Computation in Mathematics, Springer, Cham (2024)
6. Decker, W., Schreyer, F.O.: Non-general type surfaces in $\mathbf{P}^4$: some remarks on bounds and constructions. J. Symbolic Comput. **29**(4-5), 545–582 (2000). Symbolic Computation in Algebra, Analysis, and Geometry, Berkeley, CA (1998)
7. Decker, W., et al.: SINGULAR—a computer algebra system for polynomial computations, version 4.3.2-p16 (2024). http://www.singular.uni-kl.de
8. Dewar, M.: OpenMath: an overview. SIGSAM Bull. **34**(2), 2–5 (2000)
9. Ellingsrud, G., Peskine, C.: Sur les surfaces lisses de P_4. Invent. Math. **95**(1), 1–11 (1989)
10. Fateman, R.: A critique of OpenMath and thoughts on encoding mathematics, January 2001. https://people.eecs.berkeley.edu/~fateman/papers/openmathcrit.pdf
11. Fateman, R.: More versatile scientific documents (2003). https://people.eecs.berkeley.edu/~fateman/MVSD.html
12. Fieker, C., et al.: Nemo/Hecke: computer algebra and number theory packages for the Julia programming language. In: Proceedings of the 2017 ACM on International Symposium on Symbolic and Algebraic Computation, ISSAC 2017, New York, NY, USA, pp. 157–164. ACM (2017)
13. von zur Gathen, J., Gerhard, J.: Modern Computer Algebra. Cambridge University Press, New York (1999)
14. Gawrilow, E., Hampe, S., Joswig, M.: The polymake XML file format. In: Greuel, G.-M., Koch, T., Paule, P., Sommese, A. (eds.) ICMS 2016. LNCS, vol. 9725, pp. 403–410. Springer, Cham (2016). https://doi.org/10.1007/978-3-319-42432-3_50
15. Geiselmann, Z., et al.: Algebraically precise Johnson solids (2024). https://doi.org/10.5281/zenodo.10729583
16. Gleixner, A., et al.: MIPLIB 2017: Data-Driven Compilation of the 6th Mixed-Integer Programming Library. Mathematical Programming Computation (2021)
17. Paffenholz, A.: polyDB: a database for polytopes and related objects. In: Böckle, G., Decker, W., Malle, G. (eds.) Algorithmic and Experimental Methods in Algebra, Geometry, and Number Theory, pp. 533–547. Springer, Cham (2017). https://doi.org/10.1007/978-3-319-70566-8_23
18. The GAP Group: GAP – Groups, Algorithms, and Programming, Version 4.12.2 (2022). https://www.gap-system.org
19. The MaRDI Consortium: MaRDI: Mathematical Research Data Initiative Proposal, May 2022. https://doi.org/10.5281/zenodo.6552436
20. The OSCAR Team: OSCAR – Open Source Computer Algebra Research system, Version 1.0.0 (2024). https://www.oscar-system.org

21. The `SageMath` Developers: `SageMath`, version 10.2, December 2023. https://doi.org/10.5281/zenodo.10252400
22. Wright, A., Andrews, H., Hutton, B., Dennis, G.: JSON schema: a media type for describing JSON documents (2020). https://json-schema.org/draft/2020-12/json-schema-core.html

Predefined Software Environment Runtimes as a Measure for Reproducibility

Aaruni Kaushik(✉)

University of Kaiserslautern-Landau, Kaiserslautern, Germany
aaruni.kaushik@math.rptu.de

Abstract. Mathematical Research Data Initiative (MaRDI) is a consortium of the National Research Data Infrastructure (NFDI) aiming to bring FAIR data practices to mathematical research. In alignment with MaRDI, we have developed a way to preserve software packages into an easy to deploy and use sandbox environment we call a "runtime", via a program we developed called **MaPS** : **Ma**RDI **P**ackaging **S**ystem [10]. The program relies on Linux user namespaces to isolate a library environment from the host system, making the sandboxed software reproducible on other systems, with minimal effort. Moreover an overlay filesystem makes local edits persistent. This project will aid reproducibility efforts of research papers: both mathematical and from other disciplines. As a proof of concept, we provide runtimes for the OSCAR Computer Algebra System [13], polymake software for research in polyhedral geometry [9], and VIBRANT Virus Identification By iteRative ANnoTation [11]. The software is in a prerelease state: the interface for creating, deploying, and executing runtimes is final, and an interface for easily publishing runtimes is under active development. We thus propose publishing predefined, distributable software environment runtimes along with research papers in an effort to make research with software based results reproducible.

Keywords: FAIR · MaRDI · Sandbox · Runtime · Reproducibility

1 Introduction

As the capabilities of computer devices have grown, so have their use as tools in scientific research[1]. It is now possible to do wild and wonderful things by relying on computers to generate useful results in the support of both research goals as well as arguments to reach them. However, this also means that a computer program plays a more central role in the defence of a thesis. As good science requires a peer review of the results, the computer programs so central to new computational results must also be readily available to the reviewers. However, this is not straightforward with software.

---

[1] FAIR Principles https://www.go-fair.org/fair-principles/.

Supported by MaRDI, funded by the Deutsche Forschungsgemeinschaft (DFG), project number 460135501, NFDI 29/1 "MaRDI - Mathematische Forschungsdateninitiative".

A naive approach to making computer software available is to simply make the source code available to reviewers. But this poses problems with regards to ease of use, and also with regards to reproducibility. It may not be the easiest thing to get running in the first place. This may just be due to the convoluted way the prerequisites are meant to be installed. Or it may require obscure versions of libraries which may conflict with existing libraries on a reviewer's host system. It might just require a fixed version of a language which was the state of the art when the paper was written, but has been deprecated and become unsupported on modern computers. Thus, for a FAIR approach to distributing research software, a system tailor-made for the job must be created. We expect MaPS to be able to effectively preserve software in runtimes. Our work is heavily inspired by the work done in this direction by Flatpak [2] for generic Linux software, and Valve [7] for PC video games.

1.1 Organization

In Sect. 2, we describe a general overview of the system we have developed for packaging research software in an easy to distribute and easy to use way. In Sect. 3, we describe in deep technical detail how the system is implemented. Sections 4 and 5 provide some examples of using the system to run packaged software.

A casual reader may skip Sect. 3, which is purely technical, and use Sects. 4 and 5 as a rudimentary users manual. Section 3.4.4 may be of interest to users intending to create a package runtime. For documentation of the project, the wiki[2] should be consulted.

2 MaRDI Packaging System: An Overview

MaPS has only a very basic minimal set of requirements to work. It requires a reasonably recent Linux kernel (5.11+ , released February 2021), and Python (3.8+, released October 2019) pre-installed. Both of these are available in Ubuntu 22.04 LTS. The rest is grabbed automatically during installation. Root access is only required at installation time for a system wide installation, and resolving dependencies. MaPS can be installed with no root access when not installing system wide. In this case, MaPS and any missing pre requisites must be installed manually from source.

Instead of re-inventing the wheel, we make use of other free and open source[3] projects like Bubblewrap [3] and libostree [4] as mature, stable building blocks. This has a two fold advantage: firstly, that we are saved from the workload of reimplementing parts of the upstream work that we require, and secondly, that we inherit any improvements from these upstream projects for free. While we trust that these projects will stick around, we have architectured the solution

---

[2] MaPS Wiki https://github.com/MaRDI4NFDI/maps/wiki.
[3] The open source definition - Open Source Initiative https://opensource.org/osd.

in a way that in the unlikely scenario that these upstream projects are shut down, it is possible to reimplement the functionality required from scratch. The building blocks are only a time saving convenience.

In broad strokes, MaPS manages fetching and storing full runtimes from a network source onto the local disk. It deploys these runtimes in a usable format from the storage provided by libostree, and sets up the foundation of an overlay file system. When launching a runtime with MaPS, it dynamically sets up an overlay layer on top of the published runtime, so that any local changes persist on disk, without actually modifying the runtime. This allows end users to persist edits, while also leaving open the option to easily reset back to the published state for perfect reproducibility.

3 Technical Details

3.1 Bubblewrap

Bubblewrap is a low level tool to create containers using user namespaces on Linux. It is also used by other big projects like Flatpak and rpm-ostree. Using Bubblewrap, it is possible to create secure isolated containers relatively easily, and with good granularity of the exact setup. Bubblewrap prevents privilege escalation (see [3, section System Security]), and also adds a reaper process into the container to avoid the docker PID 1 problem [12].

3.2 Libostree

The libostree project (previously, just ostree) is a library for content addressed storage of arbitrary filesystem trees. One can think of it as git, but for large directories instead of source code. An ostree repository is initialized on the local disk, and a remote source is added, as a sort of "app store". Data is then fetched from this remote using standard web protocols.

Along with managing the repository, and implementing the network activity, ostree also makes efficient use of disk space. All files inside the repository are addressed by a hash of their contents. Thus, files with identical data are stored only once on disk. Checking out trees from the repository creates hardlinks on the local filesystem. Thus, file level deduplication is achieved. However, note that any changes, even just a single bitflip, will defeat this deduplication.

3.3 Fuse-Overlayfs

The fuse-overlayfs project [6] is the userspace implementation of the Overlayfs [1] filesystem. Overlayfs is a special Linux filesystem, which can present a directory overlaid on top of another directory as a single merged filesystem, with any modifications being written only to the "upper" layer. We need the userspace variant of this filesystem to maintain the rootless nature of our application. It requires at minimum four directories: a "lower directory", an "upper direcotry", a "working directory", and finally, a "merged directory". With terminology borrowed from the physical analogy, an upper directory is overlaid on top of one or more lower directories. The resulting tree is mounted onto the merged directory. A working directory is required for temporary storage while the overlay is active.

Mounting a filesystem, even inside a container, presents a potential security risk, as the contents of its superblock are executed in kernel mode. Thus, as of version 0.8.0, Bubblewrap does not support mounting any filesystems in the container, not even Overlayfs. In principle, an overlay operation is risk free as no superblocks are involved, and can safely be done as root inside a user namespace in Linux 5.11+. However, this is not currently implemented in Bubblewrap. This point serves as a first example of inheriting improvements for free: as native Overlayfs without root access is a soon to be implemented feature on the roadmap for Bubblewrap, at which point, we can stop depending on fuse-overlayfs, while still keeping all our functionality.

3.4 MaPS

At its core, MaPS is very simple and elegant. Its main task is to carefully orchestrate its building blocks to acheive desired results. It tells ostree to set up a repository for all the data in $XDG_DATA_HOME/org.mardi.maps/ostree/repo, then checks out the files in $HOME/.var/org.mardi.maps/<runtime_name>. Under this location, four subdirectories are created: rofs, rwfs, tmpfs, and live. At the time of execution, fuse-overlayfs is used to layer rwfs over rofs, and the result is exposed as a merged writeable tree at live. Finally, Bubblewrap is used to create a user namespace using live as the target filesystem tree. $HOME/Public is passed through into the namespace as /home/runtime/Public, and can be used to provide input, as well as extract output from the runtime. When the runtime is exited, i.e., when the last process in the namespace has exited, Bubblewrap will shut down the created namespace, and MaPS unmounts the filesytstem created by fuse-overlayfs.

3.4.1 Manifest File

Optionally, a runtime is allowed to contain a manifest file located at /manifest.toml in the namespace. (Note that this is in runtime, which corresponds to <runtime>/live/manifest.toml on the host.) This is a TOML[4] file that can contain arbitrary metadata relating to the runtime, and, more importantly, a custom command for MaPS invoke. This command is specified in the [Core] section of the TOML file, as a string assigned to the variable command. In the absence of this variable in the manifest file (or in the absence of a manifest file), a shell is launched by default. This behaviour can be overridden by using the command line argument --command. The Manifest file from the runtime org.oscar_system.oscar/x86_64/1.0.0 is included as an example:

```
[Core]
command = "julia -J /tmp/jl_UuXQwY/Oscar.so --banner=no"
```

---
[4] TOML: A config file format for humans. https://toml.io/en/.

```
[Meta]
Project = "OSCAR -- Open Source Computer Algebra Research
    system, Version 1.0.0, The OSCAR Team, 2024. (https://www
    .oscar-system.org)"
URL = "https://www.oscar-system.org/"
```

3.4.2 Extended Attributes

libostree requires a file system with extended attributes available and enabled to be able to manage a repository, and also to check out files from the repository. This means that both required directories ($HOME and $XDG_DATA_HOME) must be on a filesystem with extended attributes. In particular, this means that NTFS disks or NFS shares will not work. If $HOME is on such an unsupported filesystem (as is common in large organizations), but access to another, supported, filesystem is available, one may override these environment variables before invoking MaPS to point to locations on the supported filesystem.

3.4.3 Fakeroot

Even though it appears that we have root access inside the user namespace (the command whoami returns root), we don't really have access on the host. This can lead to some conflict in operations like trying to chown a file. As root in the namespace, the user expects to be able to do whatever they want in the namespace without an error. But some commands like chown would need to actually change the owner of a file on the host filesystem, which could be utilised to craft an attack to break out of the sandbox. While MaPS does not try to be a security layer, such an operation is not allowed on the kernel level, and results in an error. Fakeroot is a Linux utility which only simulates the result of such a command, and returns success to the calling program. However, the kernel does not allow even this inside a namespace, as a security measure. To get around this problem, fakeroot can be used with the environment variable FAKEROOTDONTTRYCHOWN set to 1. This makes fakeroot only pretend to execute such commands, and return a success code to the calling program. This workaround may be required in a MaPS runtime to successfully install new packages, or use tar.

3.4.4 Creating Runtimes

To create a runtime, start from a "minimal viable chroot" environment. Debian rootfs is an excellent choice for reasons of compatibility and size. Initialise the tree in a convenient location using maps package --initialise /path/to/tree. Once initialised, open a live sandbox into the tree by running maps package --sandbox /path/to/repository. Even though this is really a sandbox inside the tree, it can be treated it as if one has SSH[5]'d into a freshly installed debian box, and set up programs as normal. When done, exit the sandbox, then package the runtime using maps package --commit name_of_repository/arch/version path/to/tree. This will create a MaPS runtime out of the tree.

[5] Secure Shell https://en.wikipedia.org/wiki/Secure_Shell.

3.4.5 Publishing Runtimes

A workflow for publishing runtimes created as in the previous step is under active development. This will allow scientists to request their runtimes to be published on the "Official" repository right from the MaPS command line.

3.5 Comparison with Competing Methods

There are several methods for sharing a program for running on another machine ranging from sharing just the source code, to docker containers (via a dockerfile), or a full fat Virtual Machine (VM) disk image. We think MaPS is a superior option to these alternate methods. A MaPS runtime is more complete than just sharing source code, more light weight than sharing a VM, and more streamlined than running docker. A more in depth comparison is discussed in the full paper of this extended abstract.

3.6 Compatibility

The technology enabling MaPS deeply depends on the features provided by the Linux kernel (isolation via namespace). Technically, this means that the system and its benefits are limited to the Linux kernel. As a result the program being packaged into a runtime *MUST* work on Linux!

Packaged runtimes provided by MaPS may still be used on systems powered by other kernels, via virtual machines running Linux, or via other tightly integrated compatiblity layers. This is the same strategy used by Docker on non Linux host OSs. Wrapping the compatibility layer inside the MaPS program might be a consideration for a later date. For now, the recommended way of using MaPS on Windows is via WSL[6], and on MacOS via lima[7]. More information is available on the MaPS wiki[8].

4 Examples

4.1 AccurateArithmetic.jl

The Julia package AccurateArithmetic.jl [5] implements the algorithms described in [8] in Julia. It also contains everything needed to reproduce the results shown in the paper. Below is a full list of the actions to be taken by anyone wanting to try and reproduce the results on their own Linux system (adapted for MaPS from [8, Appendix])

Commands prefixed with a `sh>` prompt are to be entered in a shell; commands prefixed with a `julia>` prompt are to be entered in a Julia interactive session. MaPS should have been downloaded and installed beforehand by following the instructions at https://github.com/MaRDI4NFDI/maps/wiki/Installation.

---

[6] Windows Subsystem for Linux https://learn.microsoft.com/en-us/windows/wsl/.
[7] Lima: Linux Machines https://github.com/lima-vm/lima.
[8] MaPS Wiki https://github.com/MaRDI4NFDI/maps/wiki/Non-Linux-OSs.

Be aware that step (3) in this procedure might take a few hours to complete. Afterwards, all measurements should be available as JSON files in the AccurateArithmetic.jl/test directory. After step (4), all figures showed in this paper should be available as PDF files in the same directory.

1. Get and launch the runtime:

```
sh> maps -d org.juliamath.accuratearithmetic/x86_64/
    papercorrectnessv3
sh> maps -r org.juliamath.accuratearithmetic/x86_64/
    papercorrectnessv3
```

2. Start Julia and run the test suite:

```
sh> cd /AccurateArithmetic.jl/test && julia --project
julia> using Pkg
julia> pkg"test"
```

3. Run the performance tests:

```
# Additional dependencies for performance tests.
# Package testing fails if these are already included.
julia> pkg"add BenchmarkTools Plots Printf Statistics Test
"
julia> exit()
sh> julia --project -O3 -L perftests.jl -e 'run_tests()'
```

4. Plot the graphs:

```
sh> julia --project -L perfplorts.jl -e 'plot_results()'
```

4.2 OSCAR

OSCAR is a comprehensive open source computer algebra system for computations in algebra, geometry, and number theory written in Julia[9]. As a Julia program, it benefits from just in time compilation, with a speedup step of precompilation. However, this precompilation step can be lengthy, and must be done by each user **after** installation, upon first use. This can leave end users with the false first impression of OSCAR being slow. As a way to mitigate this, OSCAR can be distributed as a MaPS runtime, already pre-compiled, with a custom system image.

1. Get the runtime:

```
maps -d org.oscar_system.oscar/x86_64/latest
```

2. Start the runtime:

```
maps -r org.oscar_system.oscar/x86_64/latest
```

---

[9] The Julia Programming Language https://julialang.org/.

Starting the runtime will automatically also start a Julia REPL[10] environment and load a precompiled copy of OSCAR. To update the runtime, for the latest version of OSCAR, execute

```
maps --update org.oscar_system.oscar/x86_64/latest
```

5 Beyond Mathematical Software

We present an example which showcases non mathematical software which can benefit from MaPS.

5.1 VIBRANT

VIBRANT is a tool for automated recovery and annotation of bacterial and archaeal viruses, determination of genome completeness, and characterization of viral community function from metagenomic assemblies. VIBRANT uses neural networks of protein annotation signatures and genomic features to maximize identification of highly diverse partial or complete viral genomes as well as excise integrated proviruses. VIBRANT is a novel and useful tool which, unfortunately, requires versions of software which are not easy to install in modern operating systems. This presents a perfect opportunity for research software to be ported into a MaPS runtime, such that it can be run on current systems without having to install outdated versions of software, which might cause conflicts with, and threaten the stability of an up to date system.

As before, installing it is as simple as calling the MaPS deploy function with the runtime name :

```
maps --deploy github.anantharaman.vibrant/x86_64/1.2.1
```

6 Concluding Remarks

Packaging software for distribution always means an overhead in terms of work for a researcher. However, correctly doing so leads to great progress in the reproducibility and thus the credibility of the scientific work at hand. It also makes the research more FAIR. We have developed a software tool to simplify this packaging process, keeping in mind the long term reproducibility of the work. The runtimes we create are lighter than a VM, more straighforward than docker, and more complete than just a source distribution.

---

[10] Read-evaluate-print loop.

References

1. Overlay filesystem - the linux kernel documentation (2014). https://www.kernel.org/doc/html/latest/filesystems/overlayfs.html?highlight=overlayfs
2. Flatpak: The future of apps on linux (2015). https://flatpak.org/
3. Bubblewrap - low-level unprivileged sandboxing tool used by flatpak and similar projects (2016). https://github.com/containers/bubblewrap
4. libostree - operating system and container binary deployment and upgrades (2016). https://github.com/ostreedev/ostree
5. AccurateArithmetic.jl - calculate with error-free, faithful, and compensated transforms and extended significands (2019). https://github.com/JuliaMath/AccurateArithmetic.jl/tree/paper-correctness-2019
6. fuse-overlayfs - fuse implementation for overlayfs (2020). https://github.com/containers/fuse-overlayfs
7. Valvesoftware/steam-runtime - a runtime environment for steam applications (2020). https://github.com/ValveSoftware/steam-runtime
8. Elrod, C., Févotte, F.: Accurate and Efficiently Vectorized Sums and Dot Products in Julia (Aug 2019). https://hal.science/hal-02265534, version submitted to the Correctness2019 workshop
9. Gawrilow, E., Joswig, M.: `polymake`: a framework for analyzing convex polytopes. In: Polytopes—combinatorics and computation (Oberwolfach, 1997), DMV Sem., vol. 29, pp. 43–73. Birkhäuser, Basel (2000)
10. Kaushik, A.: Mardi packaging system (2023). https://github.com/MaRDI4NFDI/maps
11. Kieft, K., Zhou, Z., Anantharaman, K.: Vibrant: Virus identification by iterative annotation copyright (2020). https://github.com/AnantharamanLab/VIBRANT
12. Lai, H.: Docker and the pid 1 zombie reaping problem (2015). https://web.archive.org/web/20240228145942/, https://blog.phusion.nl/2015/01/20/docker-and-the-pid-1-zombie-reaping-problem
13. Oscar – open source computer algebra research system (2024). https://www.oscar-system.org

Towards a FAIR Documentation of Workflows and Models in Applied Mathematics

Marco Reidelbach[1](✉)[iD], Björn Schembera[2][iD], and Marcus Weber[3][iD]

[1] Mathematics of Complex Systems, Zuse Institute Berlin, 14195 Berlin, Germany
reidelbach@zib.de
[2] Institute of Applied Analysis and Numerical Simulation, University of Stuttgart, 70569 Stuttgart, Germany
bjoern.schembera@mathematik.uni-stuttgart.de
[3] Mathematics of Complex Systems, Zuse Institute Berlin, 14195 Berlin, Germany
weber@zib.de

Abstract. Modeling-Simulation-Optimization workflows play a fundamental role in applied mathematics. The Mathematical Research Data Initiative, MaRDI, responded to this by developing a FAIR and machine-interpretable template for a comprehensive documentation of such workflows. MaRDMO, a Plugin for the Research Data Management Organiser, enables scientists from diverse fields to document and publish their workflows on the MaRDI Portal seamlessly using the MaRDI template. Central to these workflows are mathematical models. MaRDI addresses them with the MathModDB ontology, offering a structured formal model description. Here, we showcase the interaction between MaRDMO and the MathModDB Knowledge Graph through an algebraic modeling workflow from the Digital Humanities. This demonstration underscores the versatility of both services beyond their original numerical domain.

Keywords: Mathematical Research Data · MaRDMO Plugin · MathModDB

1 Introduction

Mathematical research data holds a pivotal role in advancing scientific understanding across various disciplines, ranging from core mathematical sciences to applied fields such as Engineering, Physics, and Digital Humanities. Mathematical research data is not limited to numerical or symbolic data contained in datasets, but also to data about the model, the solution algorithm and many more appearing in the Model-Simulation-Optimization (MSO) workflow [11,17]. The intricate nature of these workflows necessitates a standardized documentation framework adhering to the FAIR (Findable, Accessible, Interoperable, and Reusable) principles [20]. They are the essential and established paradigm for good scientific practice with regard to research data and research data management should be aligned with these principles.

To foster the FAIR principles the German National Research Data Infrastructure (NFDI [8]) was founded. It is a nationwide coordinated effort, consisting of 27 consortia spanning the natural sciences, engineering, the humanities, and others, aiming to set up data infrastructure and semantic technologies. Albeit national, the NFDI also actively works to enhance research data management on an international level as a member of the Research Data Alliance[1] and participates in the European Open Science Cloud[2] and Gaia-X[3] to establish FAIR data spaces for Europe.

The Mathematical Research Data Initiative (MaRDI), a consortium within the NFDI aiming to develop a robust research data infrastructure for mathematics and bridge towards other scientific disciplines, addressed the MSO workflows through a standardized documentation template. This template provides a comprehensive workflow description, including mathematical models, methods, software, hardware, input, and output data addressing specific research objectives [3] and is related to other developments within the NFDI to capture engineering workflows [1,16].

To facilitate the widespread adoption of the documentation template, MaRDI introduced the MaRDMO Plugin [13], a tool integrated in the Research Data Management Organiser (RDMO [4]). RDMO is an open-source software designed to facilitate the creation of data management plans (DMPs). DMPs are another part of good research data management and are a prerequisite for FAIR compliance. They are a formal document that describes data handling during and after the completion of a project. Like other DMP software, e.g. DMPTool[4] from the California Digital Library, primarily addressing the funding landscape in the US, and DMPOnline[5] from the British Digital Curation Centre, tailored for the UK context, RDMO provides templates for DMP creation, focusing on the German context, which makes it the most widely used DMP software in Germany [5].

MaRDMO streamlines the process of MSO workflow documentation by fetching additional information from various sources such as Wikidata [18], swMath [7] (an information system for mathematical software) and zbMath [10] (reviewing service for articles in mathematics) and making the complete documentation accessible via the MaRDI Portal[6], a knowledge base for Mathematics. This semi-automatic approach enhances efficiency while ensuring the completeness and accuracy of the documented workflows.

At the heart of these MSO workflows lies the proper documentation of mathematical models. Recognizing this fundamental aspect, MaRDI has developed the Mathematical Models Database (MathModDB) ontology [14,15], meticulously designed to capture essential elements associated with mathematical models, including research fields, problems, formulations, quantities, and tasks.

---

[1] https://www.rd-alliance.org/.
[2] https://eosc.eu/.
[3] https://gaia-x.eu/.
[4] https://dmptool.org/.
[5] https://dmponline.dcc.ac.uk/.
[6] https://portal.mardi4nfdi.de.

Numerical models documented through this scheme will be made accessible to the public through the MathModDB Knowledge Graph (KG), ensuring widespread availability and utilization of these valuable resources.

In this paper, we aim to establish a connection between MaRDMO and the MathModDB KG, allowing researchers to access a standardized collection of mathematical models through MaRDMO. Additionally, MaRDMO can serve as a further interface for MathModDB to gather new models from scientists across disciplines via RDMO. Through a practical example of a *Logical Data Analysis* algebraic workflow from the Digital Humanities, we demonstrate how this integration facilitates comprehensive documentation and analysis, enhancing reproducibility, transparency, collaboration, and interdisciplinary innovation in the scientific community. In addition, we show the transferability of MaRDMO and MathModDB, initially developed for capturing MSO workflows and their numerical models, to algebraic modeling applications.

2 MaRDMO, MathModDB and Their Connection

The MaRDMO Plugin. builds upon the capabilities of RDMO, which enables the creation of DMPs through customizable questionnaires. By now, general questionnaires exist reflecting the requirements of individual funding organizations, as well as subject-specific catalogs[7]. With MaRDMO we introduced a questionnaire[8] initially tailored for the documentation of MSO workflows and a customized export plugin[9]. Such plugins extend the functionality of RDMO, e.g. facilitating an export to Zenodo[10]. In turn, MaRDMO enables direct export to the MaRDI Portal. As far as we know, this is the first connection of RDMO with a KG.

The MaRDMO questionnaire guides researchers through the documentation process, facilitating the capture of crucial information about their workflows. Divided into four sections, the questionnaire covers various aspects of the research process, including general aspects (1), models, variables and parameters (2), process information (3), and reproducibility considerations (4). Through a series of structured questions, researchers can input details about their research objectives, mathematical models, software and hardware used, input and output data, methods, and more (c.f. complete questionnaire[11]).

The guiding principle of MaRDMO is reusing existing information wherever possible from established data sources. This approach ensures proper integration into the existing research data landscape while minimizing duplication of effort. The latest version of MaRDMO has streamlined the retrieval of additional information, now seamlessly conducted in the background. This enhancement, coupled with dynamic on-the-fly queries of repositories enhances usability and functionality of the plugin. For instance, when the DOI of a publication related

---

[7] https://github.com/rdmorganiser/rdmo-catalog.
[8] https://github.com/MarcoReidelbach/MaRDMO-Questionnaire.
[9] https://github.com/MarcoReidelbach/MaRDMO-Export-Plugin.
[10] https://github.com/rdmorganiser/rdmo-plugins.

to the workflow is provided, MaRDMO automatically retrieves all relevant citation and author information. This information is then presented for validation and completion, ensuring proper documentation of the workflow's context.

Upon completion of the questionnaire, MaRDMO facilitates the export of the documented workflow to the MaRDI Portal. Therefore, a comprehensive summary is generated following the MaRDI template and published as wikipage. Essential details (related publication, scientific fields and mathematical areas, applied mathematical model, methods, software, input and output data) are integrated into the KG of the MaRDI Portal, making them accessible to other researchers for search and exploration. Thereby, MaRDMO empowers researchers to efficiently capture and share their MSO workflows, fostering collaboration, reproducibility, and transparency within the scientific community.

The MathModDB Ontology is designed to specifically address the intricate nature of mathematical models in the domain of MSO. Within the vast landscape of scientific reasoning, mathematical models play a fundamental role, serving as indispensable tools for abstraction, formalization, analysis, and comprehension across diverse disciplines. However, the complexity and diversity inherent in mathematical models necessitate a unified semantic framework to comprehensively capture their essence.

Recognizing this imperative, MathModDB emerges as a sophisticated ontology crafted to meet the multifaceted demands of mathematical modeling. By structuring mathematical knowledge into a coherent framework, MathModDB enhances the semantic representation of mathematical models as they appear in the MSO workflow, while fostering interoperability and accessibility across domains.

The ontology is structured around eight classes essential for describing mathematical models comprehensively, a first result of iterative (and still ongoing) development [14,15] shaped by project internal discussions and valuable feedback from the mathematics community. These classes have evolved over time to reflect the diverse needs and perspectives within the mathematical modeling domain, ensuring a comprehensive representation of mathematical models. Moreover, this iterative process underscores the ontology's adaptability, with the potential for further evolution in response to emerging trends and evolving requirements within the mathematics community.

The classes of MathModDB include the *Mathematical Model* itself, the *Research Field* within which the model operates, the *Research Problem* it addresses, the *Mathematical Formulations* formalizing the model, the *Quantities* and *Quantity Kinds* involved in the *Mathematical Formulations*, the *Computational Tasks* associated with the model, and the *Publications* inventing, studying, surveying, or using the model. The *Computational Task* class was introduced to MathModDB only recently to accommodate the diverse range of tasks or questions posed to a model, resulting in different formulations, inputs and outputs.

The Connection of MaRDMO and MathMdoDB marks a significant advancement in the documentation of workflows and mathematical models, so far only stemming from the domain of MSO. Prior to the integration, MaRDMO's documentation of mathematical models, not yet present in established repositories, was limited, as researchers could only provide basic information such as model name, description, main subject, defining formulas, and identifiers. This often fell short of providing a comprehensive understanding of the model, especially when considered in isolation from the contextual workflow.

To address this limitation, the MathModDB ontology has been integrated into MaRDMO, facilitating the generation of comprehensive and insightful documentation for the models used in MSO workflows. This integration introduces additional question sets aligned with the classes of MathModDB. Following MaRDMO's guiding principle of utilizing existing resources, these question sets leverage the MathModDB KG. Now, mathematical models can be documented by connecting documented aspects of other models using the MathModDB vocabulary. In cases where no suitable entity exists, new ones can be created, with each domain featuring mandatory and optional fields. To integrate new mathematical models and their components properly into the existing data landscape, MaRDMO is also able to establish model interconnections through the MathModDB vocabulary, e.g. using generalizations, specifications or combinations. While these interconnections are crucial for maximizing the utility of the MathModDB KG, they might pose challenges for individual researchers, leaving it an optional asset. If researchers opt not to interconnect their model, the responsibility falls on domain experts to curate the MathModDB KG. This dual approach ensures that the KG remains a valuable asset while also accommodating the varying needs and perspectives of researchers.

3 A Semantic Representation of an Algebraic Modeling Workflow

We now show a first proof of concept of how a workflow from algebraic modeling can be semantically represented by MaRDMO and MathModDB.

Introduction: The *Logical Data Analysis* algebraic modeling workflow, rooted in the Digital Humanities, particularly Egyptology, intertwines the disciplines of Mathematics and Archaeology to unearth hidden patterns within a database. Originating from the Cachette de Karnak, an archaeological repository unearthed in 1903 by G. Legrain [6], the workflow aims to discern underlying rules governing the destruction patterns observed in ancient Egyptian objects. Legrain's findings revealed common destruction patterns such as missing heads, amputated body parts, and fragmented pieces, sparking the inquiry into whether specific rules governed these occurrences.

Object Data and Encoding: An online database[11] cataloging statues discovered in the Cachette provides the foundation for the analysis. Experts from Egyptology identified 16 potentially significant properties inherent in a subset of 333 artifacts, which were then encoded into binary numbers, denoting the presence or absence of each property.

Interdisciplinary Collaboration: The encoded dataset was then passed from Egyptology to Mathematics, marking the interdisciplinary exchange of data. Mathematicians then applied an object comparison model [19] (c.f. Fig. 1), utilizing a boolean ring, to unravel the underlying rules governing the destruction patterns. Using the "Rules and Pattern" algorithm[12], written in Julia [2] and leveraging the OSCAR package [12], the generator of the ideal was computed as a comprehensible Gröbner Basis comprising 172 logical rules.

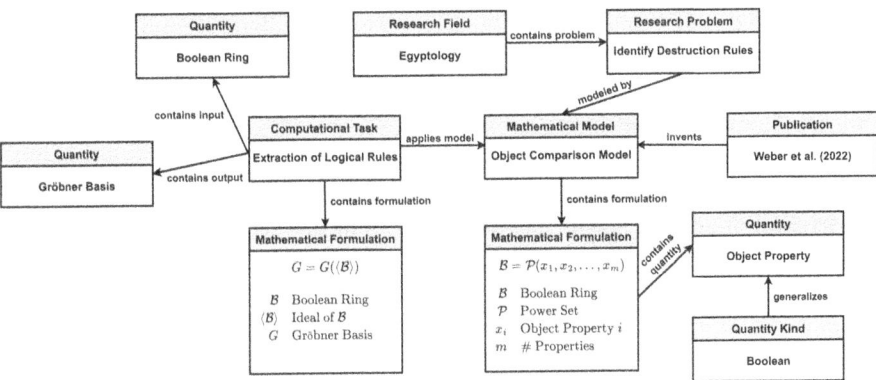

Fig. 1. Schematic representation of the object comparison model within the field of Egyptology, designed to identify destruction rules. Developed by Weber et al. (2022) [19], the model employs a boolean ring as its formulation, with object properties serving as quantities, represented as booleans. The model is associated with a task, featuring its own formulation, input, and output quantities.

Interpretation and Validation: The computational output sheds light on potential underlying rules. The validation of these expressions for scientific and statistical relevance falls upon Egyptology experts, marking the return of data to its disciplinary origin. This interdisciplinary exchange ensures a comprehensive evaluation of the identified rules.

---

[11] https://www.ifao.egnet.net/bases/cachette/.
[12] https://github.com/pynoor/The-RAP-Algorithm.

Reproducibility and Future Applications: The algebraic modeling workflow presented here follows a structured process, encompassing data selection and encoding, computational analysis, and interpretation of results. The methodology is reproducible, underscoring its reliability for future studies. Beyond Archaeology, the Object Comparison Model holds promise for solving diverse computational tasks and addressing distinct problems across various domains, e.g. literary analysis [19]. Computational tasks, other than the "Extraction of Logical Rules" include object sortings and feature extractions, broadening the applicability of the model beyond its initial context.

The documentation of the *Logical Data Analysis* workflow, following the MaRDI template, can be accessed through the MaRDI Portal[13]. The underlying "Object Comparison Model" is visualized in Fig. 1 since the MathModDB KG is not yet publicly accessible. All aspects of the algebraic modeling workflow described before could be documented through the MaRDMO Plugin and the integrated MathModDB Ontology.

4 Conclusion and Outlook

The integration of MathModDB into MaRDMO significantly enhances the quality of the workflow documentation, providing a more comprehensive and insightful portrayal of applied mathematical models. Through RDMO, we expect, once again, to reach scientists from diverse disciplines and facilitate their access to standardized and enriched model documentation.

We have shown how a research project from a seemingly completely different field, namely the humanities, can benefit from our approach. This involves the differentiation of destruction patterns in ancient objects, which can be modeled using mathematical/algebraic methods and semantically enriched and represented with the help of the tools presented (MaRDMO and MathModDB). Moreover, we have demonstrated that our solutions work beyond the domain of the classic numerical MSO workflow as they were applied to an algebraic modeling workflow. All the relevant information was captured by tools initially developed for MSO.

Looking ahead, our next objective is to connect MaRDMO with the Mathematical Algorithms Database (MathAlgoDB [9]) KG[14], also developed within MaRDI, to bolster the method component of the workflow documentation. This integration could leverage existing connections between MathAlgoDB and MathModDB, offering users a curated selection of algorithms to address specific tasks associated with particular models, thus further enhancing usability. Moreover, given the interdisciplinary nature of these workflows, we aim to collaborate

---

[13] Documentation Video, Wikipage and Knowledge Graph Entry:
https://portal.mardi4nfdi.de/wiki/MaRDMO
https://portal.mardi4nfdi.de/wiki/Logical_Data_Analysis_for_Egyptian_Objects
https://portal.mardi4nfdi.de/wiki/Item:Q6032641.
[14] https://algodata.mardi4nfdi.de/.

with the other consortia of the NFDI to explore potential connections with non-mathematical services, expanding the scope and utility of MaRDMO in the future. In addition, further research is needed to show that the connection between MaRDMO and MathModDB also works for more advanced workflows in the field of algebra.

Acknowledgement. Marco Reidelbach, Björn Schembera and Marcus Weber are supported by MaRDI, funded by the Deutsche Forschungsgemeinschaft (DFG), project number 460135501, NFDI 29/1 "MaRDI - Mathematische Forschungsdateninitiative".

Disclosure of Interests. The authors have no competing interests to declare that are relevant to the content of this article.

References

1. Arndt, S., et al.: Metadata4Ing: an ontology for describing the generation of research data within a scientific activity (2022). https://doi.org/10.5281/zenodo.7706017
2. Bezanson, J., Edelman, A., Karpinski, S., Shah, V.B.: Julia: a fresh approach to numerical computing. SIAM Rev. **59**(1), 65–98 (2017). https://doi.org/10.1137/141000671
3. Boege, T., Fritze, R., Görgen, C., Hanselman, J., Iglezakis, D., Kastner, L., et al.: Research-data management planning in the German mathematical community. Eur. Math. Soc. Mag. (2023). https://doi.org/10.4171/MAG/152
4. Engelhardt, C., Enke, H., Klar, J., Ludwig, J., Neuroth, H.: Research data management organiser. In: Proceedings of the 14th International Conference on Digital Preservation, pp. 25–29 (2017)
5. Enke, H., Hausen, D., Henzen, C., Jagusch, G., Krause, C., Schönau, S., et al.: Data management planning: concept for setting up a working group in the NFDI section common infrastructures. Zenodo (2023). https://doi.org/10.5281/zenodo.7540682
6. Legrain, G.: Les récentes découvertes de Karnak. Bull. de l'institut d'Égypte **5**, 109–120 (1904)
7. Greuel, G.-M., Sperber, W.: swMATH – an information service for mathematical software. In: Hong, H., Yap, C. (eds.) ICMS 2014. LNCS, vol. 8592, pp. 691–701. Springer, Heidelberg (2014). https://doi.org/10.1007/978-3-662-44199-2_103
8. Hartl, N., Wössner, E., Sure-Vetter, Y.: Nationale Forschungsdateninfrastruktur (NFDI). Informatik Spektrum **44**, 370–373 (2021)
9. Himpe, C., Kleikamp, H., Fritze, R., Rave, S.: MaRDI task area 2 - scientific computing @ WWU Münster. AlgoData - Algorithm Knowledge Graph - Ontology (Version 0.1) (2022). https://mardi4nfdi.de/algodata/0.1
10. Hulek, K., Teschke, O.: Die transformation von zbMATH zu einer offenen plattform für die mathematik. Mitt. der Dtsch. Mathematiker-Ver. **28**(2), 108–111 (2020). https://doi.org/10.1515/dmvm-2020-0031
11. Koprucki, T., Kohlhase, M., Tabelow, K., Müller, D., Rabe, F.: Model pathway diagrams for the representation of mathematical models. Opt. Quant. Electron. **50**, 1–9 (2018). https://doi.org/10.1007/s11082-018-1321-7

12. Oscar – open source computer algebra research system, version 0.12.1-dev (2023). https://www.oscar-system.org
13. Reidelbach, M., Ferrer, E., Weber, M.: MaRDMO Plugin - document and retrieve workflows using the MaRDI portal. In: Proceedings of the 1st Conference on Research Data Infrastructure (CoRDI) - Connecting Communities (2023). https://doi.org/10.52825/cordi.v1i.254
14. Schembera, B., et al.: Ontologies for models and algorithms in applied mathematics and related disciplines. arXiv preprint arXiv:2310.20443 (2023). https://doi.org/10.48550/arXiv.2310.20443
15. Schembera, B., et al.: Building ontologies and knowledge graphs for mathematics and its applications. In: Proceedings of the 1st Conference on Research Data Infrastructure (CoRDI) - Connecting Communities (2023). https://doi.org/10.52825/cordi.v1i.255
16. Schmitt, R.H., Anthofer, V., Auer, S., Başkaya, S., Bischof, C., Bronger, T., et al.: NFDI4Ing-the national research data infrastructure for engineering sciences. Zenodo (2020). https://doi.org/10.5281/zenodo.4015201
17. The MaRDI consortium: MaRDI: Mathematical Research Data Initiative Proposal (2022). https://doi.org/10.5281/zenodo.6552436
18. Vrandečić, D.: Wikidata: a new platform for collaborative data collection. In: Proceedings of the 21st International Conference on World Wide Web. p. 1063-1064. Association for Computing Machinery (2012). https://doi.org/10.1145/2187980.2188242
19. Weber, M., Fackeldey, K.: The mathematics of comparing objects. Zenodo (2022). https://doi.org/10.48550/arXiv.2201.07032
20. Wilkinson, M.D., Dumontier, M., Aalbersberg, I.J., Appleton, G., Axton, M., Baak, A., et al.: The FAIR guiding principles for scientific data management and stewardship. Sci. Data **3**(1), 1–9 (2016). https://doi.org/10.1038/s41597-019-0009-6

Symbolic-Numeric Methods in Algebraic Geometry

Monodromy Coordinates

Taylor Brysiewicz[✉][iD]

University of Western Ontario, London, ON N6G 2V4, Canada
tbrysiew@uwo.ca

Abstract. We introduce the concept of monodromy coordinates for representing solutions to large polynomial systems. Representing solutions this way provides a time-memory trade-off in a monodromy solving algorithm. We describe an algorithm, which interpolates the usual monodromy solving algorithm, for computing such a representation and analyze its space and time complexity.

Keywords: Monodromy · Numerical Algebraic Geometry · Random Permutations

1 Introduction

The bottleneck for numerically solving a polynomial system can lie in the space-complexity of an algorithm rather than its time-complexity: storing d points in $\mathbb{C}^n$ requires $\mathcal{O}(dn)$ bits. For instance, the solution set to the system in Example 2 requires around 100 gigabytes to store. We propose the alternative representation of a *monodromy tree* which encodes the *monodromy coordinates* of the solutions (see Fig. 1). A monodromy tree describes how to find solutions using a *monodromy solving algorithm* (Algorithm 1) and provides an iterator (Algorithm 2) for the solution set. The expected space complexity of storing a monodromy tree is $\mathcal{O}(n \ln(d))$. Using a monodromy tree, the solutions in Example 2 can be represented using less than 15 kilobytes (see Example 2).

We analyze our work under the three *strong monodromy assumptions* (see Sect. 2). These place our analysis within the realm of probabilistic group theory. Hence, our complexity estimates refer to *expected* behaviour. The memory reduction achieved through monodromy coordinates is balanced by the d queries to *monodromy oracles* required to *unpack* the solutions (see Theorem 1), which is comparable to the $\mathcal{O}(d)$ queries for solving the system in the first place. However, since a monodromy tree gives an iterator for the solution set (Algorithm 2), accumulating functions of the solutions (e.g., the number of real/complex solutions) can be achieved without ever holding all solutions in memory.

Crucially, a monodromy tree can be computed using an augmented monodromy solving algorithm (Algorithm 4) which has the same space complexity as storing a monodromy tree ($\mathcal{O}(n \ln(d))$), but comes at the cost of a time complexity increase to $\mathcal{O}(d(d/\ln(d)))$. Optionally, one may interpolate between Algorithms 1 and 4 based on a parameter $1 \leq \alpha \leq d$. This gives an algorithm with

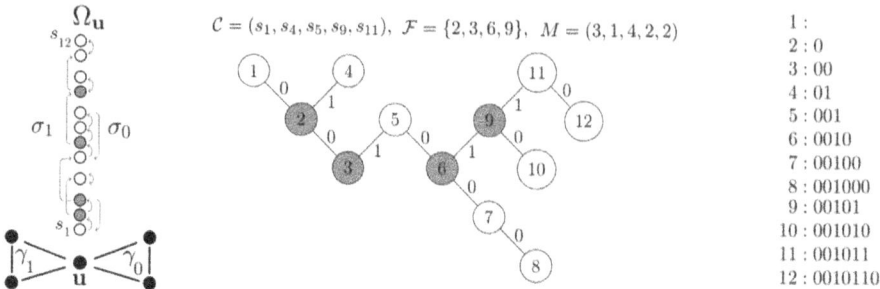

Fig. 1. A depiction of monodromy (left) a monodromy tree of type IV (center) and monodromy coordinates (right), for a system with twelve solutions.

space complexity $\mathcal{O}(n(d/\alpha))$ and time complexity $\mathcal{O}(d\alpha)$. As a consequence, our modified monodromy solving algorithm induces a time-memory trade-off, quantified by a factor of α.

2 Monodromy Background

Fix a system in variables $\mathbf{x} = (x_1, \ldots, x_n)$ and parameters $\mathbf{p} = (p_1, \ldots, p_k)$:

$$F(\mathbf{x}; \mathbf{p}) := (f_1(\mathbf{x}; \mathbf{p}), \ldots, f_N(\mathbf{x}; \mathbf{p})) \subset \mathbb{C}[p_1, \ldots, p_k][x_1, \ldots, x_n] = \mathbb{C}[\mathbf{p}][\mathbf{x}]$$

such that the incidence variety $X = \{(\mathbf{x}, \mathbf{p}) \mid F(\mathbf{x}; \mathbf{p}) = 0\} \subseteq \mathbb{C}^n \times \mathbb{C}^k$ is irreducible and the image of the projection $\pi : X \to \mathbb{C}^k$ has the same dimension as X. Let $Y := \overline{\pi(X)}$ be the Zariski closure of this image. Then $\pi : X \to Y$ is a branched cover of degree d for some $d \in \mathbb{N}$ and there exists a Zariski open subset $\mathcal{U} \subseteq Y$ of *regular values* for which $\pi|_{\pi^{-1}(\mathcal{U})}$ is a d-to-one covering space.

Given a regular value $\mathbf{u} \in \mathcal{U}$, any loop $\gamma : [0,1] \to \mathcal{U}$ based at $\mathbf{u}$ lifts to d paths $\pi^{-1}(\gamma([0,1]))$. These paths induce a permutation σ_γ on the fibre $\pi^{-1}(\mathbf{u}) = \Omega_\mathbf{u}$ mapping start points to end points. Write $G_\mathbf{u}$ for the image of the homomorphism $\gamma \mapsto \sigma_\gamma$ from the fundamental group of $\mathcal{U}$ to the symmetric group $\mathfrak{S}_{\Omega_\mathbf{u}}$. The loop γ is a *(monodromy) loop*, the permutation σ_γ is a *(monodromy) permutation*, and the group $G_\mathbf{u}$ is the *monodromy group* of π based at $\mathbf{u}$.

We analyze our work under the *strong monodromy assumptions*:

- **Full Symmetry:** $G_\mathbf{u} = \mathfrak{S}_{\Omega_\mathbf{u}} \cong \mathfrak{S}_d$
- **Uniform Sampling:** We can sample elements of $G_\mathbf{u}$ uniformly at random
- **Evaluation Oracle:** Given $\sigma \in G_\mathbf{u}$ and $s \in \Omega_\mathbf{u}$ we can evaluate $\sigma(s)$

These assumptions are reasonable. (Full-Symmetry) Many branched covers have full-symmetric monodromy groups [7,12]. (Uniform Sampling) Given a distribution on $\mathfrak{S}_d$ with no zero probabilities, taking convolutions gives distributions converging to the uniform distribution [3, Ch 4. Thm 3.]. (Evaluation Oracle)

The numerical method of *homotopy continuation* is incredibly reliable and functions as a practical evaluation oracle for monodromy. For details about homotopy continuation and numerical algebraic geometry, see [1,11].

The strong monodromy assumptions are often made when analyzing monodromy algorithms because they place the analysis of these algorithms directly within the realm of *probabilistic group theory*. To that end, we recall some facts regarding random permutations.

The first result, due to Dixon [4], states that the probability p_d that two random permutations $\sigma_0, \sigma_1 \in \mathfrak{S}_d$ generate a transitive subgroup is $1 - d^{-1} + \mathcal{O}(d^{-2})$. For reference, $p_{1000} > 0.998998$. The second fact, is that the expected number of i-cycles in a random permutation $\sigma \in \mathfrak{S}_d$ is $\frac{1}{i}$. Consequently, the expected number of cycles in a permutation is given by the d-th Harmonic number $H_d = \sum_{i=1}^{d} \frac{1}{i} = \frac{1}{1} + \frac{1}{2} + \cdots + \frac{1}{d} \approx \ln(d)$. Finally, the expected largest cycle in a random permutation is asymptotically $\lambda \cdot d$ where $\lambda \approx 0.6243$ is the *Golomb-Dickman constant* [6].

3 Monodromy Solving

Monodromy plays a central role in the algorithms and applications of modern numerical algebraic geometry [5,8,12]. A *monodromy solving algorithm* computes the d solutions $\Omega_\mathbf{u}$ to a system $F(\mathbf{x}; \mathbf{u}) = 0$ from three inputs: (a) a *seed solution* $s_1 \in \Omega_\mathbf{u}$, (b) evaluation oracles σ for D monodromy permutations $\sigma_0, \ldots, \sigma_{D-1} \in \mathfrak{S}_{\Omega_\mathbf{u}}$, and (c) a stopping criterion.

A monodromy solving algorithm succeeds as long as $\langle \sigma_0, \ldots, \sigma_{D-1} \rangle$ is a transitive permutation group. This is incredibly likely for $D = 2$ and $d \gg 0$ by Dixon's result. Hence, we consider only two random permutations $\sigma_0, \sigma_1 \in \mathfrak{S}_{\Omega_\mathbf{u}}$ and give a deterministic monodromy solve algorithm: Algorithm 1. For a more versatile framework for monodromy algorithms, see [5].

Algorithm 1. MonodromySolve

Input: • A seed solution $s_1 \in \Omega_\mathbf{u}$ • Evaluation oracles σ for $\sigma_0, \sigma_1 \in \mathfrak{S}_{\Omega_\mathbf{u}}$ which generate a transitive permutation action   • A stopping criterion
Output: The d points of $\Omega_\mathbf{u}$

1: `initialize` $S = \{s_1\}$ and $j = 1$ ▷ Initialize the set of found solutions
2: `while stopping criterion = false do`
3:    `if` $q := \sigma_0(\text{last}(S))$ `is new then`   ▷ Check if a new point is found via σ_0
4:      $S \leftarrow q$
5:    `else`
6:      `while` $q = \sigma_1(s_j)$ `is not new do`   ▷ If not, try finding new points via σ_1
7:        $j = j + 1$
8:      $S \leftarrow q;\ j = j + 1$ ▷ Such j are called founder indices
9: `return` S ▷ The order of S gives the monodromy ordering of $\Omega_\mathbf{u}$

Algorithm 1 greedily discovers new solutions by applying σ_0 to known solutions, until the known cycles of σ_0 have been saturated. It then applies σ_1 to the known solutions, in the order they have been found, until a new cycle of σ_0 is discovered, and repeats until the stopping criterion is met. For now, we take the stopping criterion in Algorithm 1 to be $|S| = d$.

Algorithm 1 induces an ordering on $S = \Omega_{\mathbf{u}}$ called the *monodromy ordering* of $\Omega_{\mathbf{u}}$ (with respect to σ_0, σ_1, s_1). We write $\mathcal{C} := (c_1, \ldots, c_r)$ for the minimal elements of each cycle of σ_0 with respect to this order, called *initial cycle solutions*, and $M := (m_1, \ldots, m_r)$ for the corresponding cycle sizes. In particular, r is the number of cycles of σ_0. Each cycle is *found* by some solution s_j via $\sigma_1(s_j)$ in line 8 in which case we call j a *founder index*. We denote the set of $r - 1$ founder indices by $\mathcal{F}$, whose largest element we denote by j^*. For $\alpha \in \mathbb{N} \cup \{\infty\}$, we write $\mathcal{C}_\alpha$ for the tuple of initial cycle solutions, along with every α-th solution in that cycle (e.g. $\mathcal{C}_\infty = \mathcal{C}$ and $\mathcal{C}_1 = \Omega_{\mathbf{u}}$).

Example 1. Consider $\Omega_{\mathbf{u}} = \{s_1, \ldots, s_{12}\}$ along with the permutations

$$\sigma_0 = (s_1, s_2, s_3)(s_4)(s_5, s_6, s_7, s_8)(s_9, s_{10})(s_{11}, s_{12})$$
$$\sigma_1 = (s_1, s_3, s_5, s_2, s_4)(s_6, s_9, s_{11})(s_7)(s_8, s_{10})(s_{12}) \in \mathfrak{S}_{\Omega_{\mathbf{u}}}$$

shown in Fig. 2.

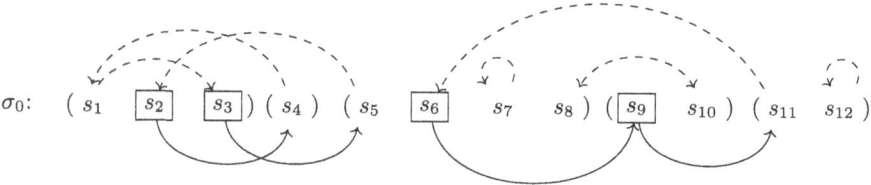

Fig. 2. Two permutations σ_0, σ_1 on twelve solutions $\Omega_{\mathbf{u}}$. In the middle, σ_0 is written in cycle notation, whereas σ_1 is represented via arrows. A solid arrow from a boxed founder indicates an instance of line 6 in Algorithm 1 proceeding to line 8.

The solutions $\{s_1, \ldots, s_{12}\}$ are already indexed with respect to their monodromy order and Algorithm 1 finds them via $20 = 11_{\sigma_0} + 9_{\sigma_1}$ queries. Running Algorithm 1 on $(\sigma_0, \sigma_1, s_1)$ gives the values $\mathcal{F} = \{2, 3, 6, 9\}$, $j^* = 9$, $\mathcal{C} = (s_1, s_4, s_5, s_9, s_{11}), \mathcal{C}_2 = (s_1, s_3, s_4, s_5, s_7, s_9, s_{11})$, and $M = (3, 1, 4, 2, 2)$.

We write $\rho(\Box)$ for the number of bits used to store an object $\Box$. For example, $\rho(\mathbb{Z}) = 64$ for moderate sized integers, and $\rho(\mathbb{C}^n) = \rho(\mathbb{R}^{2n}) = 2n\rho(\mathbb{R})$. With this notation, we describe the time and space complexity of Algorithm 1.

Proposition 1. *Algorithm 1 has time complexity $\mathcal{O}(d)$ and space complexity $\mathcal{O}(nd)$ as measured in oracle queries and bits respectively. Specifically, Algorithm 1 queries a monodromy oracle $d + j^* - 1$ times and requires $d\rho(\mathbb{C}^n)$ bits to store the solutions.*

Proof. Index $\Omega_{\mathbf{u}}$ via the monodromy ordering. Evaluation oracles are queried in lines 3 and 6. The former is evaluated $d-1$ times, once for every solution other than s_d. Line 6 is evaluated for $j = 1, \ldots, j^*$. This totals to $d + j^* - 1$ as stated. The space required to perform the algorithm is negligibly more than the cost of storing d points in $\mathbb{C}^n$ (e.g. q and j must be stored).

4 Monodromy Coordinates: Compression

Each solution $s \in \Omega_{\mathbf{u}}$ found via Algorithm 1 occurs as a sequence of permutations, each either σ_0 or σ_1, applied to s_1. Identify this sequence with a word $\omega = \omega_1 \cdots \omega_\ell$ in $\{0,1\}$, that is, $s = (\sigma_{\omega_k} \circ \cdots \circ \sigma_{\omega_1})(s_1) \longleftrightarrow \omega = \omega_1 \cdots \omega_\ell$. We say that ω gives *monodromy coordinates* for s with respect to $(\{\sigma_0, \sigma_1\}, s_1)$. By construction, the monodromy coordinates of s with respect to $(\{\sigma_0, \sigma_1\}, s_1)$ describes the lexicographically smallest $(\sigma_0 < \sigma_1)$ path from s_1 to s in the Cayley-graph of $\langle \sigma_0, \sigma_1 \rangle$ acting on $\Omega_{\mathbf{u}}$. The induced monodromy ordering of $\Omega_{\mathbf{u}}$ agrees with the lexicographic ordering on the monodromy coordinates.

Monodromy coordinates can be as long as d letters and so directly storing may require quadratic memory in d. A much more efficient encoding is via a binary tree as depicted in Fig. 3, whose structure is given by the monodromy coordinates: an edge labelled 0 from $j \to j'$ occurs if $s_{j'}$ is found via $\sigma_0(s_j)$ in line 4 of Algorithm 1, and an edge labelled 1 occurs if $s_{j'}$ is found via $\sigma_1(s_j)$ in line 8. The resulting labelled rooted binary tree $\mathcal{T}$ is the *monodromy tree* of $\Omega_{\mathbf{u}}$, with respect to $(\{\sigma_0, \sigma_1\}, s)$. See Fig. 3 for the monodromy tree/coordinates of Example 1.

Fig. 3. (Left) The monodromy tree associated to the permutations σ_0, σ_1 in Fig. 2 with founders shaded. (Right) The corresponding monodromy coordinates.

We consider five representations of a monodromy tree:

$$\underbrace{(s_1, \sigma)}_{\text{I: single seed}} \hookrightarrow \underbrace{(s_1, \sigma, \mathcal{F})}_{\text{II: minimal}} \hookrightarrow \underbrace{(s_1, \sigma, \mathcal{F}, M)}_{\text{III: with cycle sizes}} \hookrightarrow \underbrace{(\mathcal{C}, \sigma, \mathcal{F}, M)}_{\substack{\text{IV: with initial cycle} \\ \text{representatives}}} \hookrightarrow \underbrace{(\mathcal{C}_\alpha, \sigma, \mathcal{F}, M)}_{\substack{\text{V: with cycle} \\ \text{representatives}}}$$

The cost of unpacking the solution set $\Omega_{\mathbf{u}}$ from a monodromy tree $\mathcal{T}$ differs depending on how $\mathcal{T}$ is represented. Indeed, the *single seed* representation of type I is merely the input of Algorithm 1 and so one recovers $\Omega_{\mathbf{u}}$ by running Algorithm 1. Given a type II representation of $\mathcal{T}$, the solution set $\Omega_{\mathbf{u}}$ can be recovered by executing Algorithm 1 subject to the minor, efficiency-boosting, change: replace lines 6 and 7 with $q = \sigma_1(s_{\mathcal{F}_j})$. Thus, the number of oracle queries to recover all solutions from a type II representation is $d + r - 2 = (r-1) + (d-r) + (r-1)$. If cycle sizes are included (type III) then $r - 1$ evaluations of σ_0 can be saved. When initial cycle solutions are stored (type IV), evaluations of σ_1 are entirely unnecessary, saving $r - 1$ additional oracle queries. When $\lceil m_i/\alpha \rceil$-many solutions in each cycle are stored (type V), they need not be found via oracle calls.

Theorem 1. *The memory cost, more than $\rho(\sigma)$, to store a monodromy tree is*

$I: \rho(\mathbb{C}^n) \xrightarrow{\mathbb{E}} \rho(\mathbb{C}^n) \xrightarrow{\mathcal{O}} \mathcal{O}(n)$

$II: \rho(\mathbb{C}^n) + (r-1)\rho(\mathbb{N}) \xrightarrow{\mathbb{E}} \rho(\mathbb{C}^n) + (H_d - 1)\rho(\mathbb{N}) \xrightarrow{\mathcal{O}} \mathcal{O}(\max(n, \ln(d)))$

$III: \rho(\mathbb{C}^n) + (2r-1)\rho(\mathbb{N}) \xrightarrow{\mathbb{E}} \rho(\mathbb{C}^n) + (2H_d - 1)\rho(\mathbb{N}) \xrightarrow{\mathcal{O}} \mathcal{O}(\max(n, \ln(d)))$

$IV: r\rho(\mathbb{C}^n) + (2r-1)\rho(\mathbb{N}) \xrightarrow{\mathbb{E}} H_d \rho(\mathbb{C}^n) + (2H_d - 1)\rho(\mathbb{N}) \xrightarrow{\mathcal{O}} \mathcal{O}(n \ln(d))$

$V: \left(\sum_{i=1}^{r} \lceil \frac{m_i}{\alpha} \rceil \right) \rho(\mathbb{C}^n) + (2r-1)\rho(\mathbb{N}) \xrightarrow{\mathbb{E}} < ((d/\alpha) + H_d)\rho(\mathbb{C}^n) + (2H_d - 1)\rho(\mathbb{N})$

$\xrightarrow{\mathcal{O}} \mathcal{O}(\max(n(d/\alpha), n \ln(d)))$

The time cost of unpacking all solutions from a monodromy tree is $\mathcal{O}(d)$:

$I: d + j^* - 1 \quad II: d + r - 2 \quad III: d - 1 \quad IV: d - r \quad V: d - \left(\sum_{i=1}^{r} \lceil \frac{m_i}{\alpha} \rceil \right)$

Proof. The memory cost directly follows from the descriptions of the monodromy tree representations. The expectation of the memory applies (a) expectation on the number of cycles, $\mathbb{E}(r) = H_d \approx \ln(d)$, along with (b) the observation that the number of points stored in type V is at most $(d/\alpha) + r$. The asymptotics are immediate. The time cost (in oracle queries) summarizes the discussion prior to the theorem statement. Again, the expected asymptotics are immediate.

Remark 1. For monodromy solving in practice we use $\rho(\mathbb{C}^n) = 2n\rho(\mathbb{R}) = 128 \cdot n$, $\rho(\mathbb{N}) = 64$, and $\rho(\sigma) = 5\rho(\mathbb{C}^k) = 640k$, by storing a real number as a 64-bit float and an integer at most d as a 64-bit integer. For extremely large d, one would have to use larger integer types. Note that two monodromy loops can be represented by five parameter values via a bow-tie configuration (see Fig. 1). For $\alpha < d/r$ the memory storage for type V in Theorem 1 is dominated by $128n\frac{d}{\alpha}$.

Example 2. There are $d = 666841088$ quadric surfaces tangent to 9 given quadric surfaces in $\mathbb{C}^3$ (see [2]). Finding these tangent quadrics amounts to solving a polynomial system in 10 variables, over 90 parameters. Under the practical storage assumptions of Remark 1, we specialize Theorem 1 for various monodromy tree representations. We use the approximation $H_d \approx \ln(d)$ and the empirical observation that $\mathbb{E}(j^*) \approx \frac{d+1}{2}$. Note that $d/\ln(d) \approx 32820112$. To encode a monodromy tree of type I costs $58,880$ bits and requires around one-billion oracle calls to unpack. For types II – IV, the number of oracle calls is approximately d and the approximate number of bits to store are $60, 116, 61, 416$, and $112, 150$ respective. Type V with $\alpha = 100000, 10$, and 1 respectively require approximately 8-million, 85-billion, and 853-billion bits to store. To unpack these three trees of type V one needs approximately $d, \frac{9}{10}d$, and $\frac{1}{2}d$ oracle queries respectively.

5 Monodromy Solving Using Monodromy Coordinates

A key advantage of monodromy coordinates lies in the fact that the monodromy solve algorithm can be performed in such a way to produce a monodromy tree representation of type IV or V. During this procedure, the algorithm "knows" when new solutions have been found and can thus accumulate any function $\mathcal{G}(\mathbf{u}) := \sum_{s \in \Omega_{\mathbf{u}}} g(s)$ during the process, avoiding the need to store all solutions simultaneously in memory. Examples of desirable *accumulators* include

$$\#\mathbb{C}: \mathcal{G}_{\mathbb{C}} = \sum_{s \in \Omega_{\mathbf{u}}} 1, \quad \#\mathbb{R}: \mathcal{G}_{\mathbb{R}} = \sum_{s \in \Omega_{\mathbf{u}} \cap \mathbb{R}^n} 1, \quad \text{Trace: } \mathcal{G}_{\Sigma(\mathbf{v})} = \sum_{s \in \Omega_{\mathbf{v}}} s. \quad (1)$$

The trace accumulator is particularly desirable when the polynomial system corresponds to the computation of a *witness set*. In this case, the trace accumulator can be used as a stopping criterion via the *trace test* [10]. Similar trace accumulators can be designed for other trace tests, like the one in [9].

A monodromy tree of type IV gives an *iterator* for the (monodromy) ordered list $\Omega_{\mathbf{u}} = (s_1, \ldots, s_d)$, represented by a seed solution and a next function.

Algorithm 2. next

Input:
- A monodromy tree $\mathcal{T} = (\mathcal{C}, \sigma, \mathcal{F}, M)$ of type V
- j
- sj representing $s_j \in \Omega_{\mathbf{u}}$ other than s_d

Output: s_{j+1}

1: **if** j is a partial sum $\sum_{i=1}^{k-1} M_i$ **then return** $\mathcal{C}_k$ **else return** $\sigma_0(s_j)$

Algorithm 2 uses at most one oracle call and succeeds even when $\mathcal{T}$ is incomplete: for $\mathcal{T}^{(k)} := (\mathcal{C}^{(k)}, \sigma, \mathcal{F}^{(k-1)}, M^{(k)})$ where $\mathcal{C}^{(k)}$ is the truncation of $\mathcal{C}$ to those solutions only in the first k cycles, $\mathcal{F}^{(k-1)} = (\mathcal{F}_i)_{i=1}^{k-1}$, $M^{(k)} = (M_i)_{i=1}^{k}$, and j less than the sum of $M^{(k)}$, the output is correct.

Algorithm 3. inNewCycle

Input:
- An incomplete monodromy tree $\mathcal{T}^{(k)} = (\mathcal{C}^{(k)}, \sigma, \mathcal{F}^{(k-1)}, M^{(k)})$ of type IV
- $s \in \Omega_{\mathbf{u}}$

Output: `false` if s belongs to the same cycle of σ_0 as some $c \in \mathcal{C}^{(k)}$, `true` otherwise

1: **if** $s \in \mathcal{C}^{(k)}$ **then return** `false`
2: **for** i from 1 to $\min(\max(M^{(k)}), \alpha) - 1$ **do**
3:     $s = \sigma_0(s)$
4:     **if** $s \in \mathcal{C}^{(k)}$ **then return** `false`
5: **return** `true`

Algorithm 4. MonodromySolve - Monodromy Coordinates

Input:
- A seed solution $s_1 \in \Omega_{\mathbf{u}}$
- Evaluation oracles σ for $\sigma_0, \sigma_1 \in \mathfrak{S}_{\Omega_{\mathbf{u}}}$
  (where $\{\sigma_0, \sigma_1\}$ generate a transitive action)
- $\alpha \in \mathbb{N}$
- A `stopping criterion`
- (Optional:) any function $\mathcal{G} = \sum_{s \in \Omega_{\mathbf{u}}} g(s)$ to accumulate

Output: A monodromy tree representation $\mathcal{T} = (\mathcal{C}, \sigma, \mathcal{F}, M)$ of type V

1: $\mathcal{I} = \{s_1\}, \mathcal{J} = \{\}, M = \{\}$ ▷ initial cycle solutions $\mathcal{I}$, additional solutions $\mathcal{J}$, and cycle sizes M
2: $m = 1, j = 1$ ▷ current cycle size m and potential founder index j
3: $sj = s_1, s = s_1$ ▷ potential founder sj representing s_j and a placeholder s
4: $\mathcal{F} = \{\}, a = 0$ ▷ founder indices $\mathcal{F}$, and a storage counter a
5: **while** `stopping criterion` is not met **do**
6:    $\star_0$ Replace $s = \sigma_0(s)$
7:    **if** $s \neq \texttt{last}(\mathcal{I})$ **then**   ▷ Check if a new point is found via σ_0
8:      $m = m + 1, a = a + 1$ ▷ If so, increase current cycle size
9:      $\mathcal{G} = \mathcal{G} + g(s)$ ▷ Accumulate
10:      **if** $a = \alpha$ **then** $\mathcal{J} \leftarrow s; a = 0$ ▷ Store s and reset the counter a
11:    **else**   ▷ Otherwise, a cycle is complete
12:      foundNewCycle = `false`   ▷ and we must find a new one
13:      $M \leftarrow m, m = 1, a = 0$ ▷ Store cycle size in M and reset counters m and a
14:      **while** foundNewCycle = `false` **do**
15:      $\star_1$ Compute $t_* := \sigma_1(sj)$ and set $s = \texttt{copy}(t_*)$ ▷ sj is a founder if t_* is new
16:      $\star_0$ **if** inNewCycle$(\mathcal{I} \cup \mathcal{J}, \sigma, \mathcal{F}, M, s)$ **then**   ▷ If this is so
17:          foundNewCycle = `true`
18:          $\mathcal{I} \leftarrow t_*, \mathcal{F} \leftarrow j$ and $s = t_*$
19:          $\mathcal{G} = \mathcal{G} + g(s)$ ▷ Accumulate
20:      **else**
21:      $\star_0$ $sj = \texttt{next}(\mathcal{I}, \sigma, \mathcal{F}, M, j, sj)$ and $j = j + 1$
22: **return** $\mathcal{T} = (\mathcal{I}, \sigma, \mathcal{F}, M)$ ▷ Replace $\mathcal{I}$ with $\mathcal{C}_\alpha = \mathcal{I} \cup \mathcal{J}$ if desired

---

Similarly, applied to a (possibly incomplete) monodromy tree $\mathcal{T}^{(k)}$ and solution $s \in \Omega_{\mathbf{u}}$, the procedure `inNewCycle` determines whether s belongs to a cycle of σ_0 represented by an element of $\mathcal{C}^{(k)}$. We point out that Algorithm 3 uses at most $\min(\max(M), \alpha) - 1$ oracle queries.

Theorem 2. *Algorithm 4 correctly evaluates any optional accumulators $\mathcal{G}$. During which, no more than $d - 1 + j^* \cdot (\min(\max(M), \alpha) + 1)$ oracle queries are performed. The memory cost is approximately as much as a monodromy tree of type V. In expectation, Algorithm 4 has complexity*

$$\text{Space:} \ \mathcal{O}(\max(n(d/\alpha), n \ln(d))) \qquad \text{Time:} \ \mathcal{O}(d \min(\lambda d, \alpha)).$$

Proof. Algorithm 4 follows Algorithm 1 with the modifications required to ascertain whether the solution obtained in line 6 of Algorithm 1 is "new" (using Algorithm 3) and to iterate s_j (using Algorithm 2). Otherwise, the only other additional operations and storage requirements come from various book-keeping steps and accumulating the functions $\mathcal{G}$.

The lines 6, 15, 16, 21 contribute to the oracle queries σ_0 or σ_1, as indicated by $\star_0$ and $\star_1$ decorations, respectively. Line 6 queries σ_0 exactly $d - 1$ times. Lines 15, 16, and 22 are performed j^* times. Only line 16 uses more than one query: it uses at most $\min(\max(M), \alpha) - 1$, totalling

$$\underbrace{d-1}_{\sigma_0 \text{ on } 6} + \underbrace{j^*}_{\sigma_1 \text{ on } 15} + \underbrace{j^*(\min(\max(M), \alpha) - 1)}_{\sigma_0 \text{ on } 16} + \underbrace{j^*}_{\sigma_0 \text{ on } 21} = d - 1 + j^*(1 + \min(\max(M), \alpha))$$

In expectation, this is $\mathcal{O}(d \min(\lambda d, \alpha))$. If α is less than $\frac{d}{\ln(d)}$, then the expected space-time complexity estimates of Algorithm 4 become $\mathcal{O}(n(d/\alpha))$ and $\mathcal{O}(d\alpha)$ respectively, and so Algorithm 4 gives a space-time trade-off of a factor of α compared to Algorithm 1, i.e. when $\alpha = 1$.

Acknowledgement. Supported by an NSERC Discovery grant (RGPIN-2023-03551).

References

1. Bates, D.J., Breiding, P., Chen, T., Hauenstein, J.D., Leykin, A., Sottile, F.: Numerical nonlinear algebra (2023)
2. Brysiewicz, T., Fevola, C., Sturmfels, B.: Tangent quadrics in real 3-space. Matematiche (Catania) **76**, 355–367 (2021)
3. Diaconis, P.: Group Representations in Probability and Statistics. IMS Lecture Notes. Institute of Mathematical Statistics (1988)
4. Dixon, J.D.: The probability of generating the symmetric group. Math. Z. **110**(3), 199–205 (1969)
5. Duff, T., Hill, C., Jensen, A., Lee, K., Leykin, A., Sommars, J.: Solving polynomial systems via homotopy continuation and monodromy. IMA J. Numer. Anal. **39**(3), 1421–1446 (2019)
6. Golomb, S.W.: Random permutations. Bull. Amer. Math. Soc. **70**, 747 (1964)
7. Harris, J.: Galois groups of enumerative problems. Duke Math. J. **46**(4), 685–724 (1979)
8. Hauenstein, J.D., Rodriguez, J.I., Sottile, F.: Numerical computation of Galois groups. Found. Comput. Math. **18**(4), 867–890 (2018)

9. Brysiewicz, T., Burr, M.: Sparse trace tests. Math. Comp. **92**, 2893–2922 (2023)
10. Leykin, A., Rodriguez, J.I., Sottile, F.: Trace test. Arnold Math. J. **4**(1), 113–125 (2018)
11. Sommese, A.J., Verschelde, J., Wampler, C.W.: Introduction to numerical algebraic geometry. In: Solving Polynomial Equations. Algorithms and Computation in Mathematics, vol. 14, pp. 301–335. Springer, Berlin (2005). https://doi.org/10.1007/3-540-27357-3_8
12. Sottile, F., Yahl, T.: Galois groups in enumerative geometry and applications. preprint arXiv:2108.07905 (2021)

Effective Alpha Theory Certification Using Interval Arithmetic: Alpha Theory over Regions

Kisun Lee

Clemson University, Clemson, SC 29634, USA
kisunl@clemson.edu
https://klee669.github.io

Abstract. We reexamine Smale's alpha theory as a way to certify a numerical solution to an analytic system. For a given point and a system, Smale's alpha theory determines whether Newton's method applied to this point shows the quadratic convergence to an exact solution. We introduce the alpha theory computation using interval arithmetic to avoid costly exact arithmetic. As a straightforward variation of the alpha theory, our work improves computational efficiency compared to software employing the traditional alpha theory.

Keywords: Newton's method · numerical certification · analytic system · polynomial equations · alpha theory · interval arithmetic

1 Introduction

The primary focus of this paper is to *certify* a numerical solution to an analytic system $F : U \to \mathbb{C}^n$ defined in an open set $U \subset \mathbb{C}^n$. Certifying a solution is determining if a given point can be refined to an exact solution up to arbitrary precision. *Newton's method* is a widely recognized method for this task. For an analytic system F, we define the *Newton operator* $N_F(x)$ for F by

$$N_F(x) = \begin{cases} x - JF(x)^{-1}F(x) & \text{if } JF(x) \text{ is invertible,} \\ x & \text{otherwise.} \end{cases}$$

This operator is applied iteratively at a point $x \in \mathbb{C}^n$ to approximate the exact solution $x^\star$ of F. Especially, when the point x is "close" to $x^\star$, it is known that x shows the quadratic convergence to $x^\star$. Extensive research on the convergence of Newton's method has led to the development of *Smale's alpha theory* [1, Chapter 8], which establishes criteria for a point and a system to show the quadratic convergence of Newton's method.

A drawback of employing the alpha theory for certification lies in the necessity for computationally expensive exact arithmetic to ensure its rigor. Although the Krawczyk method [2,3,10–12] utilizes faster arithmetic, it does not always guarantee the quadratic convergence.

This paper introduces the *alpha theory over regions*. It executes the alpha theory computation with an interval (vector) input containing a candidate solution to enable efficient interval arithmetic. By developing the alpha theory over regions, our work introduces novel aspects; the enhancement of certification speed for numerical solutions, surpassing the method outlined in [9].

The structure of the paper is as follows. In Sect. 2, we review the concepts of Smale's alpha theory and interval arithmetic. In Sect. 3, the alpha theory over regions is introduced, which is the main result of the paper. Some remarks for implementing the alpha theory over regions are discussed in Sect. 4. The experimental results are provided in Sect. 5.

2 Preliminaries

We introduce the concepts needed for the alpha theory over regions. Initially, we discuss the alpha theory, followed by a review of interval arithmetic.

2.1 Smale's Alpha Theory

Let $F : U \to \mathbb{C}^n$ be an analytic system defined in an open set $U \subset \mathbb{C}^n$. Then, for a point $x \in \mathbb{C}^n$, the k-th Newton iteration $N_F^k(x)$ is defined by applying the Newton operator k times at x. For an exact solution $x^\star$ to the system F, suppose that we have $\|N_F^k(x) - x^\star\| \leq \left(\frac{1}{2}\right)^{2^k - 1} \|x - x^\star\|$ for every $k \in \mathbb{N}$. Then, we say that x *converges quadratically* to $x^\star$, and x is called an *approximate solution* to F with the *associated solution* $x^\star$. In other words, certifying x means proving x is an approximate solution to F associated with $x^\star$.

The alpha theory exploits three values obtained from x and F. If the Jacobian $JF(x)$ is invertible, we define

$$\alpha(F, x) := \beta(F, x)\gamma(F, x) \qquad \gamma(F, x) := \sup_{k \geq 2} \left\| \frac{JF(x)^{-1} J^k F(x)}{k!} \right\|^{\frac{1}{k-1}}$$
$$\beta(F, x) := \|x - N_F(x)\| = \|JF(x)^{-1} F(x)\|$$

where $J^k F(x)$ is a symmetric tensor whose components are k-th order partial derivative of F. The value $\beta(F, x)$ is the Euclidean norm of Newton step for F at x. The norm used in $\gamma(F, x)$ is the operator norm for $JF(x)^{-1} J^k F(x)$ which is induced from the norm on the k-fold symmetric power $S^k \mathbb{C}^n$ of $\mathbb{C}^n$. When $JF(x)^{-1}$ is not invertible, we define $\alpha(F, x) = \beta(F, x) = \gamma(F, x) = \infty$. Results of the alpha theory are like the following:

Theorem 1. *[1, Section 8.2, Theorem 2, Theorem 4 and Remark 6] Let $F : U \to \mathbb{C}^n$ be an analytic system, and x be a given point in U. Then,*

1. *if $\alpha(F, x) < \frac{13 - 3\sqrt{17}}{4}$, then x is an approximate solution to F. Moreover, $\|x - x^\star\| \leq 2\beta(F, x)$ where $x^\star$ is the associated solution to x, and*
2. *if $\alpha(F, x) < 0.03$ and $\|x - y\| < \frac{1}{20\gamma(F,x)}$ for some $y \in U$, then x and y are approximate solutions to the same associated solution to F. In addition, there is a unique solution $x^\star$ to F in the ball $B(x, \frac{1}{20\gamma(F,x)})$ centered at x with the radius $\frac{1}{20\gamma(F,x)}$.*

The first part of Smale's alpha theory checks if a point x is an approximate solution to a system F. The second part of the alpha theory identifies when two different points have the same associated solution to F, that is, it certifies distinct numerical solutions.

The most challenging part of implementing the alpha theory is the computation of the gamma value. To resolve this issue, a known approach is to find an upper limit for the gamma value. For a polynomial $f = \sum_{|\nu| \leq d} a_\nu x^\nu$ of degree d, we define the *Bombieri-Weyl norm* $\|f\|^2 = \frac{1}{d!} \sum_{|\nu| \leq d} \nu!(d-|\nu|)!|a_\nu|^2$. This norm extends to a polynomial system $F = \{f_1, \ldots, f_n\}$ with $\|F\|^2 = \sum_{i=1}^n \|f_i\|^2$. For a point $x \in \mathbb{C}^n$, we define $\|(1,x)\|^2 = 1 + \sum_{i=1}^n |x_i|^2$. Let d_i be the degree of each polynomial f_i, and set $D = \max_i \{d_i\}$. Finally, define the diagonal matrix $\Delta_F(x)$ whose diagonal entry is given by $\Delta_F(x)_{ii} := \sqrt{d_i}\|(1,x)\|^{d_i-1}$. Then, the following result provides an upper bound for $\gamma(F,x)$ for a polynomial system F:

Proposition 1. *[15, Section I-3, Proposition 3] Consider a polynomial system $F : \mathbb{C}^n \to \mathbb{C}^n$ and a point $x \in \mathbb{C}^n$ such that $JF(x)$ is invertible. We define $\mu(F,x) := \max\{1, \|F\|\|JF(x)^{-1}\Delta_F(x)\|\}$ with the operator norm for $JF(x)^{-1}\Delta_F(x)$. Then, $\gamma(F,x) \leq \frac{\mu(F,x)D^{\frac{3}{2}}}{2\|(1,x)\|}$.*

Upper bounds for the gamma value have been studied in various instances of systems of equations. In the case of systems involving polynomial-exponential equations, [7] accomplished this. Additionally, for a broader range of systems, [3] introduced an upper bound for the gamma value when dealing with systems with univariate D-finite functions.

There are known implementations of the alpha theory. For standalone software, `alphaCertified` [8] is used. On the other hand, the `Macaulay2` package `NumericalCertification` [11] provides a specialized implementation for computation in `Macaulay2` [6].

2.2 Interval Arithmetic

Interval arithmetic introduces operations between intervals to perform conservative computations to produce a certified result. For example, for two intervals $[a,b]$ and $[c,d]$ over $\mathbb{R}$, and an arithmetic operation $\odot$, we define $[a,b] \odot [c,d] := \{x \odot y \mid x \in [a,b], y \in [c,d]\}$. For explicit formulas for the standard arithmetic operations (e.g. $+, -, \cdot, /$), see [12]. Interval arithmetic can be extended to $\mathbb{C}$ by introducing an interval with real and imaginary parts, that is, $I = \Re(I) + i\Im(I)$.

For an interval I in $\mathbb{R}$, we define the minimum absolute value over the points in I by $\lfloor I \rfloor = \min_{x \in I} |x|$. Consider an interval vector $I = (I_1, \ldots, I_n)$ in $\mathbb{R}^n$. We define $\lfloor\lfloor I\rfloor\rfloor^2 = \sum_i \lfloor I_i\rfloor^2$. Note that $\lfloor\lfloor I\rfloor\rfloor$ is the minimum of 2-norms over all points in I, and it can be extended to intervals in $\mathbb{C}$ naturally. For a function $F : \mathbb{C}^n \to \mathbb{C}^m$ and an interval vector I in $\mathbb{C}^n$, we define an *interval closure* $\Box F(I)$ of F over I which is a set containing $\{F(x) \mid x \in I\}$. Usually, the smallest interval in $\mathbb{C}^m$ that contains $\{F(x) \mid x \in I\}$ is used for $\Box F(I)$. For a function F that consists of elementary functions (e.g. polynomials), the interval closure is obtained by interval arithmetic. Note that the interval closure of a function is not unique in general.

Lastly, we consider an *interval matrix*, a matrix with interval entries. For an $n \times n$-interval matrix M, we say that an $n \times n$-interval matrix is an *inverse interval matrix* of M if it contains the set $\{N^{-1} \mid N \in M\}$, which is denoted by M^{-1}. Note that M^{-1} may not be unique. We discuss how to compute the inverse interval matrix in Sect. 4.2.

3 The Alpha Theory over Regions

The goal of this section is to extend the results from Sect. 2.1 to the case when the input is given by an interval vector rather than a point. Let $F : U \to \mathbb{C}^n$ be an analytic system defined in an open set $U \subset \mathbb{C}^n$, and I be an interval vector that is contained in U. Then, we define the three values given by F and I like the following:

$$\begin{aligned} \alpha(F, I) &:= \beta(F, I)\gamma(F, I) \\ \beta(F, I) &:= \max_{x \in I} \beta(F, x) \\ \gamma(F, I) &:= \max_{x \in I} \gamma(F, x) \end{aligned}.$$

If $JF(x)$ is not invertible at some point $x \in I$, we define $\alpha(F, I) = \beta(F, I) = \gamma(F, I) = \infty$.

We state the results of the alpha theory over regions.

Theorem 2. *Let $F : U \to \mathbb{C}^n$ be an analytic system, and I be an interval vector in U. Then,*

1. *if $\alpha(F, I) < \frac{13-3\sqrt{17}}{4}$, then all points in I are approximate solutions to F with the associated solution x^* that is contained in $\bigcap_{x \in I} B(x, 2\beta(F, I))$, and*

2. *if $\alpha(F, I) < 0.03$, then all points in $\bigcup_{x \in I} B(x, \frac{1}{20\gamma(F,I)})$ are approximate solutions to the same associated solution to F.*

Proof. 1. We begin by noting $\alpha(F, x) \leq \alpha(F, I)$ for any point $x \in I$. Thus, $\alpha(F, I) < \frac{13-3\sqrt{17}}{4}$ implies that $\alpha(F, x) < \frac{13-3\sqrt{17}}{4}$, and hence, any point $x \in I$ is an approximate solution to F by Theorem 1. As $\alpha(F, I) < \infty$, we have the continuity of the k-th Newton iteration $N_F^k(x)$ over I for all k. Hence, all points in I must converge to the same associated solution x^*. Since x^* is contained in $B(x, 2\beta(F, I))$ for each point $x \in I$, we have that x^* is contained in $\bigcap_{x \in I} B(x, 2\beta(F, I))$.

2. Since $\alpha(F, I) < 0.03$, for any point $x \in I$, all points in $B(x, \frac{1}{20\gamma(F,I)})$ are approximate solutions. By the first part of this theorem, all points in I have the same associated solution to F so that the result is proved. □

Since the alpha theory over regions certifies all points in a certain region at once, the process of certifying distinct solutions is more relaxed than the known method in [9]. We introduce it as the following corollary:

Corollary 1. *Let $F : U \to \mathbb{C}^n$ be an analytic system, and I_1 and I_2 be interval vectors in U with $\alpha(F, I_1) < \frac{13 - 3\sqrt{17}}{4}$ and $\alpha(F, I_2) < \frac{13 - 3\sqrt{17}}{4}$. Then, if $I_1 \cap I_2 \neq \emptyset$, then I_1 and I_2 have the same associated solution to F. On the other hand, if $\mathrm{dist}(I_1, I_2) > 2\beta(F, I_1) + 2\beta(F, I_2)$, then I_1 and I_2 have different associated solutions to F. Here, $\mathrm{dist}(I_1, I_2) = \min\{\|x_1 - x_2\| \mid x_1 \in I_1, x_2 \in I_2\}$.*

Proof. The first part is clear by applying Theorem 2 (1) on I_1 and I_2. The second part follows from the fact that the associated solution of I_i is contained in $B(x_i, 2\beta(F, I_i))$ for any point $x_i \in I_i$. □

Note that the corollary allows a larger alpha value than that of Theorem 1 (2). Hence, it can be used for certifying distinct solutions with a more relaxed condition.

Finally, for the case with a polynomial system, we state the interval version of Proposition 1. For a polynomial system $F : \mathbb{C}^n \to \mathbb{C}^n$ and an interval vector I, we define $\mu(F, I) := \max_{x \in I} \mu(F, x)$. Then, the proposition below provides an upper bound for $\gamma(F, I)$.

Proposition 2. *Consider a polynomial system $F = \{f_1, \ldots, f_n\}$, and an interval vector $I = (I_1, \ldots, I_n)$ in $\mathbb{C}^n$. Then, $\gamma(F, I) \leq \frac{\mu(F,I) D^{\frac{3}{2}}}{2\lfloor\lfloor(1,I)\rfloor\rfloor}$ where $(1, I) = ([1, 1] + i[0, 0], I_1, \ldots, I_n)$ is the interval vector in $\mathbb{C}^{n+1}$.*

Proof. Consider any point in $x \in I$. Then,

$$\gamma(F, x) \leq \frac{\mu(F, x) D^{\frac{3}{2}}}{2\|(1, x)\|} \leq \frac{\mu(F, I) D^{\frac{3}{2}}}{2\|(1, x)\|} \leq \frac{\mu(F, I) D^{\frac{3}{2}}}{2\lfloor\lfloor(1, I)\rfloor\rfloor}.$$

Since x is an arbitrary point, the result follows. □

In the actual implementation of alpha theory over regions, we use interval closures $\Box\beta(F, I)$ and $\Box\mu(F, I)$ instead of $\beta(F, I)$ and $\mu(F, I)$. Computing these interval closures requires the computation of the interval matrix inverse due to $\Box JF(I)^{-1}$.

The expected usage of the alpha theory over regions is to replace the usual alpha theory. For a given numerical solution, we apply the alpha theory on an interval vector containing this numerical solution with preferably a small radius.

4 Implementation Details

This section points out remarks when implementing the alpha theory over regions. The first part is about the use of the interval arithmetic library MPFI [14] with arbitrary precision. Secondly, we discuss a special type of interval arithmetic for computing a tight inverse interval matrix using LU decomposition.

4.1 MPFI for Arbitrary Precision Interval Arithmetic

One of the typical ways to get input for the alpha theory over regions is by constructing an interval vector with a certain radius from a given candidate solution of a system. To execute reliable computation, one may require high precision for interval arithmetic. MPFI is a library written in C for arbitrary precision interval arithmetic using MPFR [4]. The purpose of using MPFI is to achieve guaranteed computation results without losing accuracy from the rounding error. The comparison of alpha values according to the change of precision is presented in Sect. 5.2. One possible drawback of using high precision is, however, that as the precision used in the computation increases, the speed of the calculation may decrease. The comparison of elapsed time between machine precision and MPFI is presented in Sect. 5.3.

4.2 Inverse Interval Matrix Computation via LU Decomposition

For computing the inverse interval matrix, LU decomposition may be considered for efficiency compared to other methods (e.g. cofactor expansion). We desire a tight inverse interval matrix since an unnecessarily large inverse makes the alpha value greater than what it could be. To achieve this, we introduce interval arithmetic in a special type.

For an interval I in $\mathbb{C}$, we define its *dual interval* that is denoted by $I^\star$. This dual interval $I^\star$ has the same endpoints of I with the following arithmetic:

$$I + (-I^\star) = I - I^\star = [0,0],$$
$$I \times \frac{1}{I^\star} = \frac{I}{I^\star} = [1,1] \quad \text{if } 0 \notin I.$$

Also, we define $(I^\star)^\star = I$ to make the addition and multiplication commutative. Including the dual intervals introduces *Kaucher arithmetic* with a broader collection of intervals than that of usual interval arithmetic. It has a more algebraic structure than the usual interval arithmetic while it executes the conservative computation. More specifically, the set of intervals with dual intervals is a group in addition, and the set of intervals not containing zero with dual intervals is a group in multiplication. The more general version of this interval arithmetic is introduced in [5]. The interval arithmetic with dual intervals is used for partial pivoting for LU decomposition. For other subtractions and divisions that occur except for pivoting, the usual interval arithmetic is used.

We briefly elaborate on how to compute the inverse interval matrix. For an $n \times n$ interval matrix M and an n-dimensional interval vector B, consider an interval linear system $M\mathbf{x} = B$. Solving this system using the interval LU decomposition of M, we have an interval vector $\mathbf{x}$ satisfying only $B \subseteq M\mathbf{x}$ in general (See [13, Section 4.5], for example). Using this fact, iterative solving of interval linear systems returns an inverse interval matrix M^{-1} (that is, it returns a set of interval matrices containing $\{N^{-1} \mid N \in M\}$). In particular, using interval arithmetic with dual intervals may return a tighter M^{-1} than the usual interval arithmetic.

5 Experiments

This section provides computational and experimental results as a proof of concept for the alpha theory over regions. The implementation is in C++ into two versions, one with double machine precision (in a correct rounding manner for reliable computation), and the other with arbitrary precision using MPFI [14]. It computes alpha, beta, and gamma values from a given square polynomial system and a point. All computations in this section are executed in a Macbook M2 pro 3.5 GHz, 16 GB RAM. The code and examples are available at

https://github.com/klee669/alphaTheoryOverRegions

5.1 Alpha Values According to the Radius of the Input Interval

We analyze the impact of the radius of the interval vector on alpha values. It is expected that as the size of the interval increases, the alpha value will also increase. To check this, we consider the cyclic system with 6 variables; $f_l = \sum_{j=1}^{6} \prod_{k=j}^{j+l} x_j$ for $l = 1, \ldots, 5$ and $f_6 = \prod_{j=1}^{6} x_j - 1$ with a numerical solution $a = (a_1, -a_1, a_2, a_3, -a_3, -a_2)$ where $a_1 = .782290 + .622915i$, $a_2 = .866025 - .5i$ and $a_3 = .148315 + .988940i$ whose distance from the nearest exact solution is $8.261221e - 7$. Defining the interval vector centered at a, we compute three constant values using our implementation with double precision while we change the radius of the interval (see Table 1). For reference, values computed by alphaCertified [8] with exact arithmetic are also recorded.

Table 1. The values of alpha, beta, and gamma constants for the cyclic-6 system according to the change of the radius of the input interval. All computations were conducted with double precision. The last row shows the values obtained by the software alphaCertified with exact arithmetic.

radius	alpha	beta	gamma
$1e-5$	$1.15611e+1$	$4.03387e-3$	$2.86602e+3$
$1e-7$	$1.66585e-2$	$1.26115e-5$	$1.32089e+3$
$1e-10$	$1.62002e-5$	$1.23899e-8$	$1.30754e+3$
$1e-15$	$5.75515e-11$	$4.40158e-14$	$1.30752e+3$
$1e-20$	$1.63550e-11$	$1.25084e-14$	$1.30752e+3$
$1e-30$	$1.63550e-11$	$1.25084e-14$	$1.30752e+3$
aC	$1.53040e-13$	$1.17046e-16$	$1.30752e+3$

From the result, each constant gets larger as the radius increases. The improvement in alpha and beta are more noticeable than that of gamma because the beta value is affected by the value of $F(x)$ which is more sensitive to the change of the size of the input interval than the bound for gamma value given in

Proposition 2. Once the radius gets small enough, due to the conservative computation from interval arithmetic, it shows only a slight improvement on three constant values.

5.2 Alpha Values According to Precision

In this section, we explore how precision affects the values. We consider the same cyclic-6 system, and the interval box with the radius $1e - 20$ centered at the solution a used in Sect. 5.1. The comparison of alpha, beta, and gamma values according to the change of precision is given in Table 2.

Table 2. The values of alpha, beta, and gamma constants for the cyclic-6 system according to the change of precision.

precision	alpha	beta	gamma
16	$4.44465e - 2$	$3.39918e - 5$	$1.30757e + 3$
32	$4.42851e - 7$	$3.38696e - 10$	$1.30752e + 3$
64	$2.03482e - 13$	$1.55624e - 16$	$1.30752e + 3$
128	$2.02057e - 13$	$1.54534e - 16$	$1.30752e + 3$
256	$2.02057e - 13$	$1.54534e - 16$	$1.30752e + 3$
double	$1.63550e - 11$	$1.25084e - 14$	$1.30752e + 3$

The result shows that the smaller alpha and beta values are returned as the larger precision is used. The value of gamma does not improve much since the beta value is affected by the value of $F(x)$ which is more sensitive to the change of precision. The changes in all three values become insignificant when the precision higher than 128 is used.

5.3 Time Comparison with `alphaCertified`

We provide a time comparison with the software `alphaCertified`. We experiment with the Fano problem studied in [16]. The Fano problem of type (n, k, d) where $d = (d_1, \ldots, d_l)$ is the problem of finding n-dimensional planes lying in a complete intersection of l hypersurfaces $f_1, \ldots, f_l$ in $\mathbb{P}^k$ with degrees $\deg f_1 = d_1, \ldots, \deg f_l = d_l$. Fano problems can be described as problems of solving a square polynomial system. For example, Fano problems of $(1, 5, (2, 4))$ is related to a square polynomial system of 8 variables with 1280 solutions (up to multiplicity), and $(1, 8, (2, 2, 2, 4))$ is related to a square polynomial systems of 14 variables with 47104 solutions (up to multiplicity). We find numerical solutions of these two systems with `Macaulay2` expressed in floating point arithmetic, and certify them using our implementation and `alphaCertified` by varying the number of candidate solutions. For our implementation, we compute alpha, beta, and gamma values for each candidate solution using both double precision and

MPFI with 256 precision. For `alphaCertified`, calculations are performed both using exact arithmetic and floating-point arithmetic for comparison even though floating-point arithmetic only provides soft verification. The result is recorded in Table 3.

Table 3. Elapsed time in seconds for certifying solutions for Fano problems of using the implementation of alpha theory over regions, and the software `alphaCertified`. The symbol − means that the computation does not terminate within 2 days.

$(1, 5, (2, 4))$, a square system with 8 variables.					$(1, 8, (2, 2, 2, 4))$, a square system with 14 variables.				
#sols	double	256 prec.	aC exact	aC float	#sols	double	256 prec.	aC exact	aC float
20	.06	.67	92.92	.16	20	.41	7.78	8743.73	1.51
50	.11	1.59	242.00	.27	50	.86	18.26	22500.32	2.28
200	.31	6.13	1067.02	.84	200	3.07	72.10	89455.63	6.73
1000	1.39	32.83	7649.28	3.92	1000	14.73	400.51		27.61

The result shows that the alpha theory over regions shows less elapsed time on computation than `alphaCertified` with exact arithmetic. The implementation with 256 precision may take more time than `alphaCertified` with floating point arithmetic, but it returns more reliable results than computation with floating point arithmetic. The implementation with double precision takes less time than that with MPFI.

Note that comparing the elapsed time of two software might not be fair since `alphaCertified` performs further analysis to classify distinct solutions. Nonetheless, the result shows the potential of the alpha theory over regions as it produces reliable results in a significantly shorter time than the computation with exact arithmetic.

Acknowledgement. We would like to thank Michael Burr and Thomas Yahl for helpful discussions. We also thank the anonymous referees for their constructive comments.

Disclosure of Interests. The authors have no competing interests to declare that are relevant to the content of this article.

References

1. Blum, L., Cucker, F., Shub, M., Smale, S.: Complexity and Real Computation. Springer, Berlin (2012). https://doi.org/10.1007/978-1-4612-0701-6
2. Breiding, P., Rose, K., Timme, S.: Certifying zeros of polynomial systems using interval arithmetic. ACM Trans. Math. Softw. **49**(1), 1–14 (2023)
3. Burr, M., Lee, K., Leykin, A.: Effective certification of approximate solutions to systems of equations involving analytic functions. In: Proceedings of the 2019 on International Symposium on Symbolic and Algebraic Computation, pp. 267–274 (2019)

4. Fousse, L., Hanrot, G., Lefèvre, V., Pélissier, P., Zimmermann, P.: MPFR: a multiple-precision binary floating-point library with correct rounding. ACM Trans. Math. Softw. **33**(2), 13'es (2007). https://doi.org/10.1145/1236463.1236468
5. Goldsztejn, A., Chabert, G.: A generalized interval LU decomposition for the solution of interval linear systems. In: Boyanov, T., Dimova, S., Georgiev, K., Nikolov, G. (eds.) NMA 2006. LNCS, vol. 4310, pp. 312–319. Springer, Heidelberg (2007). https://doi.org/10.1007/978-3-540-70942-8_37
6. Grayson, D.R., Stillman, M.E.: Macaulay2, a software system for research in algebraic geometry. http://www2.macaulay2.com
7. Hauenstein, J.D., Levandovskyy, V.: Certifying solutions to square systems of polynomial-exponential equations. J. Symb. Comput. **79**, 575–593 (2017)
8. Hauenstein, J.D., Sottile, F.: alphaCertified: software for certifying numerical solutions to polynomial equations (2011). https://math.tamu.edu/~sottile/research/stories/alphaCertified
9. Hauenstein, J.D., Sottile, F.: Algorithm 921: alphaCertified: certifying solutions to polynomial systems. ACM Trans. Math. Softw. (TOMS) **38**(4), 1–20 (2012)
10. Krawczyk, R.: Newton-algorithm zur Bestimmung von Nullstellen mit Fehleshranken. Computing **4**, 187–201 (1969)
11. Lee, K.: Certifying approximate solutions to polynomial systems on macaulay2. ACM Commun. Comput. Algebra **53**(2), 45–48 (2019)
12. Moore, R.E., Kearfott, R.B., Cloud, M.J.: Introduction to interval analysis. SIAM (2009)
13. Neumaier, A.: Interval Methods for Systems of Equations. No. 37, Cambridge University Press, Cambridge (1990)
14. Revol, N., Rouillier, F.: Motivations for an arbitrary precision interval arithmetic and the MPFI library. Reliab. Comput. **11**(4), 275–290 (2005)
15. Shub, M., Smale, S.: Complexity of Bézout's theorem. I. Geometric aspects. J. Am. Math. Soc. **6**(2), 459–501 (1993)
16. Yahl, T.: Computing Galois groups of Fano problems. J. Symb. Comput. **119**, 81–89 (2023)

Gröbner Degenerations of Determinantal Ideals with an Application to Toric Degenerations of Grassmannians

Fatemeh Mohammadi[✉][iD]

Department of Mathematics and Computer Science, KU Leuven, Leuven, Belgium
fatemeh.mohammadi@kuleuven.be

Abstract. The concept of the Gröbner fan for a polynomial ideal, introduced by Mora and Robbiano in 1988, provides a robust polyhedral framework where maximal cones correspond to the reduced Gröbner bases of the ideal. Within this geometric structure resides the tropical variety, a subcomplex of the Gröbner fan, utilized across various mathematical domains. Despite its significance, the computational complexity associated with tropical varieties often limits practical computations to smaller instances. In this note, we revisit a family of monomial ideals, called matching field ideals, from the context of Gröbner degenerations. We show that they can be obtained as weighted initial ideals of determinantal ideals. We explore the algebraic properties of these ideals, with a particular emphasis on minimal free resolutions and Betti numbers.

Keywords: Gröbner degeneration · Determinantal ideals · Grassmannians · Betti numbers · Cellular resolution

1 Introduction

A Gröbner degeneration refers to a process in algebraic geometry where a given ideal or variety is systematically deformed to more well-behaved ideals such as monomial or toric ideals. It produces a parametric family of ideals, preserving the geometric and algebraic properties of the original ideal. Overall, Gröbner degenerations provide a powerful tool for understanding the invariants of ideals and varieties under perturbations or deformations.

Monomial degenerations play a crucial role in simplifying the structure and analysis of ideals such as their minimal free resolutions [15,21]. For example, the semi-continuity theorem provides a bound on the Betti numbers of any ideal in terms of those of their monomial initial ideals. In this note, we will study a family of monomial degenerations of determinantal ideals [4,13,32].

Toric varieties are another family of popular objects in algebraic geometry; see e.g. [14,19]. This is mainly because there is a dictionary between their geometric properties and the combinatorial features of their polytopes. This dictionary can be extended from toric varieties to arbitrary varieties through toric degenerations. The connection between toric varieties and Newton polytopes was first developed by Khovanskii in [23].

A toric degeneration [1] of an algebraic variety X is a flat family over the affine line $\mathbb{A}^1$, where the fiber over zero is a toric variety and all the other fibers are isomorphic to X. Toric degenerations serve as useful tools for analyzing algebraic varieties, providing a means to comprehend a general variety through the geometry of its associated toric counterparts, as many geometric invariants (e.g. degree, dimension) remain invariant under such transformations. Hence, computing toric degenerations of general varieties allows us to use the robust machinery of toric varieties to understand the invariants of the original varieties. The primary focus when computing toric degenerations revolve around two pivotal questions: (1) How do we construct them? and (2) How are distinct toric degenerations of the same variety interconnected? Given the paramount importance of these questions, a plethora of methodologies have emerged across diverse fields such as algebraic geometry (e.g. [1,17,20]), representation theory [18], cluster algebras [6,30], tropical geometry [5,9,10,22,28], and combinatorics [7,8,12]. However, despite the progress made in these domains, the challenge remains: there are currently no known algorithms tailored for computing toric degenerations of a given variety or for facilitating comparisons between disparate cases from different theoretical frameworks.

In this note, we study a specific family of Gröbner degenerations, focusing on determinantal ideals and Grassmannian varieties $\mathrm{Gr}(3,n)$. We demonstrate that such degenerations are intricately tied to weight vectors and manifest either as monomial initial ideals or maximal cones within the Gröbner fan of the determinantal ideal. Moreover, we introduce a monomial map, where the kernel materializes as a binomial prime ideal, thereby signifying a toric degeneration of the Plücker ideal of $\mathrm{Gr}(3,n)$. More specifically, we identify matching field ideals as weighted monomial initial ideals of determinantal ideals, showcasing their linear quotient property. Further, we explicitly compute their Betti numbers and establish that their minimal free resolution is supported on a CW-complex, or equivalently, demonstrating their cellular resolution.

2 Gröbner Degeneration of Determinantal Ideals

Let $X = (x_{ij})$ represent a generic $d \times n$ matrix, and let I_X denote the ideal generated by the maximal minors of X. For simplicity, we focus on the case where $d = 3$ and we denote the variables corresponding to the first, second, and third rows of X as x, y, and z, respectively.

We recall the notion of block diagonal matching fields from [28]. See also [4,32]. Consider a sequence $a = (a_1, a_2, \ldots, a_r)$ of positive numbers $a_1, a_2, \ldots, a_r$ such that $\sum_{i=1}^{r} a_i = n$. For $1 \leq t \leq r$, define $I_t = \{\alpha_{t-1} + 1, \alpha_{t-1} + 2, \ldots, \alpha_t\}$, where $\alpha_t = \sum_{i=1}^{t} a_i$ and $\alpha_0 = 0$. We construct the ideal $M_{\mathbf{a}}$ as follows: For every 3-subset $\{i < j < k\}$ of $[n]$, we let s be the minimal t such that $I_t \cap \{i,j,k\} \neq \emptyset$. Then we associate the monomial

$$m_{\{ijk\}} = \begin{cases} x_j y_i z_k & \text{if } |\{i,j,k\} \cap I_s| = 1 \\ x_i y_j z_k & \text{otherwise} \end{cases}$$

The ideal $M_{\mathbf{a}}$ is generated by the monomials associated to all 3-subsets of $[n]$.

Definition 1. *To every r-block diagonal matching field of type $(a_1,\ldots,a_r)$, we associate a weight order $\prec_{w_{\mathbf{a}}}$ as follows. Let us denote $w_{\mathbf{a}}(x_i), w_{\mathbf{a}}(y_j), w_{\mathbf{a}}(z_k)$ for the weights associated to x_i, y_j, z_k for all i, j, k. Then we associate the following weights to the variables:*

- $w_{\mathbf{a}}(x_i) = w_0 \quad \text{for all } i$
- $w_{\mathbf{a}}(y_j) = \begin{cases} w_{\mathbf{a}}(y_{\alpha_{s-1}+1}) + j - (\alpha_{s-1}+1), & \text{if } j \in I_s \text{ and } j > \alpha_{s-1}+1 \\ w_{\mathbf{a}}(y_{\alpha_s+1}) + 1, & \text{if } j \in I_s, j = \alpha_{s-1}+1 \text{ and } s < r \\ w_0 + 1 & \text{if } j = \alpha_{r-1}+1. \end{cases}$
- $w_{\mathbf{a}}(z_1) = w_{\mathbf{a}}(z_2) = w_0$ and $w_{\mathbf{a}}(z_i) = w_{\mathbf{a}}(y_{\alpha_1}) + (n-2)(i-2)$ for all $3 \leq i \leq n$.

We let $\prec_w$ be the monomial order such that for two monomials m, m' we have that $m < m'$ if the associated weight of m is less than that of m' or they have the same weight and $m < m'$ with respect to the revlex order induced by:

$$z_n > \cdots > z_3 > \underbrace{y_{\alpha_1} > \cdots > y_{\alpha_0+1}}_{\text{coming from } I_1} > \cdots > \underbrace{y_{\alpha_r} > \cdots > y_{\alpha_{r-1}+1}}_{\text{coming from } I_r} > z_2 > z_1 > x_1 > \cdots > x_n.$$

Example 1. Let $n = 7, \mathbf{a} = [12|345|67], w_0 = 1$. The order on the variables is

$$z_7 > z_6 > \cdots > z_3 > y_2 > y_1 > y_5 > y_4 > y_3 > y_7 > y_6 > z_2 > z_1 > x_1 > \cdots > x_7$$

The weight on the variables (represented in a matrix) is given as:

$$w_{\mathbf{a}} = \begin{bmatrix} 1 & 1 & 1 & 1 & 1 & 1 & 1 \\ 7 & 8 & 4 & 5 & 6 & 2 & 3 \\ 1 & 1 & 13 & 18 & 23 & 28 & 33 \end{bmatrix}.$$

This induces $\prec_{w_{\mathbf{a}}}$ weight order and $M_{\mathbf{a}} = \text{in}_{\prec_{w_{\mathbf{a}}}}(I)$.

Theorem 1. *$M_{\mathbf{a}}$ is the initial ideal of the determinantal ideal I_X w.r.t. $\prec_{w_{\mathbf{a}}}$. In particular, the maximal minors form a Gröbner basis for I_X w.r.t. $\prec_{w_{\mathbf{a}}}$.*

Proof. We first show that $M_{\mathbf{a}} \subseteq \text{in}_{\prec_{w_{\mathbf{a}}}}(I_X)$. For every 3-subset $\{i < j < k\}$ of $[n]$, let s be the minimal t such that $I_t \cap \{i, j, k\} \neq \emptyset$. Consider the minor $[ijk]$ of X on the columns i, j, k. Note that:

$$[ijk] = \sum_{\sigma \in \text{Sym}(i,j,k)} \text{sign}(\sigma) x_{\sigma(i)} y_{\sigma(j)} z_{\sigma(k)} \text{ and } w_{\mathbf{a}}(x_i y_j z_k) = w_{\mathbf{a}}(x_i) + w_{\mathbf{a}}(y_j) + w_{\mathbf{a}}(z_k).$$

To show that $m_{\{ijk\}}$ is the initial term of $[ijk]$ w.r.t. $\prec_{w_{\mathbf{a}}}$, we consider two cases:

Case 1. Let $|\{i, j, k\} \cap I_s| > 1$. In this case $m_{\{ijk\}} = x_i y_j z_k$. It is enough to show that $w_{\mathbf{a}}(x_i y_j z_k) - w_{\mathbf{a}}(x_{\sigma(i)} y_{\sigma(j)} z_{\sigma(k)}) > 0$ for $\sigma \neq \text{id}$. If $\sigma(k) \neq k$, we know that $k > \sigma(k)$. Then

$$w_{\mathbf{a}}(x_i y_j z_k) - w_{\mathbf{a}}(x_{\sigma(i)} y_{\sigma(j)} z_{\sigma(k)}) \geq (n-2)(k - \sigma(k)) + w_{\mathbf{a}}(y_j) - w_{\mathbf{a}}(y_{\sigma(j)}).$$

It follows by definition of $\prec_{w_\mathbf{a}}$ that $|w_\mathbf{a}(y_j) - w_\mathbf{a}(y_{\sigma(j)})| < n-2$ and $w_\mathbf{a}(x_i y_j z_k) - w_\mathbf{a}(x_{\sigma(i)} y_{\sigma(j)} z_{\sigma(k)}) > 0$. If $\sigma(k) = k$, then $\sigma(j) = i$ as $\sigma \neq \mathrm{id}$. Hence,

$$w_\mathbf{a}(x_i y_j z_k) - w_\mathbf{a}(x_{\sigma(i)} y_{\sigma(j)} z_{\sigma(k)}) = w_\mathbf{a}(y_j) - w_\mathbf{a}(y_{\sigma(j)}) = w_\mathbf{a}(y_j) - w_\mathbf{a}(y_i).$$

As $|\{i,j,k\} \cap I_s| > 1$, we have $\{i,j\} \subset I_s$ and so $w_\mathbf{a}(y_j) - w_\mathbf{a}(y_i) = j - i > 0$.

Case 2. Let $|\{i,j,k\} \cap I_s| = 1$. Then $m_{\{ijk\}} = x_j y_i z_k$. If $\sigma(k) \neq k$, then:

$$w_\mathbf{a}(x_j y_i z_k) - w_\mathbf{a}(x_{\sigma(j)} y_{\sigma(i)} z_{\sigma(k)}) \geq (n-2)(k - \sigma(k)) + w_\mathbf{a}(y_i) - w_\mathbf{a}(y_{\sigma(i)}).$$

We know that $k > \sigma(k)$, hence $|w_\mathbf{a}(y_i) - w_\mathbf{a}(y_{\sigma(i)})| < n - 2$. (Here $|.|$ denote the absolute value.) Therefore, $w_\mathbf{a}(x_j y_i z_k) - w_\mathbf{a}(x_{\sigma(j)} y_{\sigma(i)} z_{\sigma(k)}) > 0$. If $\sigma(k) = k$, then $\sigma(i) = j$ as $\sigma \neq \mathrm{id}$. Hence,

$$w_\mathbf{a}(x_j y_i z_k) - w_\mathbf{a}(x_{\sigma(j)} y_{\sigma(i)} z_{\sigma(k)}) = w_\mathbf{a}(y_i) - w_\mathbf{a}(y_{\sigma(i)}) = w_\mathbf{a}(y_i) - w_\mathbf{a}(y_j).$$

As $|\{i,j,k\} \cap I_s| = 1$, we have that $i \in I_s$ and $j \in I_t$ for some $t > s$ and it follows from the definition of $\prec_{w_\mathbf{a}}$ that $w_\mathbf{a}(y_i) - w_\mathbf{a}(y_j) > 0$, which completes the proof.

By [13, Theorem 1.1], the maximal minors of X form a universal Gröbner basis for I_X, hence $M_\mathbf{a} = \mathrm{in}_{\prec_{w_\mathbf{a}}}(I_X)$. This also implies the second claim.

2.1 Connection to Toric Degenerations of Grassmannians

The Grassmannian $\mathrm{Gr}(3, n)$ is the space of 3-dimensional linear subspaces of $\mathbb{C}^n$, which can be embedded into a projective spaces, using Plücker embedding as follows. On the level of rings, consider the polynomial ring S on the Plücker variables p_I, corresponding to the maximal minors of X on the columns indexed by $I = \{i, j, k\}$, and the polynomial ring R on the variables x_i, y_j, z_k of the matrix X. The Plücker embedding is obtained by the map:

$$\phi : S \to R \quad \text{such that} \quad p_I \mapsto \det(X_I).$$

A good candidate for a toric degeneration of $\mathrm{Gr}(k, n)$ is given by deforming ϕ to a monomial map. This is done by sending each *Plücker variable* p_I to one of the summands of $\det(X_I)$. In particular, the monomial Gröbner degeneration of the determinantal ideal I_X provides such a degeneration. More precisely, given a matching field $\mathbf{a}$, we define the map ϕ such that $\phi(p_I) = m_I$. In [28], it is established that the kernel of this map forms a binomial and prime ideal, offering a toric degeneration of $\mathrm{Gr}(3, n)$. Furthermore, Theorem 1 further demonstrates that the matching ideal itself can be derived as a weighted Gröbner degeneration, and varying the weights $w_\mathbf{a}$, we can move through Gröbner fan and generate further toric degenerations, as the kernel of the corresponding monomial map.

For general Grassmannians and flag varieties, there are prototypic examples of toric degenerations which are related to young tableaux, Gelfand-Tsetlin integrable systems, and their polytopes. In the case of the Grassmannian $\mathrm{Gr}(2, n)$, there are many other toric degenerations generalizing this primary example. Namely, any trivalent tree with n number of labeled leaves gives rise to a toric degeneration of $\mathrm{Gr}(2,n)$. The toric variety is governed by the isomorphism type of the trivalent tree [31]. In particular, Trop $\mathrm{Gr}(2,n)$ forms the space of phylogenetic trees. In particular, all such degenerations can be obtained from the matching field ideals; see e.g. [28].

Example 2. Consider the Grassmannian $\mathrm{Gr}(2,4)$, whose corresponding Plücker ideal is $I = \langle p_{12}p_{34} - p_{13}p_{24} + p_{14}p_{23}\rangle$. A toric degeneration of $\mathrm{Gr}(2,4)$ is given by the family $I_t = \langle p_{12}p_{34} - p_{13}p_{24} + tp_{14}p_{23}\rangle$ for $t \in \mathbb{C}$. Setting $t = 1$ we obtain I, and setting $t = 0$ we get the toric ideal $I_0 = \langle p_{12}p_{34} - p_{13}p_{24}\rangle$. The Gröbner fan of I consists of seven cones (three 2-dimensional cones, three rays, and the origin). Moreover, Trop $\mathrm{Gr}(2,n)$ is the subfan of this polyhedral fan on the three rays, all leading to toric varieties; see [2].

2.2 Ordering the Generators of $M_\mathbf{a}$

Here, we further investigate the properties of the matching field ideals $M_\mathbf{a}$, in particular their minimal free resolutions and their Betti numbers. We begin by recalling the linear quotients property from [16,21]. A monomial ideal $M \subset R$ is said to have *linear quotients* if there exists an ordering of the generators $M = \langle m_1, \ldots, m_k \rangle$ such that for each j the colon ideal $\langle m_1, \ldots, m_{j-1}\rangle : m_j$ is generated by a subset of the variables $\{x_1, \ldots, x_n\}$ called the set(m_j), so that $\langle m_1, \ldots, m_{j-1}\rangle : m_j = \langle x_{j_1}, \ldots, x_{j_r}\rangle$. In particular, for each generator m_j, with $j = 1, \ldots, k$, we define set$(m_j) = \{k \in [n] : x_k \in \langle m_1, \ldots, m_{j-1}\rangle : m_j\}$.

To simplify our notation, we use (i,j,k) for the monomial $x_i y_j z_k$. (Note that in this expression we may have either $i < j < k$ or $j < i < k$). For every pair of elements (i,j,k) and (ℓ, u, v) in the ideal $M_\mathbf{a}$ with $j \in I_s$ and $u \in I_t$ we define the block ordering $\prec_\mathbf{a}$ such that

$$(i,j,k) \prec_\mathbf{a} (\ell, u, v) \quad \text{if and only if} \quad \begin{cases} k > v, & \text{or} \\ k = v,\ s < t, & \text{or} \\ k = v,\ s = t,\ j > u, & \text{or} \\ k = v,\ s = t,\ j = u,\ i < \ell \end{cases}$$

Example 3. We illustrate an ordering of the generators of $M_{(4,3)}$. The elements are arranged in a series of successive tableaux from left to right. Within each tableau, the elements are ordered from left to right and from top to bottom.

```
147 247 347 547 647      146 246 346 546
    137 237 537 637          136 236 536      145 245 345
        127 527 627               126 526         135 235       134 234
            517 617                   516             125            124      123
                567
```

For each monomial $x_\ell y_u z_v$, we define $S_\mathbf{a}(\ell, u, v)$ as the set of all monomials in $M_\mathbf{a}$ that precede (ℓ, u, v) in the block ordering $\prec_\mathbf{a}$, differing from (ℓ, u, v) by exactly one coordinate.

Lemma 1. *Consider the ordering $\prec_{\mathbf{a}}$. Then for each $(\ell, u, v) \in M_{\mathbf{a}}$ we have:*

$$|S_{\mathbf{a}}(\ell, u, v)| = \begin{cases} \alpha_s - u + \ell - 1 + n - v, & \text{if } \ell, u \in I_s \text{ for } s < r \\ \ell - 2 + n - v, & \text{if } \ell \in I_s \text{ and } u \in I_t \text{ for } t < s \\ u - 2 + n - v, & \text{if } \ell, u \in I_r. \end{cases}$$

Proof. First note that for all (ℓ, u, v), the elements $\{(\ell, u, v+1), \ldots, (\ell, u, n)\}$ come before (ℓ, u, v) which implies that the variables $z_{v+1}, \ldots, z_n$ are in the $S_{\mathbf{a}}(\ell, u, v)$. Now, we list all elements whose difference with (ℓ, u, v) is either in the first entry or in the second entry.

Case 1. Let $\ell, u \in I_s$. Then the elements coming before (ℓ, u, v) are:

$$\{(\alpha_{s-1}+1, u, v), \ldots, (\ell-1, u, v)\} \cup \{(\ell, u+1, v), \ldots, (\ell, \alpha_s, v)\} \cup \{(\ell, 1, v), \ldots, (\ell, \alpha_{s-1}, v)\}.$$

In particular, we have: $S_{\mathbf{a}}(\ell, u, v) = \{x_{\alpha_{s-1}+1}, \ldots, x_{\ell-1}\} \cup \{y_{u+1}, \ldots, y_{\alpha_s}\} \cup \{y_1, \ldots, y_{\alpha_{s-1}}\} \cup \{z_{v+1}, \ldots, z_n\}$.

Case 2. Let $\ell \in I_s$, $u \in I_t$ and $t < s$. By our construction, the elements listed below are those that precede (ℓ, u, v) (with one exception):

$$\{(\alpha_{s-1}+1, u, v), \ldots, (u-1, u, v)\} \cup \{(\alpha_s+1, u, v), \ldots, (\ell-1, u, v)\}$$

$$\cup \{(\ell, u+1, v), \ldots, (\ell, \alpha_s, v)\} \cup \{(\ell, 1, v), \ldots, (\ell, \alpha_{s-1}, v)\},$$

and we have $S_{\mathbf{a}}(\ell, u, v) = \{x_{\alpha_{s-1}+1}, \ldots, x_{u-1}\} \cup \{x_{\alpha_s+1}, \ldots, x_{\ell-1}\} \cup \{y_{u+1}, \ldots, y_{\alpha_s}\} \cup \{y_1, \ldots, y_{\alpha_{s-1}}\} \cup \{z_{v+1}, \ldots, z_n\}$.

Case 3. Let $\ell, u \in I_r$. Then

$S_{\mathbf{a}}(\ell, u, v) = \{x_{\alpha_{r-1}+1}, \ldots, x_{\ell-1}\} \cup \{y_{\ell+1}, \ldots, y_{u-1}\} \cup \{y_1, \ldots, y_{\alpha_{r-1}}\} \cup \{z_{v+1}, \ldots, z_n\}$.

Theorem 2. *$M_{\mathbf{a}}$ has the linear quotient property with respect to $\prec_{\mathbf{a}}$.*

Proof. It suffices to show that $((i, j, k) : (\ell, u, v)) \subset \langle S_{\mathbf{a}}(\ell, u, v) \rangle$ for any $(i, j, k) \prec_{\mathbf{a}} (\ell, u, v)$. Note that $((i, j, k) : (\ell, u, v)) = \langle f \rangle$ for some monomial f dividing $x_i y_j z_k$, and we need to show that this monomial lies in the ideal $\langle S_{\mathbf{a}}(\ell, u, v) \rangle$.

Case 1. Let $k \neq v$. First note that, by definition of $\prec_{\mathbf{a}}$, we have $k > v$. Thus, by Lemma 1, we have $(\ell, u, k) \in M_{\mathbf{a}}$ and $z_k \in S_{\mathbf{a}}(\ell, u, v)$. Hence, $f \in \langle z_k \rangle$.

Case 2. Suppose $k = v$, but $j \neq u$. If $f = x_i$ or y_j we are done, so assume $f = x_i y_j$. We will show in all cases either $x_i \in S_{\mathbf{a}}(\ell, u, v)$ or $y_j \in S_{\mathbf{a}}(\ell, u, v)$.

First suppose $\ell, u \in I_s$ (and so in particular $\ell < u$). If $j \in I_s$ then $j > u$ and so $\ell < j \in I_s$ and $(\ell, j, n) \prec_{\mathbf{a}} (\ell, u, v)$, hence $y_j \in S_{\mathbf{a}}(\ell, u, v)$. If $j \notin I_s$, then we have $j \in I_t$ for $t < s$ and so $j < \ell, m$. In particular, $(\ell, j, n) \prec_{\mathbf{a}} (\ell, u, v)$ so again $y_j \in S_{\mathbf{a}}(\ell, u, v)$ as desired. Now suppose $\ell \in I_s, u \in I_t$ for $s \neq t$ (so in particular $u < \ell$). If $j \in I_t$ then $j > u$ and since $\ell \in I_s$ we have $\ell > j$ and so $(\ell, j, n) \prec_{\mathbf{a}} (i, j, k)$ and again $y_j \in S_{\mathbf{a}}(\ell, u, v)$. If $j \notin I_t$ then $j < u < \ell$. If $j \notin I_s$ then $(\ell, j, n) \prec_{\mathbf{a}} (i, j, k)$ and again $y_j \in S_{\mathbf{a}}(\ell, u, v)$. Otherwise, if $j, i \in I_s$ then

$i < j$ and $(i, u, v) \prec_{\mathbf{a}} (i, j, k)$ and $x_j \in S_{\mathbf{a}}(\ell, u, v)$. Finally, if $j \in I_s$ but $i \notin I_s$ then $i > j$. Since $j, \ell \in I_s$ we also have $i > \ell > m$ and $(i, u, v) \prec_{\mathbf{a}} (\ell, u, v)$, hence $x_i \in S_{\mathbf{a}}(\ell, u, v)$.

Case 3. Suppose $k = v$ and $j = u$. Then $f = x_i$, which completes the proof.

It is shown in [21] that for any monomial ideal $M = \langle m_1, \ldots, m_k \rangle$ with linear quotient property, the Betti numbers can be explicitly described as follows.

$$\beta_\ell(M) = \sum_{j=1}^{k} \binom{|\mathrm{set}(m_j)|}{\ell}.$$

Example 4. Consider the monomial ideal $M_{\mathrm{Diag}(n)} = \langle x_i y_j z_k : 1 \leq i < j < k \leq n \rangle$. Note that each monomial is representing the diagonal term of the corresponding minor in a $3 \times n$ matrix of indeterminants. Consider a lexicographic ordering on the generators of $M_{\mathrm{Diag}(n)}$. Then

$$\mathrm{set}(x_i y_j z_k) = \{x_1, \ldots, x_{i-1}\} \cup \{y_{i+1}, \ldots, y_{j-1}\} \cup \{z_{j+1}, \ldots, z_{k-1}\}$$

and hence, $|\mathrm{set}(x_i y_j z_k)| = (i-1) + (j-1-i) + (k-1-j) = k-3$.

As an immediate corollary of Theorem 2, and Example 4 we have that:

Corollary 1. *The Betti numbers of the ideal $M_{\mathrm{Diag}(n)}$ are equal to*

$$\beta_\ell(M_{\mathrm{Diag}(n)}) = \sum_{k=3}^{n} \binom{k-1}{2} \binom{k-3}{\ell}.$$

2.3 Cellular Resolution of $M_{\mathbf{a}}$

Consider the matching field ideal $M_{\mathbf{a}}$, generated minimally by the monomials m_I for all 3-subsets I of $[n]$. We construct a CW-complex supporting the minimal free resolution of $M_{\mathbf{a}}$. Let $H_{\mathbf{a}}$ be the 3-hypergraph on the vertex set $V(H_{\mathbf{a}}) = \{x_1, \ldots, x_n, y_1, \ldots, y_n, z_1, \ldots, z_n\}$, where the edges corresponding to the monomial generators of $M_{\mathbf{a}}$. Given a 3-hypergraph $H_{\mathbf{a}}$ we define a family of 2 and 1-hypergraphs as follows:

$$z_k\text{-layer}(H_{\mathbf{a}}) = \{x_i y_j | x_i y_j z_k \in E(H_{\mathbf{a}})\},$$
$$z_k y_l\text{-layer}(H_{\mathbf{a}}) = \{x_i | x_i y_l z_k \in E(H_{\mathbf{a}})\}.$$

From the combinatorial description of the minimal generators of $M_{\mathbf{a}}$ it is easy to see that:

Lemma 2. *Assume that $x_i y_j z_k \prec_{w_{\mathbf{a}}} x_l y_m z_n$. Then z_n-layer$(H_{\mathbf{a}}) \subset z_k$-layer$(H_{\mathbf{a}})$. Moreover, if $k = n$ then $z_k y_m$-layer$(H_{\mathbf{a}}) \subset z_k y_j$-layer$(H_{\mathbf{a}})$.*

We proceed by constructing another 3-hypergraph $G_\mathbf{a}$, which is isomorphic to $H_\mathbf{a}$ and serves as a co-interval graph, as introduced in [15]. First, we order every edge of $H_\mathbf{a}$ according to the block ordering $\prec_\mathbf{a}$. Let us assume without loss of generality that we have m distinct blocks. In z_m-layer($H_\mathbf{a}$) assume that we have k distinct blocks with respect to the same ordering. Assume without loss of generality that y_{m_1} is the lowest term in this layer and $z_m y_{m_1}$-layer($H_\mathbf{a}$) has l distinct entries mainly $x_{k_1}, x_{k_2}, \ldots, x_{k_l}$.

Definition 2. *Given the 3-hypergraph $H_\mathbf{a}$ we define the graph $G_\mathbf{a}$ on the vertex set $V(G_\mathbf{a}) = [m + k + l]$ with the edge set $E(G_\mathbf{a}) = \{f(z_i)f(y_j)f(x_k) | x_k y_j z_i \in E(H_\mathbf{a})\}$, where f is the map from the set $S_\mathbf{a} = \{z_1, z_2, \ldots, z_m, y_{m_1}, y_{m_2}, \ldots, y_{m_k}, x_{k_1}, x_{k_2}, \ldots, x_{k_l}\}$ to $[m+k+l]$ as follows:*

- $f(z_i) = m - i + 1$ for $1 \leq i \leq m$
- $f(y_{m_i}) = m + i$ for $1 \leq i \leq k$
- $f(x_{k_i}) = m + k + i$ for $1 \leq i \leq l$.

We recall the notion of the v-layer and co-interval graph from [15]. Let H be a d-graph and let $v \in V(H)$ be some vertex. Then the v-layer of H is a $(d-1)$-graph on $V \setminus v$ with edge set

$$\{v_1 v_2 \ldots v_{d-1} | vv_1 v_2 \ldots v_{d-1} \in E(H) \text{ and } v < v_1, v_2, \ldots, v_{d-1}\}.$$

Definition 3. *The class of co-interval d-graphs is defined recursively as follows. Any 1-graph is co-interval. For $d > 1$, the finite d-graph H with vertex set $V(H) \subseteq \mathbb{Z}$ is co-interval if*

1. *for every $i \in V(H)$ the i-layer of H is co-interval.*
2. *for every pair $i < j$ of vertices, the j-layer of H is a subgraph of its i-layer.*

Theorem 3. *The 3-hypergraph $G_\mathbf{a}$ is a co-interval graph.*

Proof. It is enough to prove the following two claims:

1. For every $i \in V(G_\mathbf{a})$ the i-layer of $G_\mathbf{a}$ is co-interval.
2. For every pair $i < j$ of vertices, the j-layer of $G_\mathbf{a}$ is a subgraph of its i-layer.

The construction of $G_\mathbf{a}$ implies that $j\text{-layer}(G_\mathbf{a}) \subseteq i\text{-layer}(G_\mathbf{a})$ for $i < j$. More precisely, $j\text{-layer}(G_\mathbf{a})$ is the induced subgraph of the $i\text{-layer}(G_\mathbf{a})$ since in our original hypergraph $H_\mathbf{a}$ the corresponding $z_{f^{-1}(j)}$-layer($H_\mathbf{a}$) is an induced subgraph of $z_{f^{-1}(i)}$-layer($H_\mathbf{a}$). Hence, it is sufficient to show that 1-layer($G_\mathbf{a}$) is co-interval since any induced subgraph of a co-interval graph is also co-interval (see [15, Proposition 4.2]). Therefore, we only need to ensure that j-layer(1-layer($G_\mathbf{a}$)) $\subseteq i$-layer(1-layer($G_\mathbf{a}$)) for all $i < j$ which is directly implied by Lemma 2.

Corollary 2. *The polyhedral complex $X_{G_\mathbf{a}}$ supports a minimal cellular resolution of the ideal $M_\mathbf{a}$.*

Proof. The map f from Definition 2 induces a natural ring isomorphism:

$$k[z_1,\ldots,z_m,y_{m_1},\ldots,y_{m_k},x_{k_1},\ldots,x_{k_l}] \to k[t_1,\ldots,t_{m+k+l}].$$

We denote $N_\mathbf{a}$ for the image(f). The complex $X_{G_\mathbf{a}}$ supports a minimal cellular resolution of the monomial ideal $N_\mathbf{a}$ and $N_\mathbf{a} \cong M_\mathbf{a}$ (see [15, Theorem 4.4]).

Remark 1. Constructing a CW complex whose faces encode the free resolutions of the ideal is introduced in [3] by Bayer and Sturmfels, and since then, it has been established that several classes of monomial ideals arising from graphs, matroids, and posets have cellular resolutions. See, e.g. [11,24–27,29].

Example 5. Let $n = 5$ and $\mathbf{a} = [123|45]$. Then the edges of the graph $H_\mathbf{a}$ are: 123, 134, 234, 124, 135, 235, 435, 125, 425, 415. Applying the map f from Definition 2, the edges of the graph $G_\mathbf{a}$ are: 357, 247, 248, 257, 147, 148, 149, 157, 159, 169. The polyhedral complex supporting the minimal free resolution of $M_\mathbf{a}$ is shown in Fig. 1. The Betti numbers of $M_\mathbf{a}$ are: $\beta_0 = 10$, $\beta_1 = 15$, $\beta_2 = 6$.

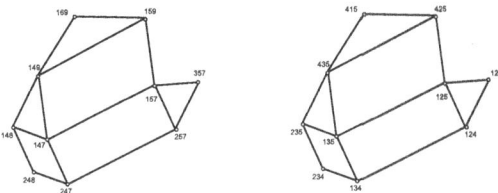

Fig. 1. Polyhedral complex $X_{G_\mathbf{a}}$ with labelling of $M_\mathbf{a}$ (right) and labelling of $H_\mathbf{a}$ (left).

Acknowledgements. The author would like to thank Oliver Clarke and Janet Page for helpful discussions. The work was partially supported by the grants G0F5921N (Odysseus programme) and G023721N from the Research Foundation - Flanders (FWO), the UiT Aurora project MASCOT and the grant iBOF/23/064 from the KU Leuven.

The authors have no competing interests to declare that are relevant to the content of this article.

References

1. Anderson, D.: Okounkov bodies and toric degenerations. Math. Annalen **356**, 1183–1202 (2013)
2. Ardila, F., Klivans, C.J.: The Bergman complex of a matroid and phylogenetic trees. J. Comb. Theory Ser. B **96**(1), 38–49 (2006)

3. Bayer, D., Sturmfels, B.: Cellular resolutions of monomial modules. J. für die reine und angewandte Mathematik **502**, 123–140 (1998)
4. Bernstein, D., Zelevinsky, A.: Combinatorics of maximal minors. J. Algebr. Combin. **2**(2), 111–121 (1993)
5. Bossinger, L., Lamboglia, S., Mincheva, K., Mohammadi, F.: Computing Toric degenerations of flag varieties. In: Smith, G.G., Sturmfels, B. (eds.) Combinatorial Algebraic Geometry. FIC, vol. 80, pp. 247–281. Springer, New York (2017). https://doi.org/10.1007/978-1-4939-7486-3_12
6. Bossinger, L., Mohammadi, F., Nájera Chávez, A. Families of Gröbner degenerations, Grassmannians and universal cluster algebras. SIGMA. Symmetry Integrability Geometry: Methods Appl. **17**, 059 (2021)
7. Clarke, O., Higashitani, A., Mohammadi, F.: Combinatorial mutations and block diagonal polytopes. Collect. Math. **12**, 1–31 (2021)
8. Clarke, O., Higashitani, A., Mohammadi, F.: Combinatorial mutations of Gelfand–Tsetlin polytopes, Feigin–Fourier–Littelmann–Vinberg polytopes, and block diagonal matching field polytopes. J. Pure Appl. Algebra, 107637 (2024)
9. Clarke, O., Mohammadi, F.: Standard monomial theory and toric degenerations of Schubert varieties from matching field tableaux. J. Symb. Comput. **104**, 683–723 (2021)
10. Clarke, O., Mohammadi, F.: Standard monomial theory and toric degenerations of Schubert varieties from matching fields. J. Symb. Comput. **104**, 683–723 (2021)
11. Clarke, O., Mohammadi, F.: Minimal cellular resolutions of powers of matching field ideals. arXiv preprint arXiv:2404.10729 (2024)
12. Clarke, O., Mohammadi, F., Zaffalon, F.: Toric degenerations of partial flag varieties and combinatorial mutations of matching field polytopes. J. Algebra **638**, 90–128 (2024)
13. Conca, A., De Negri, E., Gorla, E.: Universal Gröbner bases for maximal minors. Int. Math. Res. Not. **2015**(11), 3245–3262 (2015)
14. Cox, D., Little, J., Schenck, H.: Toric Varieties. American Mathematical Society, Providence (2011)
15. Dochtermann, A., Engström, A.: Cellular resolutions of cointerval ideals. Math. Z. **270**(1), 145–163 (2012)
16. Dochtermann, A., Mohammadi, F.: Cellular resolutions from mapping cones. J. Combin. Theory Ser. A **128**, 180–206 (2014)
17. Escobar, L., Harada, M.: Wall-crossing for Newton-Okounkov bodies and tropical Grassmannian. Int. Math. Res. Not. **230**, 09 (2020)
18. Feigin, E., Fourier, G., Littelmann, P.: PBW filtration and bases for symplectic Lie algebras. Int. Math. Res. Not. **24**, 5760–5784 (2011)
19. Fulton, W.: Introduction to Toric Varieties. Annals of Mathematics Studies, vol. 131. Princeton University Press, Princeton (2016)
20. Gross, M., Hacking, P., Keel, S., Kontsevich, M.: Canonical bases for cluster algebras. J. Am. Math. Soc. **31**(2), 497–608 (2018)
21. Herzog, J., Takayama, Y.: Resolutions by mapping cones. Homol. Homot. Appl. **4**(2), 277–294 (2002)
22. Kaveh, K., Manon, C.: Khovanskii bases, higher rank valuations, and tropical geometry. SIAM J. Appl. Algebra Geometry **3**(2), 292–336 (2019)
23. Khovanskii, A.G.: Newton polyhedra and the genus of complete intersections. Funct. Anal. Appl. **12**(1), 38–46 (1978)
24. Mohammadi, F.: Powers of the vertex cover ideals. Collect. Math. **65**(2), 169–81 (2014)

25. Mohammadi, F.: Divisors on graphs, orientations, syzygies, and system reliability. J. Algebr. Combin. **43**, 465–83 (2016)
26. Mohammadi, F.: A family of monomial Gröbner degenerations of determinantal ideals with minimal cellular resolutions. arXiv preprint arXiv:2403.17736 (2024)
27. Mohammadi, F., Shokrieh, F.: Divisors on graphs, binomial and monomial ideals, and cellular resolutions. Math. Z. **283**(1), 59–102 (2016)
28. Mohammadi, F., Shaw, K.: Toric degenerations of Grassmannians from matching fields. Algebr. Combin. **2**(6), 1109–1124 (2019)
29. Novik, I., Postnikov, A., Sturmfels, B.: Syzygies of oriented matroids. Duke Math. J. **111**(2), 287 (2002)
30. Rietsch, K., Williams, L.: Newton-Okounkov bodies, cluster duality, and mirror symmetry for Grassmannians. Duke Math. J. **168**(18), 3437–3527 (2019)
31. Speyer, D., Sturmfels, B.: The tropical Grassmannian. Adv. Geom. **4**, 389–411 (2004)
32. Sturmfels, B., Zelevinsky, A.: Maximal minors and their leading terms. Adv. Math. **98**(1), 65–112 (1993)

Polyhedral Geometry
and Combinatorics

Eigenvalue Methods for Sparse Tropical Polynomial Systems

Marianne Akian[1,2], Antoine Béreau[1,2](✉), and Stéphane Gaubert[1,2]

[1] Inria, 91120 Palaiseau, France
antoine.bereau@inria.fr
[2] CMAP, CNRS, École polytechnique, Institut polytechnique de Paris, 91120 Palaiseau, France

Abstract. We develop an analogue of eigenvalue methods to construct solutions of systems of tropical polynomial equalities and inequalities. We show that solutions can be obtained by solving parametric mean payoff games, arising to appropriate linearizations of the systems using tropical Macaulay matrices. We implemented specific algorithms adapted to the large scale parametric games that arise in this way, and present numerical experiments.

Keywords: Tropical geometry · Polynomial systems · Zero-sum games · Algorithmic complexity

1 Introduction

Motivation. We develop a method to solve systems of equations or of inequations involving multivariate tropical polynomials. Such systems arise in the study of *non-archimedean amoebas*, which are images by a non-archimedean valuation of an algebraic set [17]. The 'fundamental theorem of tropical geometry' shows that these amoebas can be described by systems of finitely many tropical polynomial equations [31].

Similarly, the solution sets of systems of tropical polynomial inequalities provide upper approximations of images by a non-archimedean valuation of semi-algebraic sets over real closed non-archimedean fields, and these approximations are exact under genericity conditions [8,27,28]. Owing to their combinatorial nature, tropical polynomial systems are often easier to grasp than their classical analogues. These ideas, which can be traced back to works of Viro [36], Gelfand, Kapranov, Zelevinsky [21], Sturmfels (see [25]), or Mikhalkin [34], are at the heart of tropical geometry, see [26,31] for background.

Systems of tropical polynomial equations and inequations also arise in specific applications, independently of the nonarchimedean interpretation. In particular, they arise in the computation of stationary behaviors of discrete event systems [13], see e.g. [7] for an application to performance evaluation. Other motivations arise from auction theory [9], or chemical reaction networks, see e.g. [14].

Grigoriev and Podolskii established in [23] a *tropical Nullstellensatz*, which states that a system of tropical polynomial equations is solvable if and only if its linearization, obtained by truncating the Macaulay matrix up to an appropriate degree, is solvable. Their results also apply to polynomial inequations. Since systems of tropical linear equations and inequations reduce to mean payoff games [1], this provides both theoretical tools (strong duality theorems) and algorithms. In [2], we provided a refinement of Grigoriev and Podolskii's tropical nullstellensatz, adapted to sparse polynomial systems, by exploiting the construction of the *Canny-Emiris set* [12,18] and its generalization by Sturmfels [35, §3], used to compute the classical resultants. We provided also an improved truncation degree for dense polynomials (which matches the optimal Macaulay bound). Our approach also applies to hybrid systems, combining tropical equalities with strict and weak inequalities. However, the approach developed in [2,23] only allows one to decide whether the solution set is empty.

Contribution. Here, we extend this approach to compute solutions. Our method may be thought of as a tropical analogue of the eigenvalue method for polynomial system solving, albeit the notion of 'eigenvalue' now has to be understood in a non-linear sense. More precisely, we construct a solution by reduction to parametric families of mean payoff games. In particular, for any coordinate index, we define a spectral function, which provides the value of a game as a function of a parameter, and show that the projection of the solution set on the given coordinate coincides with a super-level set of this spectral function. We present two algorithms building on this idea: a simple dichotomic search, allowing to construct one solution, and a path-following method, computing the graph of the spectral function, allowing one to compute the entire set of solutions when the latter is finite. We present numerical benchmarks.

Related work. The standard approach to the computation of tropical prevarieties is to exploit the duality between arrangements of tropical hypersurfaces and mixed-polyhedral subdivisions. In that way, decision problems concerning tropical prevarieties can be reduced to the enumeration of mixed cells, see [29,32]. A number of current works deal with the efficient computation of tropical varieties and prevariarieties, see [22,33] and the references therein.

In these approaches, a polyhedral complex from which all solutions can be obtained is typically constructed. In contrast, the present method allows one to obtain a more restricted information on the solution set, like a single solution, or a projection on one coordinate. Part of this work relies on ideas of parametric games, going back to [19,20]. A different approach to solve tropical polynomial systems relies on SMT solving [30].

2 The Sparse Tropical Positivstellensatz

The *tropical* (or *max-plus*) *semifield* is the semifield $(\mathbb{R} \cup \{-\infty\}, \oplus, \odot, \mathbb{0}, \mathbb{1})$ with addition $\oplus = \max$, multiplication $\odot = +$, zero element $\mathbb{0} = -\infty$ and unit element $\mathbb{1} = 0$. The operations $\oplus$ and $\odot$ are respectively refered to as the *tropical addition* and the *tropical multiplication*, and the tropical semifield is denoted by $\mathbb{R}_\infty$. The notions of formal polynomials and polynomial functions carry over $\mathbb{R}_\infty$. In particular, a polynomial function is a map $f : \mathbb{R}_\infty^n \to \mathbb{R}_\infty$ defined by an equation of the form

$$f(x) := \max_{\alpha \in \mathbb{N}^n} \left(f_\alpha + \langle x, \alpha \rangle \right) \qquad (1)$$

for all $x \in \mathbb{R}_\infty^n$, where we assume that $f_\alpha \neq -\infty$ for finitely many values of $\alpha \in \mathbb{N}^n$, and adopt the convention $(-\infty) \times 0 = 0$ when evaluating $\langle x, \alpha \rangle$. The set $\{\alpha \in \mathbb{N}^n : f_\alpha \neq -\infty\}$ is called the *support* of f and is denoted $\mathrm{supp}(f)$. Although the value $f(x)$ makes sense for any $x \in \mathbb{R}_\infty^n$, we shall restrict our attention here to vectors x with *finite* values, i.e. $x \in \mathbb{R}^n$. In particular, a *root* of the tropical polynomial f is defined as a point $x \in \mathbb{R}^n$ such that the maximum in (1) is achieved at least twice. We write $f(x) \nabla 0$ when x is a root of f, to make clear the analogy with the equation '$f(x) = 0$' defining classical roots.

A collection $f = (f_1, \ldots, f_k)$ of n-variate tropical polynomials defines a *tropical (toric) prevariety*, which is the set of common tropical roots of $f_1, \ldots, f_k$, i.e., the set of solutions $x \in \mathbb{R}^n$ of $f_i(x) \nabla 0$ for all $1 \leq i \leq k$.

We also consider tropical equations of the form $f_i^+(x) = f_i^-(x)$, $1 \leq i \leq k$, where $f_i^\pm$ are tropical polynomials, as well as systems of tropical inequalities of the form $f_i^+(x) \geq f_i^-(x)$ or $f_i^+(x) > f_i^-(x)$, for $1 \leq i \leq k$. Following [2,3], we call *basic tropical semialgebraic set* the set of solutions of a system involving any combination of weak and strict tropical polynomial inequalities. Tropical prevarieties are easily seen to be particular cases of basic tropical semialgebraic sets.

Given a collection of tropical polynomials $f = (f_1, \ldots, f_k)$, the *Macaulay matrix* $\mathcal{M}$ associated to f is defined as follows: its rows are indexed by pairs (i, α) where $1 \leq i \leq k$ and $\alpha \in \mathbb{N}^n$ and its columns are indexed by integer vectors $\beta \in \mathbb{N}^n$, and for given (i, α) and β, we set the entry $\mathcal{M}_{(i,\alpha),\beta}$ of $\mathcal{M}$ equal to the coefficient of the monomial X^β in the polynomial $X^\alpha f_i(X)$, or $-\infty$ if no such monomial exists.

Given a *nonempty* finite subset $\mathcal{E}$ of $\mathbb{N}^n$, and a collection $\mathcal{A} = (\mathcal{A}_1, \ldots, \mathcal{A}_k)$ of subsets of $\mathbb{N}^n$, we denote by $\mathcal{M}_\mathcal{E}^\mathcal{A}$ the submatrix of $\mathcal{M}$ consisting only of the columns with indices $\beta \in \mathcal{E}$, and the rows indexed by pairs (i, α) where $1 \leq i \leq k$ and $\alpha \in \mathbb{N}^n$ such that $\alpha + \mathcal{A}_i \subset \mathcal{E}$. When the polynomials f_i have their support equal to $\mathcal{A}_i$, we simply write $\mathcal{M}_\mathcal{E}$ instead of $\mathcal{M}_\mathcal{E}^\mathcal{A}$.

Set $f^+ = (f_1^+, \ldots, f_k^+)$ and $f^- = (f_1^-, \ldots, f_k^-)$ two collections of k tropical polynomials, and for $1 \leq i \leq k$, $\mathcal{A}_i^\pm = \mathrm{supp}(f_i^\pm)$, $\mathcal{A}_i = \mathcal{A}_i^+ \cup \mathcal{A}_i^-$ and $\mathcal{A} = (\mathcal{A}_1, \ldots, \mathcal{A}_k)$. We denote by $f^+(x) \geq f^-(x)$ the system

$$\max_{\alpha \in \mathcal{A}_i^+} \left(f_{i,\alpha}^+ + \langle \alpha, x \rangle \right) \geq \max_{\alpha \in \mathcal{A}_i^-} \left(f_{i,\alpha}^- + \langle \alpha, x \rangle \right) \text{ for all } 1 \leq i \leq k, \qquad (2)$$

of unknown $x \in \mathbb{R}^n$.

Moreover, we denote by $\mathcal{M}^\pm$ the pair of Macaulay matrices associated to $f^\pm$, so with entries $\mathcal{M}_{(i,\alpha),\beta}^\pm = f_{i,\beta-\alpha}^\pm$. Then, for any subset $\mathcal{E}$ of $\mathbb{N}^n$, we denote by $\mathcal{M}_{\mathcal{E}}^{\pm}$ the submatrices associated to $\mathcal{E}$ and the collection $\mathcal{A}$ defined above, i.e. $\mathcal{M}_{\mathcal{E}}^{\pm} = (\mathcal{M}^{\pm})_{\mathcal{E}}^{\mathcal{A}}$. Finally, we set for $1 \leq i \leq k$, $r_i = \dim(\text{aff}(\mathcal{A}_i^-)) + 1$.

We now call *Canny-Emiris subset* of $\mathbb{N}^n$ associated to the system $f^+(x) \geq f^-(x)$ any set $\mathcal{E}$ of the form $\mathcal{E} := (\widetilde{Q} + \delta) \cap \mathbb{N}^n$ with $\widetilde{Q} = r_1 Q_1 + \cdots + r_k Q_k$, where $Q_i = \text{conv}(\mathcal{A}_i)$ for $1 \leq i \leq k$, and δ is a generic vector in $V + \mathbb{N}^n$, with V the direction of the affine hull of $\widetilde{Q}$. Finally, for any subset $\mathcal{E}'$ of $\mathbb{N}^n$ containing a Canny-Emiris subset $\mathcal{E}$ associated to $f^+(x) \geq f^-(x)$ and $y \in \mathbb{R}^{\mathcal{E}'}$, we denote by $\mathcal{M}_{\mathcal{E}'}^+ \odot y \geq \mathcal{M}_{\mathcal{E}'}^- \odot y$ the following system of tropical linear inequalities:

$$\max_{\beta \in \mathcal{E}'} \left(m^+_{(i,\alpha),\beta} + y_\beta \right) \geq \max_{\beta \in \mathcal{E}'} \left(m^-_{(i,\alpha),\beta} + y_\beta \right) \quad \text{for all} \quad 1 \leq i \leq k \text{ and } \alpha \in \mathcal{A}_i. \quad (3)$$

The following result shows that the solvability of a system of tropical polynomial inequalities is equivalent to the solvability of a 'linearized' version obtained from the Macaulay matrices.

Theorem 1 (Sparse tropical Positivstellensatz, see [3, Theorem 4.1]).
There exists a solution $x \in \mathbb{R}^n$ to the system $f^+(x) \geq f^-(x)$ if and only if there exists a vector $y \in \mathbb{R}^{\mathcal{E}'}$ satisfying $\mathcal{M}_{\mathcal{E}'}^+ \odot y \geq \mathcal{M}_{\mathcal{E}'}^- \odot y$, where $\mathcal{E}'$ is any subset of $\mathbb{N}^n$ containing a nonempty Canny-Emiris subset $\mathcal{E}$ associated to the system $f^+(x) \geq f^-(x)$.

Theorem 4.14 of [3] provides a similar result for general *hybrid* tropical systems, allowing both strict and weak inequalities, as well as equations of the form $f_i(x) \nabla 0$. For simplicity of exposition, we consider here solutions of systems of weak inequalities, as in Theorem 1. The present method carries over to hybrid systems.

3 Deciding the Solvability of Tropical Linear Systems with Mean Payoff Games

We next recall how the linearized system of inequalities can be solved by a reduction to mean payoff games, referring to [1, Section 2] for details.

Let G be a (finite) oriented weighted bipartite graph given with its set of vertices $I \sqcup J$ and its set of arcs $E \subseteq (I \times J) \cup (J \times I)$. The vertices of G are refered to as the *states* or *positions* of the game, and the arcs of G are refered to as the *actions* or *moves*. The payments of the game are described by two matrices $A = (a_{ij})_{(i,j) \in I \times J}$ and $B = (b_{ij})_{(i,j) \in I \times J}$, where we set $a_{ij} = -\infty$ whenever the arc (j,i) does not exist in the graph G and likewise $b_{ij} = -\infty$ whenever the arc (i,j) does not exist. We assume that every node has of G has at least one successor. The associated *mean payoff game* is the zero-sum two player game defined as follows. The first player is called the *minimizer* and the second one the *maximizer*. At turn N, if the current state is $j_N \in J$, the minimizer chooses the next state $i_N \in I$ along an arc (j_N, i_N) makes a payment of $-a_{i_N j_N}$ to the maximizer. Then the maximizer, from the current state $i_k \in I$, chooses a state $j_{N+1} \in J$ such that (i_N, j_{N+1}) is an arc of G, and receives a payment of $b_{i_N j_{N+1}}$

from the minimizer. These steps repeat indefinitely. The maximizer wishes to maximize the average payment received per time unit whereas the minimizer wishes to minimize it. This game is known to have a value [16, 37], denoted by χ_{j_0} where $j_0 \in J$ is the initial state. We say that the game, from this initial state, is winning for the maximizer if this value is nonnegative.

We consider the max-plus linear operator $y \mapsto By$ defined from $(\mathbb{R} \cup \{\pm\infty\})^J$ to $(\mathbb{R} \cup \{\pm\infty\})^I$, with the convention that $(-\infty) + (+\infty) = (-\infty)$, by $(By)_i := \max_{j \in J}(b_{ij} + y_j)$ for all $i \in I$. Likewise, we consider the min-plus linear operator $z \mapsto A^\sharp z$ defined from $(\mathbb{R} \cup \{\pm\infty\})^I$ to $(\mathbb{R} \cup \{\pm\infty\})^J$, with the convention $(+\infty) + (-\infty) = (+\infty)$ this time, by $(A^\sharp z)_j := \min_{i \in I}(-a_{ij} + z_i)$ for all $j \in J$. The requirement that every node of G has at least one successor is equivalent to the following:

Assumption 2. (a) for all $j \in J$, there exists $i \in I$ such that $a_{ij} \neq -\infty$; (b) for all $i \in I$, there exists $j \in J$ such that $b_{ij} \neq -\infty$.

Finally, the *Shapley operator* of this game is the operator $T : (\mathbb{R} \cup \{\pm\infty\})^J \to (\mathbb{R} \cup \{\pm\infty\})^J$ defined by $T(y) := A^\sharp By$ for all $y \in \mathbb{R} \cup \{\pm\infty\}$. In other words, one has

$$T(y) = \left(\min_{i \in I}(-a_{ij} + \max_{k \in J}(b_{ik} + y_k)) \right)_{j \in J} \tag{4}$$

for all $y \in (\mathbb{R} \cup \{\pm\infty\})^J$. Note in particular that the tropical linear system $A \odot y \leq B \odot y$ is equivalent to the inequality $y \leq T(y)$. The vector of values $\chi(T) = (\chi_j)_{j \in J} \in \mathbb{R}^J$ only depends on the operator T, as it coincides with the limit $\lim_{N \to +\infty} T^N(0)/N$. The following non-linear spectral theorem characterizes the minimal value of an initial state.

Theorem 3 (Collatz-Wielandt property [1, Theorem 2.8 and Remark 2.10]). *Let T be the Shapley operator defined in (4). Then, the following quantities coincide and they are all equal to $\underline{\chi}(T) := \min\{\chi_j : j \in J\}$*

$$\sup\{\lambda \in \mathbb{R} : \exists u \in \mathbb{R}^J, T(u) \geq \lambda + u\} \tag{5a}$$

$$\inf\{\lambda \in \mathbb{R} \cup \{+\infty\} : \exists u \in (\mathbb{R} \cup \{+\infty\})^J, u \not\equiv +\infty, T(u) \leq \lambda + u\} \tag{5b}$$

$$\inf\{\lambda \in \mathbb{R} \cup \{+\infty\} : \exists u \in (\mathbb{R} \cup \{+\infty\})^J, u \not\equiv +\infty, T(u) = \lambda + u\} . \tag{5c}$$

Moreover, the infima and supremum in (5) are attained.

Theorem 4 ([1, Corollary 3.4]). *The tropical linear system $A \odot y \leq B \odot y$ has a solution $y \in \mathbb{R}^J$ if and only if all the initial states of the associated game have a nonnegative value, i.e. $\underline{\chi}(T) \geq 0$.*

4 Mean Payoff Games Oracles

To check the solvability of the linearized system (3) using Theorem 4, it suffices to compute the value vector of a mean payoff game. Actually, a weaker information will be enough for some of our results. We call *weak mean payoff game oracle* a procedure which takes as input two tropical matrices A, B, and decides whether $\underline{\chi}(T) \geq 0$ with $T = A^\sharp B$. We denote by w-MPG($|I|, |J|, W$) the number of arithmetic operations of a mean payoff oracle taking as input $|I| \times |J|$ matrices A, B whose entries are either relative integers of absolute values bounded by W or $-\infty$. We observe that w-MPG($|I|, |J|, W$) $\geq |I||J|$ since the input size is $\Omega(|I \times J|)$. We shall also use the notion of *strong mean payoff game oracle*, which not only decides whether $\underline{\chi}(T) \geq 0$, as a weak oracle does, but also, whenever $\chi(T) \equiv \lambda \in \mathbb{R}$, returns a vector $u \in \mathbb{R}^n$ such that $T(u) = \lambda + u$. The vector u is called a *bias*. The bias vector serves as an optimality certificate, allowing one to identify optimal policies, see [6] for background. We denote by MPG($|I|, |J|, W$) the number of arithmetic operations of a strong mean payoff oracle.

A classical algorithm to solve mean payoff games is value iteration, analyzed in [37]. It consists in computing the sequence $T^N(0)$ and inferring the limit $\lim_{N \to +\infty} T^N(0)/N$ by specializing N to an explicit sufficiently large value.

Theorem 5 (Corollary of [37, Theorem 2.4]). *Value iteration provides a weak mean payoff oracle requiring $\mathcal{O}(|J|^2 W)$ evaluations of the Shapley operator T, leading to* w-MPG($|I|, |J|, W$) $= \mathcal{O}(|I||J|^3 W)$.

The number of iterations of the method of [37] is always in $\Omega(|J|^2 W)$, which is unpracticable in our application (J will be exponentially large in the input size). We presented in [2] a refinement of value iteration, exploiting the ideas of Krasnoselskii-Mann damping with with an acceleration or 'widening' step. This accelerated version allows in practice for a much quicker check of feasibility. We also have the following result.

Theorem 6. *A strong mean payoff oracle can be implemented by making $\mathcal{O}(|J|^3 W)$ evaluations of the Shapley operator T, leading to* MPG($|I|, |J|, W$) $= \mathcal{O}(|I||J|^4 W)$.

Indeed, we first compute $\chi(T)$ by means of [37, Theorem 2.3]. Moreover, when $\chi(T) \equiv \lambda \in \mathbb{R}$ is a constant vector, we first perform the iteration $u^{k+1} = (-\lambda + T)(u^k) \wedge u^k$, starting from $u^0 = 0$, and show it converges to a vector u such that $u \leq (-\lambda + T)(u)$, in $\mathcal{O}(|J|^3 W)$ iterations, then we perform the iteration $v^{k+1} = (-\lambda + T)(v^k)$, starting from $v^0 = u$, and show it converges to a bias vector v, satisfying $v = (-\lambda + T)(v)$, again in $\mathcal{O}(|J|^3 W)$ iterations, leading to Theorem 6.

Another approach to solve mean payoff games is policy iteration, see [15, Algorithm 2], which is strongly polynomial for discounted problems with a fixed discount factor [4,24], and the mean payoff problem reduces to the discounted problem, with a discount rate in $1 - 1/\Omega(|J|^3 W)$ [37, p. 353]. This leads to pseudo-polynomial bounds, which are weaker than the one obtained for value iteration. However, policy iteration is experimentally the fastest method for our purpose.

5 Existence of Short Solutions of Tropical Polynomial Systems

We shall need an a priori bound on the solutions of a tropical polynomial system. The set of these solutions is a closed polyhedral complex. The next result shows that if this set is nonempty, there is always an element in this set with a bitsize polynomially bounded in the input size.

Theorem 7. *Let $f^\pm = (f_1^\pm, \ldots, f_k^\pm)$ be two collections of tropical polynomials and let*
$$d = \max_{1 \leq i \leq k} \deg(f_i^\pm) \quad \text{and} \quad W = \max_{1 \leq i \leq k} \|f_i^\pm\|_\infty ,$$
and for $\epsilon \in \{\pm 1\}^n$, denote by $\epsilon \mathbb{R}^n_{\geq 0}$ the orthant $\{x \in \mathbb{R}^n : \epsilon_j x_j \geq 0 \text{ for all } 1 \leq j \leq n\}$. Then:

(i) *the vertices of every polyhedral complex $\{x \in \mathbb{R}^n : f_i^+(x) \geq f_i^-(x)\} \cap \epsilon \mathbb{R}^n_{\geq 0}$ are included in a $\|\cdot\|_\infty$-ball of radius $2n(2d)^{n-1}W$ centered at point 0;*
(ii) *if moreover all the coefficients of the polynomials $f_i^\pm$ are integer, these vertices have coordinates that are rational numbers with a denominator bounded above by $(2d)^n$.*

Taking the intersection with the orthant $\epsilon \mathbb{R}^n_{\geq 0}$ is a technical convenience, which makes sures that a vertex always exists as soon as the system of inequalities has a solution in this orthant. This allows us to tackle situations in which for instance solutions sets contains affine lines. Then, the coordinates and denominator of this vertex are bounded using Hadamard's inequality for determinants, leading to the above estimate.

6 The Dichotomic Search Method

We now present a first method allowing us to construct one solution of a system of tropical polynomial inequalities. We saw in Sect. 3 that checking whether a system of weak tropical polynomial inequalities admits a solution in $\mathbb{R}^n$, using the tropical Positivstellensatz, reduces to solving a mean payoff game. We enrich this system, by adding extra inequalities of the form $a \leq x_1$ or $x_1 \leq b$. In this way, we can decide whether there is a solution such that $x_1 \in [a, b]$. Moreover, Theorem 7 provides an a priori bound for a solution, allowing us to reduce the search space to a sup-norm box centered at the origin with bounded radius, and to rational numbers with a bounded denominator. In this way, a rational number $\bar{x}_1$ which belongs to the projection of the solution set on the first variable can be obtained. Then, we substitute x_1 by the fixed value $\bar{x}_1$ in the polynomial system, and perform again a dichotomic search, now on the variable x_2, leading to a rational value $\bar{x}_2$ such that $(\bar{x}_1, \bar{x}_2)$ belongs to the projection of the solution set on the first two variables. We pursue this procedure by fixing gradually the variables $x_1, \ldots, x_n$. We call *dichotomic search* this method. Observe that dichotomic search stops at the first step if the solution set is empty. We arrive at the following complexity result.

Theorem 8. *Consider a system of weak polynomial inequalities, as in Theorem 7. Then, the dichotomic search method returns a rational solution of this system (or decides that there is none) in* $\mathcal{O}\big(\log(n(2d)^{2n-1}W)\big)$ *calls to a weak mean payoff oracle.*

7 The Path Following Method

Let ζ be a real parameter. Using Theorem 1, the feasibility of the system $f_i^+(x) \geq f_i^-(x)$ for all $1 \leq i \leq k$ with the added constraint $x_1 = \zeta$ can be expressed as a system of homogeneous tropical linear inequalities of the form $A_\zeta \odot y \leq B_\zeta \odot y$ of unknown $y \in \mathbb{R}^J$, in which the entries of the tropical matrices $A_\zeta = (a_{ij}(\zeta))_{(i,j) \in I \times J}$ and $B_\zeta = (b_{ij}(\zeta))_{(i,j) \in I \times J}$ are piecewise affine functions of the scalar ζ with non $-\infty$ coefficients bounded by $W_\zeta := W + |\zeta|d$. We consider the parametric Shapley operator $T_\zeta := A_\zeta^\sharp B_\zeta$ which, up to replacing it by the operator $u \mapsto u \wedge T_\zeta(u)$, can be assumed to send $\mathbb{R}^J$ onto itself. We define the *spectral function* of this operator to be the map $\phi : \mathbb{R} \to \mathbb{R}$ defined by $\phi(\zeta) = \chi(T_\zeta)$.

Theorem 9. *The spectral function ϕ is continuous and piecewise affine. Moreover, the projection on the first coordinate of the solution set $S = \{x \in \mathbb{R}^n : f_i^+(x) \geq f_i^-(x)$ for all $1 \leq i \leq k\}$ coincides with the super-level set $\{\zeta \in \mathbb{R} : \phi(\zeta) \geq 0\}$.*

We next show how to compute the restriction of the spectral function to an interval $[\underline{\zeta}, \overline{\zeta}]$. We define the matrix $\bar{A}_\zeta$, obtained by replacing every $-\infty$ entry of A_ζ by a number $-M$. If M is chosen larger than $4|\mathcal{E}| \max(W_{\underline{\zeta}}, W_{\overline{\zeta}})$, then $\chi(\bar{A}_\zeta^\sharp B_\zeta)$ is a constant vector whose entries coincide with $\chi(A_\zeta^\sharp B_\zeta)$. Moreover, for all $\zeta \in \mathbb{R}$, the eigenproblem $\bar{A}_\zeta^\sharp B_\zeta u = \lambda + u$ with $\lambda \in \mathbb{R}$ and $u \in \mathbb{R}^n$, depending on ζ, is solvable, and $\lambda = \lambda(\zeta) := \chi(A_\zeta^\sharp B_\zeta)$. We will construct piecewise linear functions $\lambda : [\underline{\zeta}, \overline{\zeta}] \to \mathbb{R}$ and $u : [\underline{\zeta}, \overline{\zeta}] \to \mathbb{R}^J$ and $v : [\underline{\zeta}, \overline{\zeta}] \to \mathbb{R}^I$ satisfying

$$\begin{aligned}\min_{i \in I} -\bar{a}_{ij}(\zeta) + v_i(\zeta) &= \lambda(\zeta) + u_j(\zeta) \quad \text{for all } j \in J \\ \max_{j \in J} b_{ij}(\zeta) + u_j(\zeta) &= v_i(\zeta) \quad\quad\quad\quad\;\; \text{for all } i \in I \;.\end{aligned} \quad (6)$$

Suppose that λ, u and v have been evaluated at a point ζ_0. Then, we look for a solution of (6) of the form

$$\lambda(\zeta) = \lambda(\zeta_0) + (\zeta - \zeta_0)\lambda'(\zeta_0) \;,\quad u(\zeta) = u(\zeta_0) + (\zeta - \zeta_0)u'(\zeta_0) \;,\quad v(\zeta) = v(\zeta_0) + (\zeta - \zeta_0)v'(\zeta_0) \quad (7)$$

defined on a small right neighborhood of ζ_0, where $\lambda'(\zeta_0) \in \mathbb{R}$, $u'(\zeta_0) \in \mathbb{R}^J$ and $v'(\zeta_0) \in \mathbb{R}^I$ will be computed. Denoting $I_j(\zeta_0) = \arg\min_{i \in I}(-\bar{a}_{ij}(\zeta_0) + v_i(\zeta_0))$ and $J_i(\zeta_0) = \arg\max_{j \in J}(b_{ij}(\zeta_0) + u_j(\zeta_0))$, we consider the non-linear eigenvalue problem

$$\begin{aligned}\min_{i \in I_j(\zeta_0)} -a'_{ij}(\zeta_0) + v'_i(\zeta_0) &= \lambda'(\zeta_0) + u'_j(\zeta_0) \quad \text{for all } j \in J \\ \max_{j \in J_i(\zeta_0)} b'_{ij}(\zeta_0) + u'_j(\zeta_0) &= v'_i(\zeta_0) \quad\quad\quad\quad\;\; \text{for all } i \in I \;,\end{aligned} \quad (8)$$

where a'_{ij} and b'_{ij} denote the right derivatives of the piecewise affine functions $\zeta \mapsto \bar{a}_{ij}(\zeta)$ and $\zeta \mapsto \bar{b}_{ij}(\zeta)$. This problem can be solved by calling the strong mean payoff oracle, the payment matrices being given by $a'_{ij}(\zeta_0)$ and $b'_{ij}(\zeta_0)$. We show that any solution $\lambda'(\zeta_0), u'(\zeta_0), v'(\zeta_0)$ of (8) yields a solution $\lambda(\zeta), u(\zeta), v(\zeta)$ of (6), as per (7), defined on a right neighborhood of ζ_0. We now perform a *pivoting step*, similar in its principle to a pivoting in the simplex algorithm. We denote by ζ_1 the smallest value of $\zeta > \zeta_0$ for which the Ansatz (7) is no longer a solution of (6). We reevaluate the sets of active constraints $I_j(\zeta_1)$ and $J_i(\zeta_1)$ at the new point $u(\zeta_1)$ and $v(\zeta_1)$, and solve again (8).

Whereas the eigenvalue $\lambda(\zeta)$ of the operator $\bar{A}^\sharp_\zeta B_\zeta$ is unique for all values of ζ, the eigenvector of this operator may not be unique (even up to an additive constant). In particular, the derived eigenvalue problem (8) may have several solutions $u'(\zeta), v'(\zeta)$. However, we show that if A_0 or B_0 has *generic entries*, in the sense of not belonging to an explicit finite union of hyperplanes, then the eigenvector map $\zeta \mapsto u(\zeta)$ becomes uniquely defined and then the number of pivoting steps is finite. The proof of this relies on a result of generic uniqueness of the eigenvector [5]. We resort to an unpleasant technicality for non-generic systems: we perturb explicitly the input to make it generic, at the cost of a dilation of W by a possibly large factor of at most $(2|\mathcal{E}|+1)^{|\mathcal{E}|}$ to ensure that the input remains integer. We denote by $W' \leq (2|\mathcal{E}|+1)^{|\mathcal{E}|}W$ this dilated cost.

Finally, from Theorem 7, we deduce that one can in fact retrieve the entire projection of the solution set by choosing $\underline{\zeta} = -2n(2d)^{n-1}W'$ and $\overline{\zeta} = 2n(2d)^{n-1}W'$. Denoting by $\operatorname{piv}(|I|,|J|,W)$ the maximal number of pivoting steps when A_0, B_0 ranges over the set of payment matrices with generic entries, we arrive at the following result.

Theorem 10. *The path following method computes the projection on the first coordinate of the solution set $S = \{x \in \mathbb{R}^n : f_i^+(x) \geq f_i^-(x) \text{ for all } 1 \leq i \leq k\}$ in a number of*

$$\operatorname{MPG}(\mathcal{O}(k|\mathcal{E}|), |\mathcal{E}|, 4|\mathcal{E}|(n(2d)^n+1)W') + \operatorname{MPG}(\mathcal{O}(k|\mathcal{E}|), |\mathcal{E}|, d) \operatorname{piv}(\mathcal{O}(k|\mathcal{E}|), |\mathcal{E}|, W')$$

arithmetic operations, where $\mathcal{E}$ is a Canny-Emiris set as defined in Sect. 2. Moreover one has $|\mathcal{E}| \leq \binom{n+N}{n}$ for $N = \sum_{i=1}^k r_i d_i$.

When the solution set is finite, this method allows us to recover all solutions: for each element in the projection of S on the first coordinate, we substitute x_1 to this value, and find the possible values of x_2, by considering a new spectral function, etc. We arrive at the following complexity.

Corollary 11. *A finite tropical semialgebraic set given by a collection of polynomial relations $f_i^+ \triangleright_i f_i^-$ for $1 \leq i \leq k$, and $f_i \triangledown \mathbb{0}$ for $k+1 \leq i \leq \ell$, with $\triangleright_i \in \{=, \geq\}$ for $1 \leq i \leq k$ can be computed in*

$$\prod_{j\in[n]}\left[\mathsf{MPG}\left(\mathcal{O}(kM_j),\mathcal{O}(M_j),4M_j(j(2d^{(j)})^j+1)W'\right)\right.$$
$$\left.+\mathsf{MPG}\left(\mathcal{O}(kM_j),\mathcal{O}(M_j),d^{(j)}\right)\mathsf{piv}\left(\mathcal{O}(kM_j),\mathcal{O}(M_j),W'\right)\right]$$

arithmetic operations, where $M_j = \binom{j+N_j}{j}$ with N_j a integer linear combination of the degrees $d_i^{(j)}$ in the last j variables of the polynomials $f_i^+ \oplus f_i^-$ for $1 \leq i \leq k$ and f_i for $k+1 \leq i \leq \ell$, that can be explicited using [3, Theorem 4.14] and $d^{(j)} = \max_{1 \leq i \leq \ell} d_i^{(j)}$.

Example 12. Consider the following system.

$$\begin{cases} 0 \oplus 0.x^2 y & \geq 2x \oplus 2xy \\ 2xy & \geq 1x \oplus -1y \\ 0 & \geq -3x \oplus -1y \end{cases} \quad (9)$$

Observing the zero set of the two associated spectral functions displayed on Fig. 1b, one finds that the solutions of (9) are included in $([-3,-2] \sqcup [2,3]) \times [-1,1]$, which is indeed confirmed by the representation of the solution set in Fig. 1a.

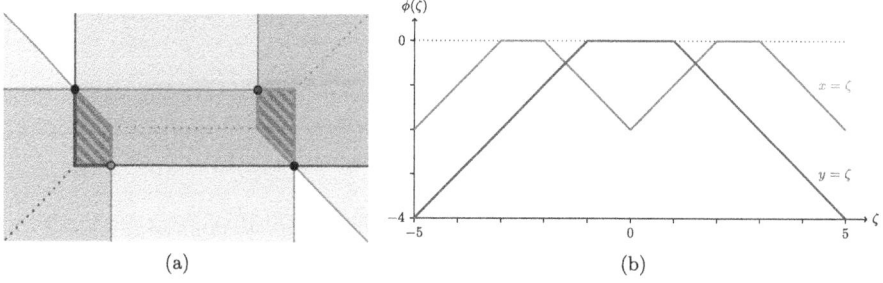

Fig. 1. The collection of tropical semialgebraic sets arising from system (9) as well as the spectral functions obtained when specializing each variable respectively.

8 Numerical Results

An implementation of algorithms to decide the feasibility of tropical polynomial systems is available at [11]. Experimental results obtained by the dichotomy method are displayed in Fig. 2. The solvability of randomized instances of systems of tropical polynomial inequalities was examined twice. First, the polynomials were treated as full polynomials, with no consideration on their support. The resulting linearized system was thus given by a Macaulay matrix simply truncated to the according degree bound. This corresponds to the curves in

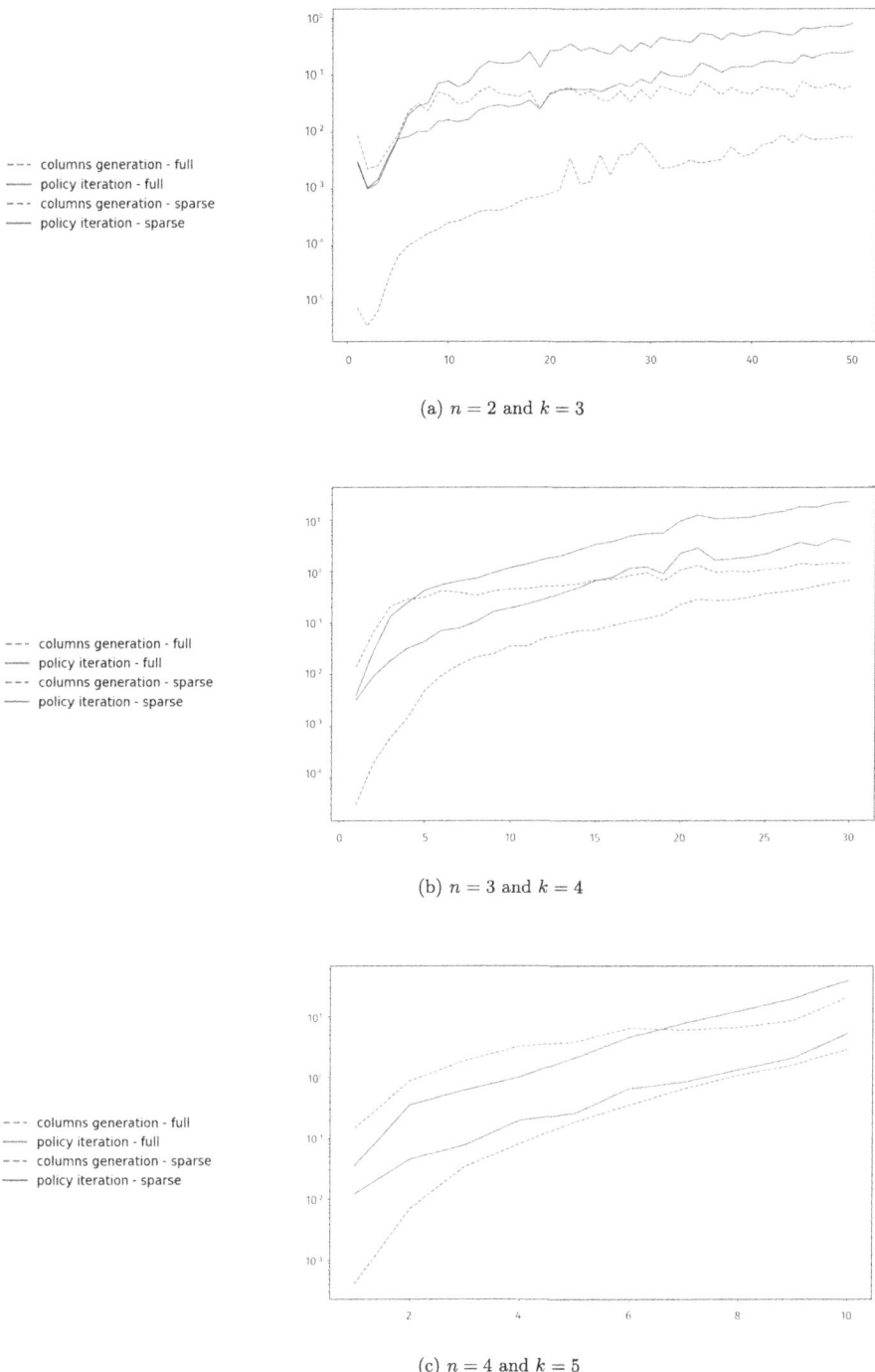

Fig. 2. Average CPU time in seconds on 10 random instances to decide the solvability of a system of k tropical polynomial inequalities in n variables in function of the degree

red. Then, the sparsity of the polynomials was used to obtain a much smaller linearized system, given by Macaulay matrices whose columns were this time obtained by computing the Minkowski sum of the supports of each polynomial. This corresponds to the curves in green.

The sheer size difference between the matrices in the first and second case is repercuted in the time difference for the policy iteration (solid line curves), which is naturally much faster in the case where the sparsity was exploited. How the bottleneck of the current implementation in this case can be observed for instances with a bigger number of variables, where the computation time of the Minkowski sum (dashed lines) to deal with sparse instances dominates the time of policy iteration.

9 Concluding Remarks

We developed an analogue of the eigenvalue method, allowing one to find solutions of systems of tropical equalities and inequalities. We focused on the problem of computing the projections of the solution set on the different coordinates. It would be interesting to extend the present methods to get additional informations on the solution set (like the number of connected components).

References

1. Akian, M., Gaubert, S., Guterman, A.: Tropical polyhedra are equivalent to mean payoff games. Int. J. Algebra Comput. **22**(1), 125001 (2012)
2. Akian, M., Béreau, A., Gaubert, S.: The tropical Nullstellensatz and Positivstellensatz for sparse polynomial systems. In: Proceedings of the 2023 International Symposium on Symbolic and Algebraic Computation, pp. 43–52. ISSAC '23, Association for Computing Machinery, New York, NY, USA (2023)
3. Akian, M., Béreau, A., Gaubert, S.: The Nullstellensatz and Positivstellensatz for sparse tropical polynomial systems (2023). arXiv:2312.05859
4. Akian, M., Gaubert, S.: Policy iteration for perfect information stochastic mean payoff games with bounded first return times is strongly polynomial (2013)
5. Akian, M., Gaubert, S., Hochart, A.: Generic uniqueness of the bias vector of finite stochastic games with perfect information. J. Math. Anal. Appl. **457**(2), 1038–1064 (2018)
6. Akian, M., Gaubert, S., Vannucci, S.: Ambitropical geometry, hyperconvexity and zero-sum games (2023)
7. Allamigeon, X., Bœuf, V., Gaubert, S.: Performance evaluation of an emergency call center: tropical polynomial systems applied to timed petri nets. In: Sankaranarayanan, S., Vicario, E. (eds.) FORMATS 2015. LNCS, vol. 9268, pp. 10–26. Springer, Cham (2015). https://doi.org/10.1007/978-3-319-22975-1_2
8. Allamigeon, X., Gaubert, S., Skomra, M.: Tropical spectrahedra. Discret. Comput. Geom. **63**, 507–548 (2020)
9. Baldwin, E., Klemperer, P.: Understanding preferences: "demand types", and the existence of equilibrium with indivisibilities. Econometrica **87**(3), 867–932 (2019)
10. Bronstein, M., Cohen, A.M., Cohen, H., Eisenbud, D., Sturmfels, B., Dickenstein, A., Emiris, I.Z. (eds.): Solving Polynomial Equations. Springer-Verlag (2005). https://doi.org/10.1007/b138957

11. Béreau, A.: Tp2s: a solver for systems of tropical polynomial equalities and inequalities (2023). python software available from https://gitlab.inria.fr/abereau/tropical-polynomial-system-solving
12. Canny, J., Emiris, I.: An efficient algorithm for the sparse mixed resultant. In: Cohen, G., Mora, T., Moreno, O. (eds.) AAECC 1993. LNCS, vol. 673, pp. 89–104. Springer, Heidelberg (1993). https://doi.org/10.1007/3-540-56686-4_36
13. Cohen, G., Gaubert, S., Quadrat, J.: Algebraic system analysis of timed Petri nets. In: Gunawardena, J. (ed.) Idempotency, pp. 145–170. Publications of the Isaac Newton Institute, Cambridge University Press (1998)
14. Desoeuvres, A., Szmolyan, P., Radulescu, O.: Qualitative dynamics of chemical reaction networks: an investigation using partial tropical equilibrations. In: Petre, I., Păun, A. (eds.) Computational Methods in Systems Biology. CMSB 2022. LNCS(), vol. 13447. Springer, Cham (2022). https://doi.org/10.1007/978-3-031-15034-0_4
15. Dhingra, V., Gaubert, S.: How to solve large scale deterministic games with mean payoff by policy iteration. In: Valuetools '06: Proceedings of the 1st International Conference on Performance Evaluation Methodologies and Tools, p. 12. ACM Press, New York, NY, USA (2006)
16. Ehrenfeucht, A., Mycielski, J.: Positional strategies for mean payoff games. Internat. J. Game Theory **8**(2), 109–113 (1979)
17. Einsiedler, M., Kapranov, M., Lind, D.: Non-archimedean amoebas and tropical varieties. Journal für die reine und angewandte Mathematik (Crelles Journal) **2006**(601) (2006)
18. Emiris, I.Z.: Toric resultants and applications to geometric modelling. In: [10], pp. 269–300 (2005)
19. Gaubert, S., Sergeev, S.: The level set method for the two-sided max-plus eigenproblem. J. Disc. Event Dyn. Syst. **23**(2), 105–134 (2013)
20. Gaubert, S., Katz, R.D., Sergeev, S.: Tropical linear-fractional programming and parametric mean payoff games. J. Symb. Comput. **47**(12), 1447–1478 (2012)
21. Gelfand, I.M., Kapranov, M.M., Zelevinsky, A.V.: Discriminants, Resultants, and Multidimensional Determinants. Birkhäuser (1994)
22. Görlach, P., Ren, Y., Zhang, L.: Computing zero-dimensional tropical varieties via projections. Comput. Complex. **31**(1), 5 (2022)
23. Grigoriev, D., Podolskii, V.: Tropical effective primary and dual Nullstellensätze. Discrete Comput. Geom. **59**, 507–552 (2018)
24. Hansen, T., Miltersen, P., Zwick, U.: Strategy iteration is strongly polynomial for 2-player turn-based stochastic games with a constant discount factor. In: Innovations in Computer Science 2011, pp. 253–263. Tsinghua University Press (2011)
25. Huber, B., Sturmfels, B.: A polyhedral method for solving sparse polynomial systems. Math. Comput. **64**(212), 1541–1555 (1995)
26. Itenberg, I., Mikhalkin, G., Shustin, E.: Tropical algebraic geometry, Oberwolfach Semin., vol. 35. Birkhäuser, Basel, second edn. (2009)
27. Itenberg, I., Viro, O.: Patchworking algebraic curves disproves the Ragsdale conjecture. Math. Intelligencer **18**(4), 19–28 (1996)
28. Jell, P., Scheiderer, C., Yu, J.: Real tropicalization and analytification of semialgeaic sets. Int. Math. Res. Not. **2022**(2), 928–958 (2020)
29. Jensen, A.N.: Tropical homotopy continuation (2016). arXiv:1601.02818

30. Lüders, C.: Computing tropical prevarieties with satisfiability modulo theories (SMT) solvers. In: Fontaine, P., Korovin, K., Kotsireas, I.S., Rümmer, P., Tourret, S. (eds.) Proceedings of SC2'20: Fifth International Workshop on Satisfiability Checking and Symbolic Computation, July 05, 2020, Paris, France. CEUR Workshop Proceedings (CEUR-WS.org) (2020)
31. Maclagan, D., Sturmfels, B.: Introduction to Tropical Geometry. Graduate Studies in Mathematics, American Mathematical Society (2015)
32. Malajovich, G.: Computing mixed volume and all mixed cells in quermassintegral time. Found. Comput. Math. **17**(5), 1293–1334 (2016)
33. Markwig, T., Ren, Y.: Computing tropical varieties over fields with valuation. Found. Comput. Math. **20**(4), 783–800 (2019)
34. Mikhalkin, G.: Enumerative tropical algebraic geometry in $\mathbb{R}^2$. J. Amer. Math. Soc. **18**, 313–377 (2005)
35. Sturmfels, B.: On the newton polytope of the resultant. J. Algebraic Combin. **3**(2), 207–236 (1994)
36. Viro, O.Y.: Real plane algebraic curves: constructions with controlled topology. Algebra i Analiz **1**(5), 1–73 (1989)
37. Zwick, U., Paterson, M.: The complexity of mean payoff games on graphs. Theoret. Comput. Sci. **158**(1–2), 343–359 (1996)

Dynamic Decomposition of Tropical Prevarieties for Celestial Mechanics

Anders Nedergaard Jensen[✉]

Aarhus University, Aarhus, Denmark
jensen@math.au.dk

Abstract. Every tropical hypersurface is the union of finitely many polyhedra. By the distributive law, intersecting a list of tropical hypersurfaces amounts to doing a huge computation of intersections of polyhedra that may be naturally organised in an enumeration tree. The idea of *dynamic enumeration* proposed by Mizutani, Takeda and Kojima (2007) is extended to *dynamic decomposition* where tropical hypersurfaces are not just chosen dynamically but also split dynamically into disjoint half-open polyhedra. This has the advantage of making the enumeration tree even thinner. We present an implementation of this enumeration algorithm in the C++ library gfanlib and do a comparison against an implementation of an existing algorithm. Finally we give an example of how the software has been used in a celestial mechanics problem.

Keywords: Tropical prevarieties · Celestial mechanics

1 Introduction

It has been almost 20 years since Hampton and Moeckel [7] used their "big Minkowski" computation to prove finiteness of relative equilibria in the 4-body problem. With the almost simultaneous introduction of tropical geometry their argument became simpler to state using the language of tropical prevarieties, i.e. intersections of tropical hypersurfaces. This makes prevarieties interesting to not only to tropical geometry but more generally to polynomial system solving. We refer to [12] for a general introduction to tropical geometry.

Definition 1. *A finite formal maximum* $f = \text{Max}_{u \in S}(c_u + u \cdot x)$ *with* $S \subseteq \mathbb{Z}^n$ *finite,* x *a vector of variables and* $c_u \in \mathbb{Q}$ *is called a* tropical polynomial *and defines a piecewise linear function of* x.

$$T(f) := \{x \in \mathbb{R}^n : \text{Max}_{u \in S}(c_u + u \cdot x) \text{ is attained at least twice}\}$$

is the tropical hypersurface *of* f. *It naturally has the structure of a polyhedral complex also denoted* $T(f)$. *The tropical prevariety of a list of polynomials* $f_1, \ldots, f_m$ *is the complex obtained by common refinement* $\bigwedge_i T(f_i)$ *or its support.*

Here the refinement of two complexes is $F \wedge G := \{\sigma \cap \tau | (\sigma, \tau) \in F \times G\}$, making the support of $\bigwedge_i T(f_i)$ the intersection $\cap_i T(f_i)$ of the supports $T(f_1), \ldots, T(f_m)$.

To keep prevariety computations lighter one wishes to avoid non-maximal faces when possible and would cover each $T(f_i)$ with for example codimension 1 polyhedra. The support of the prevariety is then computable by a naive tree enumeration using the distributive law of union and intersection, with each level corresponding to a particular $T(f_i)$. It is not uncommon that the $T(f_i)$ are fans and all intersections therefore would contain 0. Hence it is better to write $T(f_i)$ as a disjoint union of *half-open* cones, so that the tree can be pruned at vertices giving an empty intersection. If the $T(f_i)$ are not fans it is still an advantage splitting $T(f_i)$ disjointly into half-open polyhedra as the output would then consist of disjoint polyhedra, avoiding redundancy, although non-maximal polyhedra may still exist (after taking closures).

These considerations can be completely avoided in the application to mixed volume computation, where a generic choice of coefficients can be used [5,6]. In [14,15] *dynamic enumeration* was introduced, where the tropical hypersurface considered at level i may be different in different parts of the tree. However, for mixed volume computation it is better to use *tropical homotopy continuation* [11, 13].

For computing tropical prevarieties, an implementation using dynamic enumeration, relation tables [5] and half-open cones and relying on cddlib [4] has been present in gfan [9] for a long time. In [8] the ideas were further refined and combined with the Parma Polyhedra Library (PPL) [3]. An observation is that adjacency information should not be forgotten when passing from edge graphs to normal tropical hypersurfaces.

The contribution of this article is the idea of dynamic decomposition, where we do not make a predefined decomposition of $T(f_1), \ldots, T(f_m)$ into half-open polyhedra. Instead, *dynamically*, at each vertex of the enumeration tree, a *decomposition* is chosen in an attempt to keep the number of required half-open polyhedra low. We report on the implementation of this algorithm in gfanlib.

The algorithm described here is not compatible with the Osserman-Payne trick [16] which says that if an intersection of some polyhedra gets too low-dimensional, then it can be discarded because it does not contribute a maximal polyhedron to the prevariety. An other drawback of our approach is that it does not currently incorporate the situation of *positive tropical hypersurfaces* where some faces of the hypersurfaces are missing [18, Definition 3.2].

2 Representing Polyhedra and Polyhedral Complexes

Imagine the situation where $T(f_1), \ldots, T(f_m)$ are fans. If they are covered by closed cones, every cone would contain the origin, and every intersection would be non-empty potentially contributing to the prevariety. To avoid this half-open polyhedra will be used, and the support of a polyhedral complex is represented as a *disjoint* union of half-open polyhedra.

Since both open and closed half-spaces are half-open cones and our class of polyhedra should be closed under intersection, a half-open polyhedron will be any set described by a finite number of strict and non-strict linear inequalities - including $\mathbb{R}^n$ and $\emptyset$.

A nice feature of working with half-openness is that it is *no restriction* considering only cones – as opposed to the closed case, where things get messy at infinity. Indeed, a half-open polyhedron in $\mathbb{R}^n$ is *represented* by the cone over its embedding into $\{1\} \times \mathbb{R}^n$ inside $\mathbb{R}^{n+1}$, which is a half-open cone. Further restricting all computations to the half-open cone $\mathbb{R}_{>0} \times \mathbb{R}^n$ makes things such as f-vectors match up between a polyhedron and its cone. *Therefore, from now on in the dual space we shall only consider half-open cones and not general polyhedra.* (In particular, coefficients of polynomials are ignored in the next section.)

A last thing to notice is that while the vectors of the support set S of a polynomial f typically has single digit entries, to fit the coefficients into $\mathbb{Z}$ after appropriate scaling, the coordinate of an inequality where these entries end up needs much higher precision.

3 Face Enumeration and Decomposition

In this section we describe how to obtain a tropical hypersurface as a disjoint union of half-open cones given the Newton polytope of the defining polynomial.

A typical method is to direct the edge graph according to a generic direction to get a unique sink orientation and letting outgoing edges define strict inequalities. Alternatively, we define the half-open shift of a closed cone C:

$$\tilde{C} = \{x \in \mathbb{R}^n | (x + (\varepsilon^1, \ldots, \varepsilon^n) \in C \text{ for } \varepsilon > 0 \text{ sufficient small}\}.$$

For any polytope P the (dual) ambient space is a disjoint union of shifts of the maximal normal cones of P.

We pass from the disjoint union of half-open shifts to a disjoint description of $T(f)$ by applying [8, Algorithm 1] having the following specifications:

Algorithm 1
Input: *A full-dimensional half-open cone C.*
Output: *A collection of disjoint relatively open cones with union equal $C \cap \partial C$.*

3.1 Dynamic Enumeration

Computing a tropical prevariety can now be done (in parallel) in an enumeration tree. Each level of the enumeration tree would then correspond to a hypersurface. As for the mixed volume computation in [15] it is an advantage to choose the ordering of hypersurfaces differently in different parts of the enumeration tree. The advantage of this for prevariety computation in terms of the required number of cone intersections was documented for the Albouy-Chenciner equations in [8]. As in [8] we use relation tables to guide the choice of the next hypersurface to consider. This is less useful for dynamic decomposition, since the actual

4 The Dynamic Decomposition Algorithm

The idea of dynamic decomposition is illustrated in Fig. 1. Suppose we wish to compute the blue tropical hypersurface $T(f)$ restricted to the grey half-open polyhedron C. Clearly a decomposition that includes the blue center vertex in one of the two higher dimensional blue polyhedra will be advantageous since, if used in an enumeration tree, only two children of the current vertex are needed.

A way to find a good decomposition is to imagine a generic point w in C. Either we can use w to direct the edge graph of Newton polytope of f, or consider the half-open shifts $K^w := \{x \in \mathbb{R}^n | x + \varepsilon w \in K \text{ for } \varepsilon > 0 \text{ sufficiently small}\}$ for any full-dimensional normal cone K of the Newton polytope of f.

In general the situation can be more complicated, since $T(f)$ and C may not intersect transversely and we could have $\dim(T(f) \cap C) = \dim(C)$. In that case a good choice of w would be in $T(f) \cap C$.

A global (i.e. static) choice of w has some of the same advantages, but using different choices in different parts of the enumeration tree is better. This is what we refer to as *dynamic decomposition*.

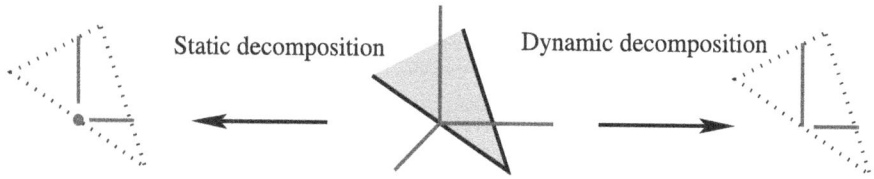

Fig. 1. A schematic drawing of a half-open polyhedron C and a blue tropical hypersurface $T(f)$ (center). The hypersurfaces can be decomposed in several ways. The dynamic decomposition will try to minimise the number of components in the decomposition (Color figure online)

4.1 Restricted Tropical Hypersurface Computation

The core part of the dynamic decomposition is Algorithm 2 presented below. We do not present the higher level algorithm in more detail. We let $\mathrm{NF}(f)$ denote the normal fan of the Newton polytope of f and use supp for taking supports.

Algorithm 2 (Restricted Tropical Hypersurface)
Input: *A tropical polynomial f and inequalities defining a half-open cone C.*
Output: *A collection $\mathcal{F}$ of disjoint half-open polyhedral cones in $\mathbb{R}^n$ such that*

$$\bigcup_{\sigma \in \mathcal{F}} \sigma = \mathrm{supp}(T(f)) \cap C.$$

1. *Compute a minimal set $S \subseteq \overline{C} \wedge \mathrm{NF}(f)$ of closed cones covering $\overline{C}$.*
2. *Create the dual graph G having S as vertices and two cones connected iff they share a facet.*
3. *Colour each vertex of G arising from multiple terms of f.*
4. *Pick a vertex R of G to become a sink (giving priority to coloured vertices of high degree).*
5. *Choose a generic w in the relative interior of R.*
6. *Use w to orient the edges of G, making R the unique sink.*
7. *Let $\mathcal{F} = \{K^w | \text{coloured } K \in V(G)\} \cup \bigcup_{\text{uncoloured } K \in V(G)} K^w \wedge \partial K$.*

Step 1 can be performed by doing a graph traversal on the maximal cones of $C \wedge NF(f)$, at the same time obtaining adjacency information for step 2. To get a starting cone for the traversal a perturbed relative interior point of C can be used to find a vertex with normal cone intersecting $\overline{C}$ in dimension $\dim(C)$.

In step 3 the purpose of the colouring is to be able to figure out whether it is a cone or its boundary contributing to $\mathrm{supp}(T(f)) \cap C$.

In step 4 K^w is the half-open shift described above. With a slight misuse of notation $K^w \wedge \partial K$ is a covering of $K^w \cap \partial K$ by half-open cones obtainable with Algorithm 1.

5 Implementation Details

The dynamic decomposition prevariety algorithm has been implemented as a C++ template that could in theory be instantiated with any implementation of an ordered integral domain as coefficient type. The implementation follows the ideas described in [11], in particular concerning integer arithmetic and parallelisation. Generic points are represented as symbolic perturbations.

The library contains an implementation of a cone class for V-represented cones. Generators are stored as columns of a matrix and simplex type pivoting is done. By the observation at the end of Sect. 2, before pivoting this matrix has only one row with big entries. Therefore machine integers are often sufficient for tableau entries as they are bounded by subdeterminants of the starting matrix. H-cones are represented by their duals and half-open H-cones by lifts.

The source code of gfanlib will be available in a beta version from the gfan webpage [9]. At the moment a problem is that overflow checks are not fully implemented.

5.1 Stack Allocators

Because of the needed intermediate data, the memory management is more involved than in [11], where tableau memory could be allocate for all threads at initialisation. A new feature in C++17 is that of polymorphic allocators and accompanying container classes (such as pmr::vector). We make an allocator that is a specialisation of memory_resource and allocates from a stack. This allows fast allocation and deallocation, efficient use of data cache and avoids fragmentation. Each threads then uses two such stack allocators. A typically function will take a primary allocator and secondary allocator as arguments. The returned (pass-by-value) result is supposed to be have its internals allocated with the primary allocator and the efficiency of this is ensured by move semantics. If nested calls to function are needed the two allocators can be swapped in such calls, with the old primary allocator acting now as a temporary stack. Objects collected for the final output (i.e. the computed half-open cones covering the prevariety) will have their internal data allocated with a separate allocator, typically requiring a deep copy.

6 Experiments

We report on experiments with our dynamic decomposition implementation on cyclic-n examples and some celestial mechanics problems. Since the implementation does not beat Sommars' implementation [8] with PPL [3] by much, the main contribution of this section is actually the experiments supporting the claim that dynamic decomposition leads to fewer cones than static decomposition. This observation is independent of how well optimised the implementation is and suggests that dynamic decomposition has some merit and a smaller enumeration tree and might be faster with an efficient implementation. Note that it is reported in [8] that gmp integers were used in the experiments. On the other hand our templates were instantiated with 32 or 64 bit machine integers in tests.

6.1 Data Supporting Claim of Fewer Cones

In Table 1 the columns "#cones in static" and "#cones in dynamic" list the number of half-open polyhedra in the computed disjoint unions in the static and dynamic decomposition algorithm respectively. It is worth noting that the first column therefore depends only on the initially chosen decomposition of $T(f_1), \ldots, T(f_m)$ including the order in which the facets are processed in Algorithm 1. Besides using the same order for directing all edges of the Newton polytopes no attempt has been made to reduce the numbers in this column. The second column could be highly sensitive to the implementation choices of the dynamic decomposition. For example not attempting to (heuristically) minimize the number of produced cones could lead to higher counts. We observe that a 30–40 percent reduction in the number of produced cones is not uncommon when passing from static to dynamic, but sometimes the reduction is negligible (~ 10 percent).

6.2 Running Time

We make a comparison against ([8], Table 1). Involved are two systems: a dual Intel Xeon E2670 CPU and a 2.2 GHz Intel Xeon E5-2699. The latter is according to https://cpubenchmark.net approximately 30% faster in single thread performance and has 37.5% more cores than the first, resulting in the latter being 61% faster. These numbers are of course unreliable as they highly depend on the actual task being measured. Both systems were used in [8] and are outdated. The present experiments were performed on the first with comparison against Sommars running on the second. We restrict ourselves to the cyclic n-roots example class to illustrate the assymptotics, while the bigger celstial mechanics problem of Sect. 6.3 was done with a slightly older version of the software.

Sommars managed to compute the cyclic-16 prevariety with his software. We have made no attempt to redo that computation with dynamic decomposition. In the table the number of cones in the static decomposition was computed with an old prevariety implementation in gfan that cannot compute cyclic-14 and cyclic-15, explaining why the numbers are missing for those examples.

Table 1. Comparison of half-open cones in decompositions and algorithms for some cyclic n-root systems. The three columns attributed to Sommars are for static decomposition (dynamic enumeration) and were taken from [8]. For comparison, the time for computing the mixed volume of Cyclic-15 was 461.3 s with one thread and 35.8 s for 16 threads as reported in [11]. The appearance of non-monotonicity in the number of cones in the static decomposition comes as no surprise since $2^2|12$, see [2]

Example	# cones in static	# cones in dynamic	Sommars' [8] 1 thr.	10 thr.	20 thr.	Dynamic decomposition 1 thr.	10 thr.	20 thr.
Cyclic-9	520	389	2.8 s	1.2 s	1.4 s	2.1 s	0.9 s	0.9 s
Cyclic-10	3532	2022	9.8 s	4.4 s	3.7 s	10.0 s	2.1 s	2.1 s
Cyclic-11	6536	5819	50 s	16.8 s	20.3 s	1 m 15 s	10.0 s	7.2 s
Cyclic-12	111473	54679	5 m 02 s	1 m 05 s	1 m 03 s	6 m 47 s	47.6 s	31.5 s
Cyclic-13	92492	78360	46 m 59 s	8 m 30 s	6 m 20 s	63 m 28 s	7 m 02 s	4 m 24 s
Cyclic-14		1143350	383 m	67 m 31 s	46 m 37 s	374 m	40 m 31 s	25 m 19 s
Cyclic-15		3467340		626 m	464 m	2892 m	312 m	194 m

The new implementation is faster than the old one, but not convincingly. In particular not when taking into account that overflow checks are not fully implemented in the new implementation. Moreover, there is some uncertainty as to whether the Parma Polyhedra Library [3] was used optimally in [8]. While PPL does support computations with machine integers it is unclear if this was attempted by the used PPL class NNC_Polyhedron. Profiling data shows that a major part of the computation takes place in gmp. If there was no attempt to use machine integers there is much room for improvement since gmp numbers are easily between 1–100 times slower than with machine integers. The required memory management for gmp integers would explain why the new algorithm performs better with more threads.

It would be interesting to investigate

- whether our simplex implementation could be improved and
- whether Sommars' implementation works well with fixed precision.

Alternatively, one could try implementing dynamic decomposition using PPL.

6.3 Finiteness Computation for Generic Masses N=5

In [10] our implementation was used to provide a computer proof of the main result of [1], that for generic masses the Newtonian 5-body problem has only finitely many relative equilibria up to symmetry. We briefly describe the argument here.

We are given 35 polynomial equations in 10 variables and 5 parameters and need to prove that for almost all choices of parameters the system has a finite number of solutions. Making particular choices of valuations for the parameters the prevariety of the tropicalisation of the system can be computed with our implementation. The computation took 5 min. The recession cone is pointed, implying that the tropical variety is 0 dimension. It is concluded that for almost all choices of parameters the system has only finitely many solutions in $(\mathbb{C}^*)^{10}$.

An inconclusive computation for $N = 6$ took 3 days. There is a lot of room for experimenting with different defining equations, but the time consuming prevariety computation is the limiting factor. A successful computation would settle the generic answer in the next instance of Smale's 6th problem.

7 Conclusion

Tropical prevarieties can be a powerful tool for arguing in polynomial system solving. While the proposed dynamic decomposition method for their computation disjointly covers the prevariety with fewer cones than conventional methods, the algorithm is considerably more complicated and may not be worth the savings. On the implementation side, we almost have a tie in performance against [8]. Both implementations have potential for improvement. It should be investigated whether Sommars' implementation [8] can be adjusted to working with machine integer precision. On the other hand our dynamic decomposition implementation would benefit from a more mature simplex implementation.

Aknowledgements. The idea of dynamic decomposition was discussed with Jeff Sommars. This is why the same term appears as future work in his thesis [17]. Bjarne Knudsen contributed the abstract parallel tree traverser to the gfan project. The experiments with celestial mechanics in Sect. 6.3 is joint work with Anton Leykin and was reported in [10].

References

1. Albouy, A., Kaloshin, V.: Finiteness of central configurations of five bodies in the plane. Ann. Math. **176**(1), 535–588 (2012)
2. Backelin, J.: Square multiples n give infinitely many cyclic n-roots. Reports, Matematiska Institutionen 8, Stockholms universitet (1989)
3. Bagnara, R., Hill, P.M., Zaffanella, E.: The Parma Polyhedra Library: toward a complete set of numerical abstractions for the analysis and verification of hardware and software systems. Sci. Comput. Program. **72**(1–2), 3–21 (2008)
4. Fukuda, K.: cddlib reference manual, cddlib Version 094m. Swiss Federal Institute of Technology, Lausanne and Zürich, Switzerland (2021), https://people.inf.ethz.ch/fukudak/cdd_home/
5. Gao, T., Li, T.: Mixed volume computation for semi-mixed systems. Discr. Comput. Geometry **29**(2), 257–277 (2003)
6. Gao, T., Li, T.Y., Wu, M.: Algorithm 846: MixedVol: a software package for mixed-volume computation. ACM Trans. Math. Softw. **31**(4), 555 560 (2005)
7. Hampton, M., Moeckel, R.: Finiteness of relative equilibria of the four-body problem. Invent. Math. **163**(2), 289–312 (2006)
8. Jensen, A., Sommars, J., Verschelde, J.: Computing tropical prevarieties in parallel. In: Proceedings of the International Workshop on Parallel Symbolic Computation. PASCO 2017, Association for Computing Machinery, New York, NY, USA (2017)
9. Jensen, A.N.: Gfan, a software system for Gröbner fans and tropical varieties. https://math.au.dk/~jensen/software/gfan/gfan.html
10. Jensen, A.N., Leykin, A.: Smale's 6th problem for generic masses (2024), submitted to Journal of Experimental Mathematics
11. Jensen, A.N.: An implementation of exact mixed volume computation. In: Proceedings, Mathematical Software - ICMS 2016 - 5th International Conference, Berlin, Germany, 11–14 July 2016, pp. 198–205 (2016)
12. Maclagan, D., Sturmfels, B.: Introduction to Tropical Geometry, Graduate Studies in Mathematics, vol. 161. American Mathematical Society, Providence, RI (2015)
13. Malajovich, G.: Computing mixed volume and all mixed cells in quermassintegral time. Found. Comput. Math. **17**(5), 1293–1334 (2017)
14. Mizutani, T., Takeda, A.: DEMiCs: a software package for computing the mixed volume via dynamic enumeration of all mixed cells. In: Stillman, M., Takayama, N., Verschelde, J. (eds.) Software for Algebraic Geometry, The IMA Volumes in Mathematics and Its Applications, vol. 148, pp. 59–79. Springer-Verlag (2008). https://doi.org/10.1007/978-0-387-78133-4_5
15. Mizutani, T., Takeda, A., Kojima, M.: Dynamic enumeration of all mixed cells. Discr. Comput. Geometry **37**(3), 351–367 (2007)
16. Osserman, B., Payne, S.: Lifting tropical intersections. Doc. Math. J. DMV **18**, 121–175 (2013)
17. Sommars, J.C.: Algorithms and Implementations in Computational Algebraic Geometry (2018), PhD thesis, University of Illinois at Chicago
18. Speyer, D., Williams, L.K.: The positive Dressian equals the positive tropical Grassmannian. Trans. Am. Math. Soc. Ser. B **8**, 330–353 (2021)

Regular Flips in mptopcom

Lars Kastner

Technische Universität Berlin, Chair of Discrete Mathematics/Geometry,
Berlin, Germany
kastner@math.tu-berlin.de

Abstract. A triangulation of a point configuration is regular if it can be given by a height function, that is every point gets lifted to a certain height and projecting the lower convex hull gives the triangulation. Checking regularity of a triangulation usually is done by solving a linear program. However when checking many flip-connected triangulations for regularity, one can instead ask which flips preserve regularity. When traversing the flip graph for enumerating all regular triangulations, this allows for vast reduction of the linear programs needing to be solved. At the same time the remaining linear programs will be much smaller.

Keywords: triangulation · reverse search · mptopcom

1 Introduction

Enumerating regular triangulations is an important problem in many areas of mathematics, such as tropical geometry [10,12], optimization [5, Section 1.2], and mathematical physics [4]. There are two main software systems for enumerating triangulations, namely TOPCOM [13] and mptopcom [8] which is a fusion of the former and polymake [7]. This paper will provide details on recent developments of mptopcom.

The algorithm of mptopcom is reverse search [2]. We will leave out most details, but essentially reverse search on the flip graph associates to every triangulation a unique predecessor. In our case this is the flip neighbor with lexicographically largest GKZ-vector. Taking only the edges from such a predecessor relation in the flip graph results in the so-called *reverse search tree*.

The two main advantages of reverse search are that reverse search is output-sensitive, meaning it only consumes a constant amount of memory and that it can be parallelized, mptopcom uses the *budgeted reverse search* [3]. Further explanation of mptopcom's internal workings can be found in [8] and on the mptopcom webpage[1]. Note that parallelization has also been adopted in TOPCOM.

To enumerate regular triangulations exclusively, one can restrict to regular triangulations, but still allowing all flips in between. Second, we could restrict to regular flips only. The second approach results in a different reverse search

---
[1] https://polymake.org/mptopcom.

tree, as it considers the subgraph of regular flips of the flip graph. Thus the goal of this paper is to mitigate the main bottleneck of solving regularity linear programs (LP): First, we want to reduce the size of the LPs appearing. Second, we want to reduce the number of LPs that need to be solved altogether.

2 Regular Flips

In this section we will carve out a description of regular flips between regular triangulations in terms of GKZ-vectors. We follow the notation and conventions of the triangulation bible [5].

Start with a point configuration $A = \{a_1, \ldots, a_n\}$ and let T be a triangulation of A. Recall that the *GKZ-vector* $\operatorname{gkz}(T) \in \mathbb{R}^n$ of T is defined as

$$\operatorname{gkz}(T)_i = \sum_{s \ni a_i, s \in T} \operatorname{vol}(s).$$

Now for a flip $f : T \rightsquigarrow T'$ we define $\operatorname{gkz}(f) := \operatorname{gkz}(T') - \operatorname{gkz}(T)$.

Example 1. Consider the two triangulations of the standard square $[0,1]^2$ and the flip between them.

$$f: \quad \begin{array}{c} 3 2 \\ \boxed{\diagup} \\ 0 1 \end{array} \quad \rightsquigarrow \quad \begin{array}{c} 3 2 \\ \boxed{\diagdown} \\ 0 1 \end{array}$$
$$(2,1,2,1) \quad (-1,1,-1,1) \quad (1,2,1,2)$$

The second row contains the GKZ-vectors of the triangulations and the flip GKZ-vector between them.

To decide whether f is regular from $\operatorname{gkz}(f)$ it is important that two flips $f_0 : T \rightsquigarrow T'$ and $f_1 : T \rightsquigarrow T''$ have different GKZ-vectors. For an idea why this could be true, take the following proposition.

Proposition 2. *Let* $f : T \rightsquigarrow T'$ *be a flip. Then* $\operatorname{gkz}(f) \neq 0$.

Proof. It is already clear that $\operatorname{gkz}(f)$ will have zero entries in all coordinates, but those that belong to the corresponding circuit. Next Lemma 2.4.2 of [5] states that the two different triangulations of a circuit are regular. But two distinct regular triangulations cannot have the same GKZ-vector.

Every flip arises from a *corank-one configuration* [5, 2.4.1], that is a subset $J \subseteq \{1, \ldots, n\}$ such that $\operatorname{conv}(\{a_j | j \in J\})$ is full-dimensional and there exists a unique, non-trivial affine dependence relation $\sum_{j \in J} \lambda_j a_j = 0$ with $\sum_{j \in J} \lambda_j = 0$, of course up to scalar multiple. The set J can now be divided into three parts:

$$J_+ = \{j \in J \mid \lambda_j > 0\}, \ J_0 = \{j \in J \mid \lambda_j = 0\}, \ J_- = \{j \in J \mid \lambda_j < 0\}.$$

A corank-one configuration has exactly two triangulations [5, Lem. 2.4.2], namely

$$T_+ = \{J \setminus \{j\} \mid j \in J_+\}, \text{ and } T_- = \{J \setminus \{j\} \mid j \in J_-\}.$$

We say that a triangulation T of A is *compatible* with a corank-one configuration J if it contains the simplices of the triangulation T_+. Replacing these by the simplices of T_- is exactly what it means to apply a flip. The pair (J_+, J_-) is (the *Radon partition* of) a circuit. Note that a circuit can give rise to at most one flip:

Lemma 3. *Assume that J is a corank-one configuration with circuit (J_+, J_-) compatible with the triangulation T. Then there cannot be a second corank-one configuration compatible with T and the same circuit (J_+, J_-).*

Proof. Write $J = J_+ \cup J_0 \cup J_-$ and assume that there is $J' = J_+ \cup J'_0 \cup J_-$ with $J_0 \neq J'_0$. Denote the (sub-)triangulations associated to J by T_+ and T_-, and those of J' by T'_+ and T'_-. Each of these corank-one configurations give rise to a flip we can apply to T. By assumption, no maximal simplex of T_+ is part of T'_+ and vice versa, since every maximal simplex of T_+ contains J_0 and of T'_+ contains J'_0. Now if we apply the flip associated to J, the maximal simplices of T'_+ remain untouched. In particular, every simplex of T'_+ will still have a simplex corresponding to J_- as a face. On the other hand, T_- got inserted and here every maximal simplex has a face corresponding to J_+. However

$$\mathrm{conv}(\{a_j \mid j \in J_+\}) \cap \mathrm{conv}(\{a_j \mid j \in J_-\}) \neq \emptyset,$$

stemming from the affine dependence equation, while $J_+ \cap J_- = \emptyset$. That means that this intersection of faces does not form a face of both and hence is invalid. Thus such J and J' cannot simultaneously exist.

Let us investigate the GKZ-vector of a flip. We already deduced that it can only have non-zero entries in the indices corresponding to the corank-one configuration J. Thus we can write the subvector of $\mathrm{gkz}(f)$ in the indices of J as

$$\mathrm{gkz}(f)_J = \mathrm{gkz}(T_-) - \mathrm{gkz}(T_+).$$

Next, one deduces that the points a_j for $j \in J_0$ are part of every simplex for both T_+ and T_-, hence for these points we have $\mathrm{gkz}_j(f) = 0$. Now take $j \in J_+$. Then we have

$$\mathrm{gkz}_j(f) = \mathrm{gkz}_j(T_-) - \mathrm{gkz}_j(T_+) = \sum_{a_j \in S \in T_-} \mathrm{vol}(S) - \sum_{a_j \in S \in T_+} \mathrm{vol}(S).$$

By construction though, a_j is part of every simplex of T_-, so the first part is just the volume of the convex hull of the corank-one configuration, write $VJ = \mathrm{vol}(\mathrm{conv}\{a_i \mid i \in J\})$. There is exactly one maximal simplex in T_+ that a_j is not part of, namely stemming from $J \setminus \{j\}$. Thus we get

$$\mathrm{gkz}_j(f) = VJ - (VJ - \mathrm{vol}(\mathrm{conv}\{a_i \mid i \in J \setminus \{j\}\}))$$
$$= \mathrm{vol}(\mathrm{conv}\{a_i \mid i \in J \setminus \{j\}\}) > 0.$$

Similarly we get $\mathrm{gkz}_j(f) < 0$ for $j \in J_-$. Thus $\mathrm{gkz}(f)$ uniquely determines the corresponding circuit and using Lemma 3 a unique corank-one configuration.

Hence given gkz(f) we can uniquely identify the corresponding flip f. In particular, for two flips f and f' of T, gkz(f) cannot be a scalar multiple of gkz(f').

Note that for a fixed regular triangulation T, all the vectors gkz(f) for $f : T \rightsquigarrow T'$ form the edge cone of the secondary polytope at the vertex gkz(T). The dual of this cone is sec(T). With this in mind we can state the following theorem.

Theorem 4. *Let $f : T \rightsquigarrow T'$ be a flip between two triangulations, with T being regular. If gkz(f) forms a facet of sec(T), then T' is regular. Equivalently, if gkz(f) generates an extremal ray of the edge cone of the secondary polytope at T, then T' is regular.*

Proof. Assume gkz(f) forms a facet, then dually it corresponds to an edge of the secondary polytope. By the discussions above it is the unique flip with that edge direction. Hence it is the unique flip to the vertex neighboring gkz(T) in direction gkz(f). But vertices of the secondary polytope correspond to regular triangulations. Hence T' must be regular.

Usually the linear program we need to solve to determine whether gkz(f) forms a facet of sec(T) is much smaller than the LP for determining whether T is regular, finishing our goals of reducing the size of the LPs we need to solve.

3 Caching

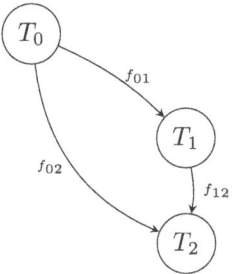

Fig. 1. Flips between three triangulations

At this point we want to illustrate how the new approach interferes with our previous caching approach [8, Sect. 7.4], where we would cache whether a triangulation was regular, and consider a flip valid whenever the target triangulation is regular. Using this caching approach can result in mixing in non-regular flips with the regular flips that we want to restrict to. For a flip $f_{ij} : T_i \rightsquigarrow T_j$ denote the reverse flip as $f_{ji} : T_j \rightsquigarrow T_i$.

To further explain this example, we need the notion of up- and downflips, which for the sake of brevity will just be the vertical direction of the flip in Fig. 1.

A flip is part of the reverse search tree if its reverse is the largest regular upflip from the target triangulation, see [8] for further details.

In Fig. 1, assume that the T_i are regular triangulations, and that f_{01} and f_{12} are regular flips, while f_{02} is not. Furthermore assume that regularity of T_2 was not cached when we were in T_0. Then f_{02} was correctly identified as non-regular and thus not used. But when we arrive in T_2 via f_{12}, f_{20} is considered valid, since caching determines that T_0 is regular and actual regularity of f_{20} is never checked. Since f_{02} is the largest upflip from T_2, the algorithm incorrectly determines that f_{12} is not part of the reverse search tree and T_2 is never visited.

4 Avoiding Linear Programs

The edge cone of the secondary polytope at $\text{gkz}(T)$ is spanned by the $v^1, \ldots, v^n$, with $v^i = \text{gkz}(f_i)$, where f_i is the i-th flip of the triangulation T. We assume that T is regular, hence this cone is pointed. None of the v^i can be scalar multiples of each other, due to Lemma 3. Both conditions are necessary for the following tricks to work. According to Theorem. 4, we need to find the extremal rays among the v^i, translating to solving the following linear program for every $i = 1, \ldots, n$:

$$v^i = \sum_{j \neq i} x_j \cdot v^j, \text{ with } x_j \geq 0.$$

Thus, for every regular triangulation T we have to solve n LPs, where n is the number of flips from T.

One could try to compute the rays of this cone directly, but due the high dimension this has its own set of challenges [1]. Nevertheless, considering all flips at the same time allows us to preprocess this system. As remarked in the proof of Proposition 2, the only non-zero entries of a flip GKZ-vector corresponds to

	0	1	2	3	4	5	6	7	8	9	10	11	12	13	14	15	16	17
v^1	0	0	0	0	0	0	0	0	-3	3	0	0	0	0	3	-3	0	0
v^2	0	0	0	0	0	0	-1	0	0	0	0	1	1	0	0	0	0	-1
v^3	-1	0	0	0	1	0	1	0	0	0	-1	0	0	0	0	0	0	0
v^4	0	0	1	0	-1	0	0	0	-1	0	1	0	0	0	0	0	0	0
v^5	-1	1	0	0	0	0	1	0	0	0	0	-1	0	-1	0	0	0	1
v^6	0	0	0	0	-1	1	0	0	0	0	0	0	0	0	0	0	1	-1
v^7	0	0	-3	3	0	0	0	0	3	-3	0	0	0	0	0	0	0	0
v^8	0	0	0	0	0	0	0	1	0	0	-1	0	0	-1	0	0	1	0
v^9	0	0	0	-1	1	0	0	0	0	1	0	-1	0	0	0	0	-1	1
v^{10}	0	0	0	0	0	0	1	0	0	-1	0	0	-1	0	0	1	0	0
v^{11}	0	-1	1	0	0	0	0	0	0	0	0	0	1	-1	0	0	0	0
v^{12}	1	-1	0	0	0	0	-1	0	0	0	1	0	0	1	0	0	-1	0

Fig. 2. An example set of flip GKZ-vectors for a regular triangulations of $\Delta_2 \times \Delta_5$. The corresponding triangulation can be found at our GitHub repo

the circuit giving rise to the flip. Thus most entries of v^i will be zero, unless in boundary case point configurations. Hence the system we are considering is quite sparse and there are a few simple tricks we can apply to spot some rays immediately.

We present these in a simplified form, where v^1 is the pivot. Note that by a convex combination inside of a cone we mean a positive linear combination of the generators. We will use the term ray to mean both ray and ray generator.

Proposition 5. *Assume that $v^1[0] > 0$ and $v^i[0] = 0$ for all other $i \neq 1$. Then v^1 is a ray and the subsystem $v^2, \ldots, v^n$ can be solved without including v^1.*

Proof. Every convex combination using v^1 will have first coordinate > 0, so it cannot give $v^2, \ldots, v^n$. Similarly convex combinations of $v^2, \ldots, v^n$ result in the first entry being zero. Thus all vectors v^i with $v^i[0] = 0$ form a face of the cone.

Example 6. One example is v^6 in Fig. 2, it is the only vector with $v^6[5] > 0$.

Remark 7. The above proposition can be generalized such that we have

$$v^1[0], \ldots, v^m[0] > 0 \text{ and } v^{m+1}[0], \ldots, v^n[0] = 0.$$

However then we cannot deduce that $v^1, \ldots, v^m$ are rays, but the linear programs for $v^{m+1}, \ldots, v^n$ become much smaller.

Next we will reduce to smaller systems by including some modified vectors that are needed to solve the smaller system correctly, but that are already known not to be rays, so we do not need to solve a LP for these.

Proposition 8. *Assume that $v^1[0] > 0$ and that $v^2[0] < 0$, while $v^i[0] = 0$ for all other $i > 2$. Then v^1 and v^2 are rays and it suffices to solve the smaller system $v^3, \ldots, v^n, v^1[0] \cdot v^2 - v^2[0] \cdot v^1$, where the last vector is not a ray of the larger system.*

Proof. Considering convex combinations it is again obvious that v^1 and v^2 are rays. Now for a convex combination resulting in one of the vectors $v^3, \ldots, v^n$, we need the first entries of v^1 and v^2 to cancel, resulting in the last vector. However this last vector is already a convex combination of v^1 and v^2.

Remark 9. Just as with Proposition 5, we can generalize the previous proposition: Assume that $v^1[0] > 0$, $v^2[0], \ldots, v^m[0] < 0$ and $v^i[0] = 0$ for all other $i > m$. Then v^1 is a ray and it suffices to solve the smaller system

$$v^{m+1}, \ldots, v^n, \ v^1[0] \cdot v^2 - v^2[0] \cdot v^1, \ \ldots, \ v^1[0] \cdot v^m - v^m[0] \cdot v^1,$$

where the last $m-1$ vectors are not rays of the larger system.

Example 10. There are several examples for this situation in Fig. 2, e.g. the columns 0, 1, 2, 3, 4, 7, 8 and 9 work for Remark 9. On the other hand, column 6 does not work.

Given a column index it is clear which operation to apply and which subsystem to consider. Take the following sequence of column indices: $[5, 7, 16, 0, 1, 2, 8, 9, 6]$. After applying these operations the only rows left to consider are v^3, v^5, v^9, v^{12}, all other rows have now been confirmed to be rays.

Remark 11. In the case that $n = 2$, where we need to determine whether v^1 is a ray and we already know that v^2 is not a ray of the larger system, it is enough to determine whether v^2 is a scalar multiple of v^1.

All of these tricks can be applied recursively to the smaller subsystems. They are much cheaper than solving actual LPs.

Example 12. In Table 1 we let mptopcom 1.4 output the number of LPs it actually solved for different examples. We chose the four dimensional cube I^4 since it has become the standard first benchmark for enumeration of triangulations, P_5 from [11] due to its relevance in optimization, and the products of simplices due to their connection to matroid theory [14] and tropical geometry [6,9]. The entry 0 for $\Delta_2 \times \Delta_5$ means that all LPs were of the type from in Remark 11. An interpretation of a small number of LPs is that non-regular triangulations have very few flips, i.e. have high flip-deficiency [5, Sec. 7.2].

Table 1. Number of linear programs actually solved by mptopcom 1.4 (single threaded) for some examples. (with --regular --orbit-cache 40000 --flip-cache 40000)

Example	I^4	P_5	$\Delta_2 \times \Delta_5$	$\Delta_2 \times \Delta_6$
# orbits of regular triangulations	235 277	27 248	13 621	531 862
Linear programs solved	6 642	638	0	261

5 Experimental Results

Table 2. Example timings for enumerating regular triangulations

	Example	I^4	P_5	$\Delta_2 \times \Delta_5$	$\Delta_2 \times \Delta_6$
1	mptopcom 1.1 (20 threads)	1159.15	793.61	67.44	15786.31
2	mptopcom 1.2 (20 threads)	1380.39	601.68	67.61	15379.18
3	mptopcom 1.3 (20 threads)	991.45	170.66	52.39	2670.98
4.	mptopcom 1.4 (20 threads)	82.79	31.22	9.44	405.05
5	TOPCOM 1.1.2 (single threaded)	709.97	–	50.17	5178.04
6	mptopcom 1.4 (single threaded)	958.92	250.55	30.22	5327.87
7	mptopcom 1.4 (single threaded)	495.98	83.17	20.20	3641.78
8	mptopcom 1.4 (single threaded)	339.08	78.78	20.04	2688.20

All timings are measured in seconds

Table 2 contains a few example runs for different settings, all run on the cluster of the mathematics department of TU Berlin. The examples are the same as in Table 1. The runs 1 and 2 are almost the same except for small outliers caused by other jobs running in parallel on the cluster. We ran TOPCOM with the option --affinesymmetries, since affine symmetries are assumed for mptopcom anyway. For P_5 we only counted *central* regular triangulations, i.e. triangulations such that all simplices have the first point of the input as a vertex. This option does not exist for TOPCOM to the best of our knowledge. The default cache settings in run 6 of mptopcom are not enough to beat TOPCOM, thus for run 7 we increased the cache size of all three caches to 100 000 and for run 8 to 200 000.

Remark 13. Our GitHub repository contains some valgrind profiles for of mptopcom 1.2, 1.3, and 1.4 running on I^4. While mptopcom 1.2 and 1.3 spend 88% of the runtime checking regularity, this number is only 43% for mptopcom 1.4. Furthermore mptopcom 1.4 spends only 1% of the time solving LPs, while mptopcom 1.3 spends 79% and mptopcom 1.2 spends 32% of the time on LPs. The latter difference is due to mptopcom 1.2 having to assemble much larger LPs for regularity of triangulations, while mptopcom 1.3 only solves flip regularity LPs, but a much larger number of these.

Supplementary material, such as cluster scripts, our input files used and the benchmark logs, is available on GitHub in the repository

https://github.com/dmg-lab/regular_flips_in_mptopcom.

Aknowledgements. We are very grateful to Michael Joswig, Benjamin Lorenz, Marta Panizzut, and Francisco Santos Leal for many helpful discussions. Furthermore we are thankful to Jörg Rambau for the friendly competition. The author has been supported by MaRDI [15] under DFG project ID 460135501.

References

1. Assarf, B., et al.: Computing convex hulls and counting integer points with polymake. Math. Program. Comput. **9**(1), 1–38 (2017). https://doi.org/10.1007/s12532-016-0104-z
2. Avis, D., Fukuda, K.: Reverse search for enumeration. Discrete Appl. Math. **65**(1), 21–46 (1996). https://doi.org/10.1016/0166-218X(95)00026-N, first International Colloquium on Graphs and Optimization
3. Avis, D., Jordan, C.: mplrs: A scalable parallel vertex/facet enumeration code. Math. Program. Comput. **10**(2), 267–302 (2018). https://doi.org/10.1007/s12532-017-0129-y
4. Bies, M., Cvetič, M., Donagi, R., Liu, M., Ong, M.: Root bundles and towards exact matter spectra of F-theory MSSMs. J. High Energy Phys. **2021**(9), 1–65 (2021). https://doi.org/10.1007/JHEP09(2021)076
5. De Loera, J.A., Rambau, J., Santos, F.: Triangulations, Algorithms and Computation in Mathematics: Structures for Algorithms and Applications, vol. 25. Springer-Verlag, Berlin (2010). https://doi.org/10.1007/978-3-642-12971-1,
6. Develin, M., Sturmfels, B.: Tropical convexity. Doc. Math. **9**, 1–27 (2004)

7. Gawrilow, E., Joswig, M.: polymake: a framework for analyzing convex polytopes. In: Polytopes—combinatorics and Computation (Oberwolfach, 1997), DMV Sem., vol. 29, pp. 43–73. Birkhäuser, Basel (2000). https://doi.org/10.1007/978-3-0348-8438-9_2
8. Jordan, C., Joswig, M., Kastner, L.: Parallel enumeration of triangulations. Electron. J. Combin. **25**(3), Paper 3.6, 27 (2018). https://doi.org/10.37236/7318
9. Joswig, M.: Essentials of tropical combinatorics. Grad. Stud. Math. **219**. Providence, RI: American Mathematical Society (AMS) (2021). https://doi.org/10.1090/gsm/219
10. Joswig, M., Panizzut, M., Sturmfels, B.: The Schläfli fan. Discrete Comput. Geom. **64**(2), 355–381 (2020). https://doi.org/10.1007/s00454-020-00215-x
11. Joswig, M., Schröter, B.: Parametric shortest-path algorithms via tropical geometry. Math. Oper. Res. **47**(3), 2065–2081 (2022). https://doi.org/10.1287/moor.2021.1199
12. Panizzut, M., Vigeland, M.D.: Tropical lines on cubic surfaces. SIAM J. Discrete Math. **36**(1), 383–410 (2022). https://doi.org/10.1137/20M136520X
13. Rambau, J.: TOPCOM: Triangulations of point configurations and oriented matroids. In: Cohen, A., Gao, X.S., Takayama, N. (eds.) Mathematical Software — ICMS 2002, pp. 330–340. World Scientific (2002). https://doi.org/10.1142/9789812777171_0035
14. Schröter, B.: Multi-splits and tropical linear spaces from nested matroids. Discrete Comput. Geom. **61**(3), 661–685 (2019). https://doi.org/10.1007/s00454-018-0021-1
15. The MaRDI consortium: MaRDI: mathematical research data initiative proposal (2022). https://doi.org/10.5281/zenodo.6552436

A Framework for Generalized Tropical Homotopy Continuation

Oliver Daisey and Yue Ren

Department of Mathematical Sciences, Durham University, Durham DH1 3LE, UK
{oliver.j.daisey,yue.ren2}@durham.ac.uk

Abstract. We present a framework for generalizing tropical homotopy continuation for computing stable intersections of tropical hypersurfaces, tropical linear spaces, and tropical inverted linear spaces. We also report on beginning implementations in OSCAR.

Keywords: Tropical geometry · Stable intersections · OSCAR

1 Introduction

In [9], Jensen develops a tropical analogue of homotopy continuation for computing the stable intersection of n tropical hypersurfaces in $\mathbb{R}^n$. Jensen's main motivation was the computation of mixed volumes and their application to polyhedral homotopies [6]. The algorithm was initially implemented in GFAN [7,8], where it is used in the mixed volume computation to date. It has since been adopted by polynomial system solvers like HOMOTOPYCONTINUATION.JL [3].

In this extended abstract, we describe a general framework that allows us to extend Jensen's approach to stable intersections of hypersurfaces, linear spaces, and inverted linear spaces. All three types of tropical varieties have in common that they belong to large parametrized families of balanced polyhedral complexes which have a prominent dual picture. For example:

1. tropical hypersurfaces are dual to Newton subdivisions [11, §3.1] [10, §1], and the space of tropical hypersurfaces with a fixed monomial support is parametrized by the corresponding coefficient vectors.
2. tropical linear spaces are dual to matroid subdivisions [11, §4.4] [10, §10], and the space of tropical linear spaces with a fixed matroid is parametrized by the corresponding Plücker vectors.

This duality is essential for Jensen's approach, and, similar to Jensen, our motivation also partially stems from polynomial system solving: The stable intersection of linear spaces and hypersurface is relevant for determining the number of steady states of chemical reaction networks [4, Proposition 6.9 + Example 6.11],

---

Yue Ren is supported by the UKRI Future Leaders Fellowship "Computational Tropical Geometry and its Applications" (MR/Y003888/1).

© The Author(s), under exclusive license to Springer Nature Switzerland AG 2024
K. Buzzard et al. (Eds.): ICMS 2024, LNCS 14749, pp. 331–339, 2024.
https://doi.org/10.1007/978-3-031-64529-7_32

and the intersection of linear spaces and inverted linear spaces is useful for computing the beta invariant of affine matroids [2].

The purpose of this extended abstract is to present a unified framework and its basic implementation in OSCAR [1]. Like Jensen, our approach tracks mixed cells individually as much as possible, which avoids the complexities of working with the secondary fan. Details on starting point construction, path creation, and path tracking are not aspects we can properly treat in this manuscript, and are left to a future in-depth paper. The same goes for generalizations to even more families of tropical varieties. The code is openly available at

https://github.com/oliverdaisey/TropicalHomotopyContinuation.jl.

Note that, to date, our software package does not construct paths or starting data manually. For the sake of brevity, we will refrain from showing any code in this extended abstract and rather refer to the documentation on github, which contains code snippets for all examples in this manuscript.

2 Tropical Varieties

In this section, we assume some basic familiarity with

(1) tropical hypersurfaces and Newton subdivisions [11, §3.1] [10, §1],
(2) tropical linear spaces and matroid subdivisions [11, §4.4] [10, §10],

Both (1) and (2) are classes of balanced polyhedral complexes with prominent dual pictures. We will unify them and inverted tropical linear spaces under a common name and notation, which allows us to work agnostically in subsequent sections. Here, an inverted tropical linear space is the image of a tropical linear space under the negation map $\mathbb{R}^n \to \mathbb{R}^n$, $w \mapsto -w$.

Convention 1. For the remainder of the article, let $\mathbb{T} := (\mathbb{R} \cup \{\infty\}, \oplus, \odot)$ denote the min-plus tropical semiring and fix a multivariate (Laurent) polynomial ring $\mathbb{T}[x^{\pm}] := \mathbb{T}[x_1^{\pm}, \ldots, x_n^{\pm}]$. We will use $[n]$ to denote the set $\{1, \ldots, n\}$.

Dual supports and weight spaces
A *dual support* is a finite set $S \subseteq \mathbb{Z}^n$ that is either:

(Hypersurface) of cardinality at least two, representing the support of a non-monomial tropical polynomial.
(LinearSpace) of the form $S := \{-e_B \mid B \in M\} \subseteq \mathbb{Z}^n$, where M is a matroid on $[n+1]$ and $e_B := \sum_{i \in B} e_i \in \mathbb{Z}^n$ is the indicator vector of basis B,
(InvertedLinearSpace) a set of the form $S := \{e_B \mid B \in M\} \subseteq \mathbb{Z}^n$, where M is a matroid on $[n]$ and $e_B \in \mathbb{Z}^n$ is as before.

We assume that a dual support S knows which class it belongs to. In our code, they are objects of type DualSupport{Hypersurface}, DualSupport{Linear} or DualSupport{InvertedLinear}.

Every dual support S has an associated *dual weight space* $\mathcal{M}_S \subseteq \mathbb{T}^S$ that is, depending on the class of S, either

[(Hypersurface)] The coefficient space of tropical Laurent polynomials with support S and non-trivial tropical vanishing set:
$$\mathcal{M}_S := \{(c_\alpha)_{\alpha \in S} \in \mathbb{T}^S \mid c_{\alpha_1} \neq \infty \neq c_{\alpha_2} \text{ for at least two } \alpha_1, \alpha_2 \in S\}.$$

[(LinearSpace) & (InvertedLinearSpace)] The space of tropical Plücker vectors with support S:
$$\mathcal{M}_S := \mathrm{Dr}(S) \subseteq \mathbb{T}^S,$$

The reader may know this as the local Dressian [12], though we consider the ambient space of $\mathcal{M}_S$ indexed by S and not indexed by the bases of a matroid.

Points $c \in \mathcal{M}_S$ are referred to as *dual weights* and they can be regarded as coefficient vectors of tropical Laurent polynomials with support S:
$$f_S(c) := \bigoplus_{\alpha \in S} c_\alpha \odot x^{\odot \alpha}.$$

Our package contains no features for constructing these weight spaces, as for example computing Dressians is incredibly challenging even for small matroids [5].

Dual cells and tropical varieties

Let $S \subseteq \mathbb{Z}^n$ be a dual support. We refer to a subset $s \subseteq S$, as a *dual cell candidate* if, depending on the class of S,

[(Hypersurface)] s is of cardinality at least 2,
[(LinearSpace) & (InvertedLinearSpace)] s is loopless, i.e., whenever either $\bigcup_{-e_B \in s} B = [n]$ or $\bigcup_{e_B \in s} B = [n]$ respectively.

Let further $c \in \mathcal{M}_S$ be a dual weight. We call a dual cell candidate $s \subseteq S$ a *dual cell* induced by c, if $s = S \cap \delta$ for some cell δ in the regular subdivision on S induced by c. The *dual subdivision* on S induced by c is set of all dual cells induced by c:
$$D_S(c) := \{s \subseteq S \mid s \text{ is a dual cell induced by } c\}.$$

Any dual cell $s \in D_S(c)$ defines a *tropical polyhedron* (here, tropical refers to the polyhedron being part of a tropical variety in $\mathbb{R}^n$, and not to it being tropically convex):
$$\sigma_S(c, s) := \overline{\{w \in \mathbb{R}^n \mid \text{the minimum in } f_S(c)(w) \text{ is attained exactly at } s\}},$$

where $\overline{(\cdot)}$ denotes the closure under the Euclidean topology. Note that $s \subseteq s'$ implies $\sigma_S(c, s) \supseteq \sigma_S(c, s')$ and vice versa, hence maximal $\sigma_S(c, s)$ arise from minimal s.

The set of all such polyhedra form a polyhedral complex, which we refer to as the *tropical variety* with dual support S induced by weight c:
$$\Sigma_S(c) := \{\sigma_S(c, s) \mid s \in D_S(c)\}.$$

Moreover, $\Sigma_S(c)$ can be made into a balanced polyhedral complex with the following weights on maximal $\sigma_S(c, s)$:

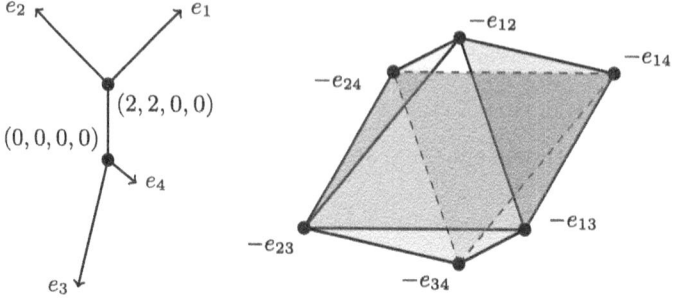

Fig. 1. Dual subdivision of a tropical linear space

[(Hypersurface)] $\text{mult}_{\Sigma_S(c)}(\sigma_S(c,s))$ is the lattice length of the edge s,
[(LinearSpace) & (InvertedLinearSpace)] $\text{mult}_{\Sigma_S(c)}(\sigma_S(c,s)) := 1$,

Example 2. Figure 1 shows the tropical linear space and its dual subdivision, with linear dual support $S_{\text{lin}} := \{-e_{12}, -e_{13}, -e_{14}, -e_{23}, -e_{24}, -e_{34}\}$ and dual weight $(2,0,0,0,0,0) \in \mathbb{T}^{S_{\text{lin}}}$. The tropical linear space is drawn modulo its lineality space $\mathbb{R} \cdot (1,1,1,1) \subseteq \mathbb{R}^4$, while the dual subdivision lives in $\{(w_1, w_2, w_3, w_4) \in \mathbb{R}^4 \mid \sum_{i=1}^4 w_i = -2\} \subseteq \mathbb{R}^4$. The dual subdivision is the usual matroid subdivision of the bipyramid [11, Figure 4.4.1], but with vertices e_B replaced by their negation $-e_B$. The dual cells consist of one middle quadrilateral corresponding to a bounded tropical edge, and four triangles corresponding to four unbounded rays.

3 Stable Intersections

The duality between points in the stable intersection of tropical hypersurfaces and the mixed cells in the induced subdivision of the Minkowski sum of their Newton polytopes [11, §4.6] [10, §4] naturally extends to our setting in Sect. 2.

First note that the dimension of $\Sigma_S(c)$ is independent of $c \in \mathcal{M}_S$, namely:

[(Hypersurface)] $\dim(\Sigma_S(c)) = n - 1$,
[(LinearSpace) & (InvertedLinearSpace)] $\dim(\Sigma_S(c)) = |B|$ for any $-e_B \in S$ or $e_B \in S$ respectively.

Moreover, the lineality space of $\Sigma_S(c)$ is also independent of $c \in \mathcal{M}_S$, namely

$$\text{Lineality}(\Sigma_S(c)) = \text{AffineSpan}(S)^\perp.$$

We therefore refer to the dual supports $S_1, \ldots, S_k$ as being of *complementary dimension (modulo lineality)*, if

$$\sum_{i=1}^k \text{codim}\left(\Sigma_{S_i}(c_i)\right) + \dim\left(\bigcap_{i=1}^k \text{Lineality}(\Sigma_{S_i}(c_i))\right) = n$$

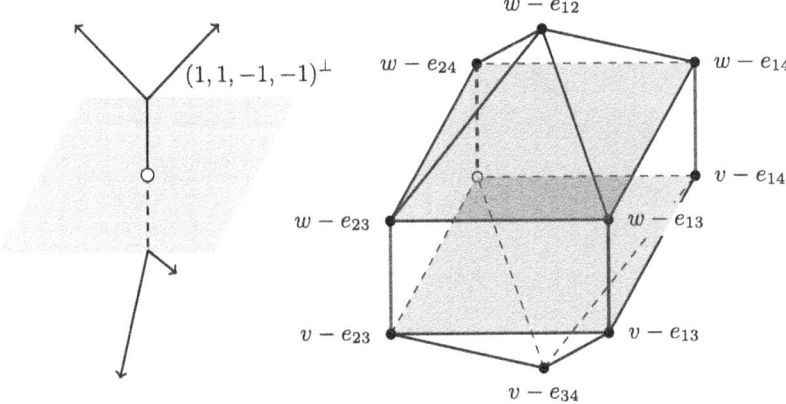

Fig. 2. Linear space and a hypersurface with one intersection point and mixed cell

Geometrically, $S_1, \ldots, S_k$ being of complementary dimension implies the stable intersection $\Sigma_{S_1}(c_1)) \cap_{\text{st}} \cdots \cap_{\text{st}} \Sigma_{S_k}(c_k)$ is either empty or a finite union of affine translates of $\bigcap_{i=1}^{k} \text{Lineality}(\Sigma_{S_i}(c_i))$.

A *mixed support* $\mathbf{S} := \coprod_{i=1}^{k} S_i = \{(i, \alpha) \mid \alpha \in S_i\}$ is a disjoint union of dual supports of complementary dimension. The *mixed weight space* $\mathcal{M}_{\mathbf{S}} := \mathcal{M}_{S_1} \times \cdots \times \mathcal{M}_{S_k} \subseteq \mathbb{T}^{\mathbf{S}}$ is the product of their dual spaces. Elements $\mathbf{c} = (c_1, \ldots, c_k) \in \mathcal{M}_{\mathbf{S}}$ are referred to as *mixed weights*, and they define tropical Laurent polynomials $f_{\mathbf{S}}(\mathbf{c}) := \bigodot_{i=1}^{k} f_{S_i}(c_i)$ with support $S_1 + \cdots + S_k$.

A *mixed cell candidate* is a subset $\mathbf{s} = \coprod_{i=1}^{k} s_i \subseteq \mathbf{S}$, where $s_i \subseteq S_i$ is a dual cell candidate. We call a mixed cell candidate $\mathbf{s}$ a *mixed cell* induced by $\mathbf{c}$, if $s_1 + \cdots + s_k = (S_1 + \cdots + S_k) \cap \delta$ for some cell δ in the dual mixed subdivision on $S_1 + \cdots + S_k$ induced by the coefficients of $f_{\mathbf{S}}(\mathbf{c})$. We use $D_{\mathbf{S}}(\mathbf{c})$ to denote the set of all mixed cells induced by $\mathbf{c}$:

$$D_{\mathbf{S}}(\mathbf{c}) := \{\mathbf{s} \subseteq \mathbf{S} \mid \mathbf{s} \text{ is a mixed cell induced by } \mathbf{c}\}.$$

We say $D_{\mathbf{S}}(\mathbf{c})$ is *transverse*, if all mixed cells $\mathbf{s} = (s_1, \ldots, s_k) \in D_{\mathbf{S}}(\mathbf{c})$ are comprised of minimal dual cells $s_i \in D_{S_i}(c_i)$.

Any mixed cell $\mathbf{s} = \coprod_{i=1}^{k} s_i \in D_{\mathbf{S}}(\mathbf{c})$ defines a tropical polyhedron $\sigma_{\mathbf{S}}(\mathbf{c}, \mathbf{s}) \subseteq \mathbb{R}^n$:

$$\sigma_{\mathbf{S}}(\mathbf{c}, \mathbf{s}) := \overline{\{w \in \mathbb{R}^n \mid \text{the minimum in } f_{\mathbf{S}}(\mathbf{c})(w) \text{ is attained exactly at } s_1 + \cdots + s_k\}}.$$

The polyhedral complex $\Sigma_{\mathbf{S}}(\mathbf{c})$ is the set of all such polyhedra:

$$\Sigma_{\mathbf{S}}(\mathbf{c}) := \{\sigma_{\mathbf{S}}(\mathbf{c}, \mathbf{s}) \mid \mathbf{s} \in D_{\mathbf{S}}(\mathbf{c})\}.$$

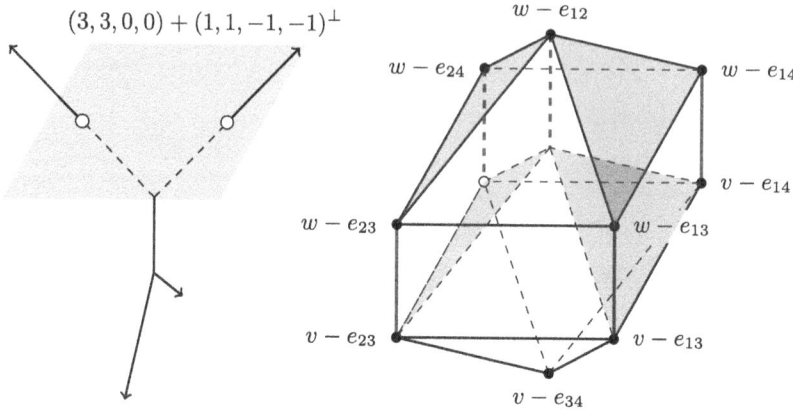

Fig. 3. Linear space and hypersurface with two intersection points and mixed cells

Using the notation above, we have

- $\sigma_{\mathbf{S}}(\mathbf{c}, \mathbf{s}) = \sigma_{S_1}(c_1, s_1) \cap \cdots \cap \sigma_{S_k}(c_k, s_k)$ as polyhedra in $\mathbb{R}^n$, and
- $\Sigma_{\mathbf{S}}(\mathbf{c}) = \Sigma_{S_1}(c_1) \wedge \cdots \wedge \Sigma_{S_k}(c_k)$ as polyhedral complexes in $\mathbb{R}^n$, where $\wedge$ is the common refinement.

Moreover, the latter intersection of balanced polyhedral complexes is transverse if and only if $D_{\mathbf{S}}(\mathbf{c})$ is transverse.

Example 3. Figures 2 and 3 each shows the intersections and mixed cells of a hypersurface and a linear space. In both figures, the linear space is from Example 2, while the hypersurface has dual support $S_{\text{Hyper}} = \{v, w\}$ with $v := e_1 + e_2$ and $w := e_3 + e_4$.

In Fig. 2, the hypersurface has dual weight $(0, 2) \in \mathbb{T}^{S_{\text{Hyper}}}$, intersecting the linear space in the bounded edge at point $(1, 1, 0, 0) \in \mathbb{R}^n$ modulo the common lineality space $\mathbb{R} \cdot (1, 1, 1, 1)$. The corresponding mixed cell consists of the red edge and the blue quadrilateral. In Fig. 3, the hypersurface has dual weight $(0, 5) \in \mathbb{T}^{S_{\text{Hyper}}}$, intersecting the linear space in two unbounded rays at points $(3, 2, 0, 0)$ and $(2, 3, 0, 0)$. The two mixed cells each consist of the red edge and one of the two top blue triangles.

4 Tropical Homotopy Continuation

The idea of tropical homotopy continuation [9] can be extended naturally to encompass linear and inverted linear spaces using our notation from Sects. 2 and 3. Given a target support $\mathbf{S}_{\text{target}}$ and weight $\mathbf{c}_{\text{target}} \in \mathcal{M}_{\mathbf{S}_{\text{start}}}$, computing the stable intersection of interest $\Sigma_{\mathbf{S}_{\text{target}}}(\mathbf{c}_{\text{target}})$ consists of two major steps:

Step 1: Finding a starting support $\mathbf{S}_{\text{start}} \supseteq \mathbf{S}_{\text{target}}$ weight $\mathbf{c}_{\text{start}} \in \mathcal{M}_{\mathbf{S}_{\text{start}}}$ as well as a path $\mathbf{h}\colon [0,\infty] \to \mathcal{M}_{\mathbf{S}_{\text{start}}}$ connecting $\mathbf{h}(0) = \mathbf{c}_{\text{start}}$ to $\mathbf{h}(\infty) = \mathbf{c}_{\text{target}}$ such that
1. $D_{\mathbf{S}}(\mathbf{c}_{\text{start}})$ is known and transverse,
2. $D_{\mathbf{S}}(\mathbf{h}(t))$ is transverse for all but finitely many t.

Step 2: Tracking the mixed cells along $\mathbf{h}$.

In Step 1, we regard $\mathcal{M}_{\mathbf{S}_{\text{target}}}$ as a subset of $\mathcal{M}_{\mathbf{S}_{\text{start}}}$ under the inclusion $\mathbb{T}^{\mathbf{S}_{\text{start}}} \cong \mathbb{T}^{\mathbf{S}_{\text{start}}} \times \{\infty\}^{\mathbf{S}_{\text{target}} \setminus \mathbf{S}_{\text{start}}} \subseteq \mathbb{T}^{\mathbf{S}_{\text{target}}}$. This enlargement of the support is for example done in Jensen's total degree homotopy [9, § 7.1], where the starting polynomials have more monomials than the target polynomials. There remains open problems on how to execute Step 1 in the general setting, as the choice of starting data strongly depends on the classes involved in the mixed support $\mathbf{S}_{\text{target}}$. As such, we leave it to a future in-depth mathematical paper.

Step 2 boils down to two key tasks:

Task 1: Identifying at which points $\mathbf{h}(t)$ the mixed cell changes,
Task 2: Describing how the mixed cell changes around $\mathbf{h}(t)$.

Mixed cell cones

Let $\mathbf{S}$ be a mixed support with a mixed weight space $\mathcal{M}_{\mathbf{S}} \subseteq \mathbb{T}^{\mathbf{S}}$. Let $\mathbb{R}^{\mathbf{S}} \subseteq \mathbb{T}^{\mathbf{S}}$ denote the real vector space with coordinate indexed by $\mathbf{S}$. Then any mixed cell candidates $\mathbf{s} \subseteq \mathbf{S}$ gives rise to a *mixed cell cone*:

$$C_{\mathbf{S}}(\mathbf{s}) := \overline{\{\mathbf{c} \in \mathbb{R}^{\mathbf{S}} \mid \mathbf{s} \text{ is a mixed cell induced by } \mathbf{c}\}}.$$

One can adapt the proof in [9, Lemma 4.4] to show that $C_{\mathbf{S}}(\mathbf{s})$ is a closed convex polyhedral cone with inequalities induced by elements in $\mathbf{S} \setminus \mathbf{s}$.

By definition, Task 2 now entails finding out when $\mathbf{h}$ leaves $C_{\mathbf{S}}(\mathbf{s})$, which is straightforward if $\mathbf{h}$ is (piecewise) linear. This is why in our software package objects of type `MixedPath` are always piecewise linear path.

Inflation and deflation of dual cells

Describing how a mixed cell $\mathbf{s} \subseteq \mathbf{S}$ changes along $\mathbf{h}$ in our more general setting is a bit more involved than Jensen's bistellar flips [9, § 5].

We will leave the exact definition to the future in-depth paper, but it reduces to describing the individual changes of each dual cell $s_i \subseteq S_i$. Suppose $\mathbf{h}(t_0)$ lies on the boundary of a mixed cell cone $C_{\mathbf{S}}(\mathbf{s})$.

If s_i remains a dual cell induced by $h_i(t_0)$, then nothing changes. The dual cell candidate s_i is a dual cell induced by $h_i(t_0 \pm \varepsilon)$ for $\varepsilon > 0$ sufficiently small, and the tropical intersection point $\sigma_{\mathbf{S}}(\mathbf{h}(t_0 \pm \varepsilon), \mathbf{s})$ remains in the relative interior of $\sigma_{S_i}(h_i(t_0 \pm \varepsilon), s_i)$.

If s_i is no dual cell induced by $h_i(t_0)$, then the tropical intersection point $\sigma_{\mathbf{S}}(\mathbf{h}(t_0-\varepsilon), \mathbf{s})$ moves onto the boundary of $\sigma_{S_i}(h_i(t_0-\varepsilon), s_i)$ as $\varepsilon \searrow 0$. Describing how the support changes involves

1. identifying the dual cell candidate $s_i' \subseteq S_i$ with $s_i \subseteq s_i'$, such that $\sigma_{S_i}(h_i(t_0), s_i')$ is the lower-dimensional tropical face containing $\lim_{\varepsilon \to 0} \sigma_{\mathbf{S}}(\mathbf{h}(t_0-\varepsilon), \mathbf{s})$,

2. identifying all dual cell candidates $s_i'' \subseteq S_i$ with $s_i'' \subseteq s_i'$, such that the limit of the tropical intersection point $\lim_{\varepsilon \to 0} \sigma_\mathbf{S}(\mathbf{h}(t_0 + \varepsilon), \mathbf{s}'')$ is contained in $\sigma_{S_i}(h_i(t_0), s_i')$.

In our code, we refer to s_i' as the inflation of s_i, while the s_i'' are deflations of s_i'.

Example 4. In the transition from Figs. 2 to 3, where the dual weight of the hypersurface goes from $(2,0) \in \mathbb{T}^{S_{\text{Hyper}}}$ to $(5,0) \in \mathbb{T}^{S_{\text{Hyper}}}$, the dual cell of the hypersurface $s_{\text{Hyper}} = \{e_1 + e_2, e_3 + e_4\} \subseteq S_{\text{Hyper}}$ remains unchanged. The quadrilateral dual cell of the tropical linear space $s_{\text{lin}} = \{-e_{13}, -e_{14}, -e_{23}, -e_{24}\}$ first inflates to the upper bipyramid $s_{\text{lin}}' = \{-e_{12}, -e_{13}, -e_{14}, -e_{23}, -e_{24}\} \supseteq s_{\text{lin}}$ which corresponds to the vertex $(2,2,0,0)$ on the tropical linear space, and then deflates to the two upper triangles $s_{\text{lin},1}'' = \{-e_{12}, -e_{13}, -e_{14}\} \subseteq s_{\text{lin}}'$ and $s_{\text{lin},2}'' = \{-e_{12}, -e_{23}, -e_{24}\} \subseteq s_{\text{lin}}'$. All of this happens while the tropical intersection point moves from the bounded edge $\text{conv}((0,0,0,0), (2,2,0,0))$, over vertex $(2,2,0,0)$, onto the unbounded rays $(2,2,0,0) + \mathbb{R}_{\geq 0} \cdot e_1$ and $(2,2,0,0) + \mathbb{R}_{\geq 0} \cdot e_2$ on the tropical linear space.

References

1. Oscar - open source computer algebra research system, version 1.0 (2024). https://www.oscar-system.org
2. Ardila-Mantilla, F., Eur, C., Penaguiao, R.: The tropical critical points of an affine matroid. Sém. Lothar. Combin. **89B**, Art. 28, 12 (2023)
3. Breiding, P., Timme, S.: HomotopyContinuation.jl: a package for homotopy continuation in Julia. In: Davenport, J.H., Kauers, M., Labahn, G., Urban, J. (eds.) ICMS 2018. LNCS, vol. 10931, pp. 458–465. Springer, Cham (2018). https://doi.org/10.1007/978-3-319-96418-8_54
4. Helminck, P.A., Ren, Y.: Generic root counts and flatness in tropical geometry (2022)
5. Herrmann, S., Joswig, M., Speyer, D.E.: Dressians, tropical Grassmannians, and their rays. Forum Math. **26**(6), 1853–1881 (2014). https://doi.org/10.1515/forum-2012-0030
6. Huber, B., Sturmfels, B.: A polyhedral method for solving sparse polynomial systems. Math. Comput. **64**(212), 1541–1555 (1995). https://doi.org/10.2307/2153370
7. Jensen, A.N.: Gfan, a software system for Gröbner fans and tropical varieties. http://home.imf.au.dk/jensen/software/gfan/gfan.html
8. Jensen, A.N.: An implementation of exact mixed volume computation. In: Greuel, G.M., Koch, T., Paule, P., Sommese, A. (eds.) Mathematical Software - ICMS 2016, pp. 198–205. Springer International Publishing, Cham (2016). https://doi.org/10.1007/978-3-319-42432-3_25
9. Jensen, A.N.: Tropical homotopy continuation (2016)
10. Joswig, M.: Essentials of Tropical Combinatorics, vol. 219. Graduate Studies in Mathematics. Providence, RI: American Mathematical Society (AMS) (2021). https://doi.org/10.1090/gsm/219

11. Maclagan, D., Sturmfels, B.: Introduction to Tropical Geometry, vol. 161. Providence, RI: American Mathematical Society (AMS) (2015)
12. Olarte, J.A., Panizzut, M., Schröter, B.: On local Dressians of matroids. In: Algebraic and Geometric Combinatorics on Lattice Polytopes. Proceedings of the Summer Workshop on Lattice Polytopes, Osaka, Japan, July 23 – August 10, 2018, pp. 309–329. Hackensack, NJ: World Scientific (2019). https://doi.org/10.1142/9789811200489_0020

General Session

Integrating GeoGebra with React and WebAssembly: A Web-Based Approach for Mathematical Software Development

Mitsushi Fujimoto[✉]

University of Teacher Education Fukuoka, 1-1 Akamabunkyo-machi,
Munakata 811-4192, Japan
fujimoto@fukuoka-edu.ac.jp
https://staff.fukuoka-edu.ac.jp/fujimoto/

Abstract. A method for developing a GeoGebra-based web app is introduced, utilizing React for the user interface and WebAssembly for computation. React enhances UI components and improves error detection, overcoming the limitations of GeoGebra's script input. Additionally, WebAssembly facilitates the use of C functions within web browsers, thereby enhancing GeoGebra's computational capabilities. The effective collaboration between React and WebAssembly is demonstrated through the implementation of the 'Lights Out' puzzle.

Keywords: GeoGebra · React · WebAssembly · Lights Out

1 Introduction

A dynamic mathematics software, GeoGebra [1], can be embedded on the web, and a JavaScript API [2] is provided for interacting with the embedded GeoGebra. By utilizing web platforms, GeoGebra can collaborate with various web technologies. In this article, I would like to introduce a method for creating a GeoGebra-based web app using React [3] for the user interface (UI) and WebAssembly [4] for the computational engine.

One of the limitations of GeoGebra is its script input dialog for JavaScript, which has the following drawbacks:

- Creating an object is needed to access the script input dialog.
- Code formatting and autocompletion features are insufficient.
- Lack of script error detection.

React is a JavaScript library developed by Meta and the community for creating UIs. It provides rich UI components like switch buttons and toasts, and you can write clean and well-organized code using JSX, a syntax extension for JavaScript. Additionally, React can detect code errors during the transpilation of JSX to regular JavaScript. The use of React can address the limitations of GeoGebra's script input dialog mentioned earlier.

GeoGebra provides numerous calculation functions, including functions from a powerful Computer Algebra System (CAS). However, using external calculation engines created in C or C++ is not possible. WebAssembly (Wasm) is a type of binary format that can be executed in web browsers. Although it is possible to write WebAssembly code directly, it is designed to compile and generate code from programming languages such as C/C++/Rust. Converting C functions to WebAssembly makes them accessible in web browsers. The web app version of GeoGebra can access external computational engines created in C using WebAssembly.

The following steps can be taken to implement an app with GeoGebra/React for the UI and C language/WebAssembly for the computational engine:

1. Create a React component for GeoGebra using react-geogebra [5], and build a UI with components such as switch buttons.
2. Develop a computational engine in C, then convert it to a Wasm binary using Emscripten [6].
3. Execute the C functions converted to Wasm binary from the React side and update the UI dynamically.

Some considerations are required for the collaboration between React and WebAssembly. I will explain the details through the implementation of a mathematical puzzle, 'Lights Out' [7].

2 Creating a UI

2.1 Overview of the UI Design

Lights Out is a mathematical puzzle in which the goal is to turn off all the lights in a 5×5 grid, some of which are initially lit. Clicking on a cell inverts the lighting state of that cell and its adjacent cells (up, down, left, and right). However, if a cell on the edge of the grid is clicked, the missing adjacent cells are ignored. Whether a cell remains lit or not is determined by the parity of the overlay of changes made by clicks. Therefore, each cell needs to be clicked at most once, and the result is the same regardless of the order of clicking the cells.

Here, we will create a UI for Lights Out using React. The main UI components used are `Geogebra` component by react-geogebra, along with `Switch`, `Button`, and `Snackbar` components from Material UI [8]. react-geogebra is a component package that embeds GeoGebra into React, and Material UI is a library for using Google's Material Design in React.

In the `Geogebra` component, the 5×5 grid is generated using the GeoGebra JavaScript API, and an event listener is used to enable the changing of cell lighting states through clicks. The `Switch` component is used to implement the mode switching between 'Setup' for problem creation and 'Play' for gameplay. In 'Setup' mode, only the clicked cell is inverted. The Judge and Solve buttons execute C functions using WebAssembly and display the results in a snackbar (Fig. 1).

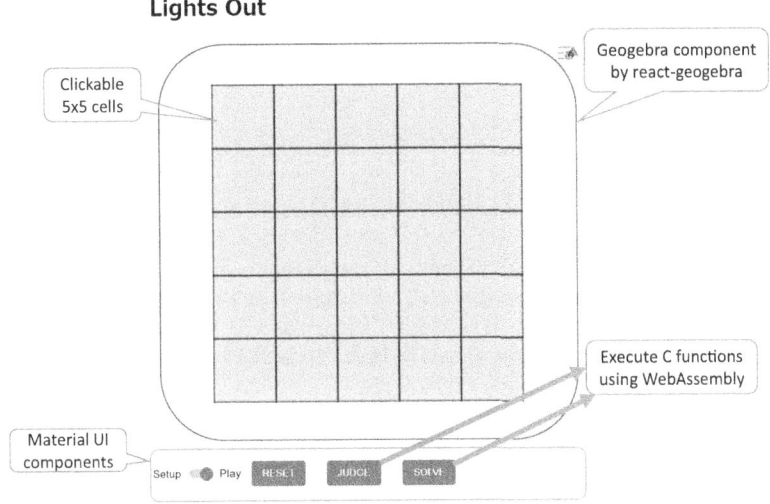

Fig. 1. UI of the Lights Out app

2.2 Preparing React Components

After generating a React template in the my-app directory using a project generation tool create-react-app, install react-geogebra and Material UI using a package manager npm.

```
$ npm install --legacy-peer-deps react-geogebra
$ npm install @mui/material @emotion/react @emotion/styled
```

--legacy-peer-deps option is used to automatically resolve issues with package dependencies.

2.3 Implementation of the UI

Then, we describe the code for the UI in the file src/App.js.

Code 1. App.js

```
1  import Geogebra from 'react-geogebra';
2  import React from 'react';
3  import { Stack, Typography, Switch, Button, Snackbar, Alert } from
       '@mui/material';
4
5  function App() {
6    const [mode, setMode] = React.useState(true);
7    const handleChange = (event) => {
8      setMode(event.target.checked);
9    };
```

```
10    const stateMode = React.useRef(); // Create an object to obtain
         state
11    stateMode.current = mode; // Keep the state of mode in current
12    return (
13      <div className="App">
14        <h1>Lights Out</h1>
15        <Geogebra
16          width="650" height="600" showToolBar="false" showMenuBar="
               false"
17          showAlgebraInput="false" allowStyleBar appletOnLoad={
               ggbOnInit}
18        />
19        <Stack direction="row" alignItems="center">
20          <Typography>Setup</Typography>
21          <Switch checked={mode} onChange={handleChange} />
22          <Typography>Play</Typography>
23          <Button variant="contained" onClick={init}>Reset</Button>
24          <Button variant="contained" onClick={judgement}>Judge</
               Button>
25          <Button variant="contained" onClick={solution}>Solve</
               Button>
26        </Stack>
27        <Snackbar open={open} autoHideDuration={5000} onClose={
             handleClose}>
28          <Alert icon={false} onClose={handleClose} severity="success
               ">
29            <strong>{message}</strong>
30          </Alert>
31        </Snackbar>
32      </div>
33    );
```

The necessary React components are imported in lines 1–3. The `Geogebra` component is set up in lines 15–18, the toggle switch in lines 20–22, the three buttons of Reset/Judge/Solve in lines 23–25, and the `Snackbar` for displaying the execution result of the Judge button in lines 27–31.

The `React.useState` function in line 6 is used to manage the state of a React component, and it returns a pair of the current state and a function to update it. The argument of the `React.useState` function is the initial value of the state. Lines 10–11 are necessary to refer to the latest state value within the cell click event listener, and the `current` property of `stateMode` holds the mode of the toggle switch. The `handleChange` function in lines 7–9 is an event handler that is executed when the toggle switch is used, and it changes the state value of `mode` using the `setMode` function.

The `ggbOnInit` function specified in line 17 creates the 5 × 5 grid with the GeoGebra command `Polygon` and sets up a listener to invert the lighting state of cells on click events. This function is automatically executed when the Geogebra component is loaded by passing it to the appletOnLoad property.

3 Creating a Computational Engine

Let 1 be a lit cell and 0 an unlit cell. Consider the following system of linear equations using a 25 × 25 matrix, where the i-th column represents the state of the grid after clicking the i-th cell (scanning from top-left to right).

$$\begin{pmatrix} 1 & 1 & 0 & 0 & 0 & 1 & 0 & 0 & 0 & 0 & 0 & 0 & 0 & 0 & 0 & 0 & 0 & 0 & 0 & 0 & 0 & 0 & 0 & 0 & 0 \\ 1 & 1 & 1 & 0 & 0 & 0 & 1 & 0 & 0 & 0 & 0 & 0 & 0 & 0 & 0 & 0 & 0 & 0 & 0 & 0 & 0 & 0 & 0 & 0 & 0 \\ 0 & 1 & 1 & 1 & 0 & 0 & 0 & 1 & 0 & 0 & 0 & 0 & 0 & 0 & 0 & 0 & 0 & 0 & 0 & 0 & 0 & 0 & 0 & 0 & 0 \\ 0 & 0 & 1 & 1 & 1 & 0 & 0 & 0 & 1 & 0 & 0 & 0 & 0 & 0 & 0 & 0 & 0 & 0 & 0 & 0 & 0 & 0 & 0 & 0 & 0 \\ 0 & 0 & 0 & 1 & 1 & 0 & 0 & 0 & 0 & 1 & 0 & 0 & 0 & 0 & 0 & 0 & 0 & 0 & 0 & 0 & 0 & 0 & 0 & 0 & 0 \\ 1 & 0 & 0 & 0 & 0 & 1 & 1 & 0 & 0 & 0 & 1 & 0 & 0 & 0 & 0 & 0 & 0 & 0 & 0 & 0 & 0 & 0 & 0 & 0 & 0 \\ 0 & 1 & 0 & 0 & 0 & 1 & 1 & 1 & 0 & 0 & 0 & 1 & 0 & 0 & 0 & 0 & 0 & 0 & 0 & 0 & 0 & 0 & 0 & 0 & 0 \\ 0 & 0 & 1 & 0 & 0 & 0 & 1 & 1 & 1 & 0 & 0 & 0 & 1 & 0 & 0 & 0 & 0 & 0 & 0 & 0 & 0 & 0 & 0 & 0 & 0 \\ 0 & 0 & 0 & 1 & 0 & 0 & 0 & 1 & 1 & 1 & 0 & 0 & 0 & 1 & 0 & 0 & 0 & 0 & 0 & 0 & 0 & 0 & 0 & 0 & 0 \\ 0 & 0 & 0 & 0 & 1 & 0 & 0 & 0 & 1 & 1 & 0 & 0 & 0 & 0 & 1 & 0 & 0 & 0 & 0 & 0 & 0 & 0 & 0 & 0 & 0 \\ 0 & 0 & 0 & 0 & 0 & 1 & 0 & 0 & 0 & 0 & 1 & 1 & 0 & 0 & 0 & 1 & 0 & 0 & 0 & 0 & 0 & 0 & 0 & 0 & 0 \\ 0 & 0 & 0 & 0 & 0 & 0 & 1 & 0 & 0 & 0 & 1 & 1 & 1 & 0 & 0 & 0 & 1 & 0 & 0 & 0 & 0 & 0 & 0 & 0 & 0 \\ 0 & 0 & 0 & 0 & 0 & 0 & 0 & 1 & 0 & 0 & 0 & 1 & 1 & 1 & 0 & 0 & 0 & 1 & 0 & 0 & 0 & 0 & 0 & 0 & 0 \\ 0 & 0 & 0 & 0 & 0 & 0 & 0 & 0 & 1 & 0 & 0 & 0 & 1 & 1 & 1 & 0 & 0 & 0 & 1 & 0 & 0 & 0 & 0 & 0 & 0 \\ 0 & 0 & 0 & 0 & 0 & 0 & 0 & 0 & 0 & 1 & 0 & 0 & 0 & 1 & 1 & 0 & 0 & 0 & 0 & 1 & 0 & 0 & 0 & 0 & 0 \\ 0 & 0 & 0 & 0 & 0 & 0 & 0 & 0 & 0 & 0 & 1 & 0 & 0 & 0 & 0 & 1 & 1 & 0 & 0 & 0 & 1 & 0 & 0 & 0 & 0 \\ 0 & 0 & 0 & 0 & 0 & 0 & 0 & 0 & 0 & 0 & 0 & 1 & 0 & 0 & 0 & 1 & 1 & 1 & 0 & 0 & 0 & 1 & 0 & 0 & 0 \\ 0 & 0 & 0 & 0 & 0 & 0 & 0 & 0 & 0 & 0 & 0 & 0 & 1 & 0 & 0 & 0 & 1 & 1 & 1 & 0 & 0 & 0 & 1 & 0 & 0 \\ 0 & 0 & 0 & 0 & 0 & 0 & 0 & 0 & 0 & 0 & 0 & 0 & 0 & 1 & 0 & 0 & 0 & 1 & 1 & 1 & 0 & 0 & 0 & 1 & 0 \\ 0 & 0 & 0 & 0 & 0 & 0 & 0 & 0 & 0 & 0 & 0 & 0 & 0 & 0 & 1 & 0 & 0 & 0 & 1 & 1 & 0 & 0 & 0 & 0 & 1 \\ 0 & 0 & 0 & 0 & 0 & 0 & 0 & 0 & 0 & 0 & 0 & 0 & 0 & 0 & 0 & 1 & 0 & 0 & 0 & 0 & 1 & 1 & 0 & 0 & 0 \\ 0 & 0 & 0 & 0 & 0 & 0 & 0 & 0 & 0 & 0 & 0 & 0 & 0 & 0 & 0 & 0 & 1 & 0 & 0 & 0 & 1 & 1 & 1 & 0 & 0 \\ 0 & 0 & 0 & 0 & 0 & 0 & 0 & 0 & 0 & 0 & 0 & 0 & 0 & 0 & 0 & 0 & 0 & 1 & 0 & 0 & 0 & 1 & 1 & 1 & 0 \\ 0 & 0 & 0 & 0 & 0 & 0 & 0 & 0 & 0 & 0 & 0 & 0 & 0 & 0 & 0 & 0 & 0 & 0 & 1 & 0 & 0 & 0 & 1 & 1 & 1 \\ 0 & 0 & 0 & 0 & 0 & 0 & 0 & 0 & 0 & 0 & 0 & 0 & 0 & 0 & 0 & 0 & 0 & 0 & 0 & 1 & 0 & 0 & 0 & 1 & 1 \end{pmatrix} \begin{pmatrix} x_1 \\ x_2 \\ x_3 \\ x_4 \\ x_5 \\ x_6 \\ x_7 \\ x_8 \\ x_9 \\ x_{10} \\ x_{11} \\ x_{12} \\ x_{13} \\ x_{14} \\ x_{15} \\ x_{16} \\ x_{17} \\ x_{18} \\ x_{19} \\ x_{20} \\ x_{21} \\ x_{22} \\ x_{23} \\ x_{24} \\ x_{25} \end{pmatrix} = \begin{pmatrix} b_1 \\ b_2 \\ b_3 \\ b_4 \\ b_5 \\ b_6 \\ b_7 \\ b_8 \\ b_9 \\ b_{10} \\ b_{11} \\ b_{12} \\ b_{13} \\ b_{14} \\ b_{15} \\ b_{16} \\ b_{17} \\ b_{18} \\ b_{19} \\ b_{20} \\ b_{21} \\ b_{22} \\ b_{23} \\ b_{24} \\ b_{25} \end{pmatrix}$$

The variable x_i stores the click information for the i-th cell. If the i-th cell is clicked, then $x_i = 1$, otherwise $x_i = 0$. The variable b_i represents the final lit state of the i-th cell. To find the solution of Lights Out, we substitute the lit state of the grid into b_i and solve this system of linear equations over the binary field $GF(2)$.

When we apply the forward elimination of Gaussian elimination method to this system of linear equations, all the elements in the 24th and 25th rows of the coefficient matrix become 0. Therefore, the system is solvable if $b'_{25} = b'_{24} = 0$, and unsolvable otherwise[1]. In the solvable case, if we fix the values of x_{25} and x_{24}, then x_{23} to x_1 can be obtained sequentially by the back substitution. Since the possible values of (x_{25}, x_{24}) are $(0,0), (0,1), (1,0)$ and $(1,1)$, there are exactly 4 solutions in the solvable case.

We implement this algorithm in C language. The following two functions are called by the Judge and Solve button to determine solvability and to find the solution, respectively.

int solvable(int *num, int n): A function that determines the solvability of Lights Out. The arguments are an array of integers num and an integer n.

---

[1] b'_{25} and b'_{24} are the 25th and 24th entries of the 26th column of the augmented matrix obtained after the forward elimination.

num consists of the indices (from 0 to 24) of lit cells. n is the number of lit cells. It returns 1 if solvable, and 0 otherwise.

int *solve(int *num, int n): A function that finds the solution of Lights Out. In the solvable case, it returns the solution as an integer array of length 25 in the form [0,1,0,0,0,1,1,1,0,0,...]. The cells to be clicked are represented by 1, and those not to be clicked by 0. This function outputs one of the four possible solutions at random.

4 Utilizing WebAssembly

4.1 Creating a Wasm Module

Emscripten is a compiler toolchain that converts C/C++ code into Wasm binaries. To execute C functions from the React side, we use the Emscripten compiler frontend emcc for the C source file lightsout.c, which includes the functions from the previous section, as follows:

```
$ emcc -O3 --no-entry lightsout.c -o lightsout.mjs \
  -s ENVIRONMENT='web' \
  -s SINGLE_FILE=1 \
  -s EXPORT_NAME='createModule' \
  -s EXPORTED_FUNCTIONS='["_solvable","_solve","_free"]' \
  -s EXPORTED_RUNTIME_METHODS='["ccall"]'
```

The meanings of each option are as follows:

--no-entry   Ignore the main function in the C source file.
-s ENVIRONMENT='web'   Use only in a web browser.
-s SINGLE_FILE=1   Combine the .mjs and .wasm files into a single file.
-s EXPORT_NAME   Set the name for the module creation function.
-s EXPORTED_FUNCTIONS   Set the C functions to be used from JavaScript.
-s EXPORTED_RUNTIME_METHODS   Set the runtime functions that execute C functions.

The above command generates the Wasm module file lightsout.mjs.

4.2 Importing the Wasm Module File

Some preparations are required before importing the Wasm module file. First, copy the Wasm module file lightsout.mjs to the my-app/src directory. Then, modify the file package.json located in the root directory my-app of the React template as follows:

Code 2. package.json (addition)

```
24    "eslintConfig": {
25      "extends": [
26        "react-app",
27        "react-app/jest"
28      ],
29      "ignorePatterns": [
30        "src/lightsout.mjs"
31      ]
32    },
```

ESLint [9] is a syntax checker used when transpiling JSX code, but it cannot check the Wasm code in `lightsout.mjs`. The above modification makes ESLint ignore `lightsout.mjs`, allowing the transpilation to proceed.

Next, we add an `import` statement to the head of `App.js` to use the module creation function `createModule`.

Code 3. App.js (addition)

```
1   import Geogebra from 'react-geogebra';
2   import React from 'react';
3   import { Stack, Typography, Switch, Button, Snackbar, Alert } from '@mui/material';
4   import createModule from './lightsout.mjs';
5
6   function App() {
```

4.3 Event Handling for the Solve Button

When the Solve button is clicked, it uses the C function `solve` to find the solution of Lights Out, and adds white circles on the cells that are part of the solution. To achieve this, the following `solution` function is added to `App.js`.

Code 4. App.js (addition)

```
1    // Find the solution of Lights Out
2    function solution() {
3      const ggbApplet = window.ggbApplet;
4      const numlist = []; // Array to store the indices of the lit cells
5      for (let i = 0; i < ggbApplet.getObjectNumber(); i++) {
6        const obj = ggbApplet.getObjectName(i);
7        if (ggbApplet.getObjectType(obj) === "quadrilateral" && ggbApplet.getColor(obj) !== "#000000") {
8          numlist.push(parseInt(obj.substring(1))); // Store the index of q_i in numlist
9        }
10     }
```

```
11      // Prepare the array for the C function
12      const numarray = new Uint8Array(new Uint32Array(numlist).buffer);
13      createModule().then(mod => {
14        // Execute the C function 'solve'
15        const ptr = mod.ccall('solve','number',['array','number'],[numarray
            ,numlist.length]);
16        // Convert the result to a JavaScript array
17        const rslt = new Int32Array(mod.HEAP32.buffer, ptr, 25);
18        if (rslt[0] === -1) { // Unsolvable case
19          setMessage('Unsolvable!');
20          setOpen(true);
21        } else { // Solvable case
22          for (let i = 0; i < 25; i++) {
23            if (rslt[i] === 1){ // Add white circles on the cells for the
                solution
24              const x = ggbApplet.getXcoord("P"+(i))+1;
25              const y = ggbApplet.getYcoord("P"+(i))-1;
26              ggbApplet.evalCommand("AP"+(i)+"=("+(x)+","+(y)+")");
27              ggbApplet.setLabelVisible("AP"+(i),false);
28              ggbApplet.setFixed("AP"+(i),true,false);
29              ggbApplet.evalCommand("SetDynamicColor(AP"+(i)+",1,1,1,0.2)")
                ;
30            }
31          }
32        }
33        mod._free(ptr);
34      });
35    }
```

The indices of the lit cells are stored in the array numlist in lines 5–10. Line 12 performs preprocessing to pass the array to the C function. In line 15, the C function solve is executed using the Emscripten runtime function ccall[2]. In line 17, the array returned by solve is converted to a JavaScript array. Finally, white circles are added on the cells corresponding to the solution of Lights Out in lines 22–31.

5 Building and Running the App

To build the app, execute the command 'npm run build' in the root directory my-app of the React template. This command transpiles the JSX code within App.js to regular JavaScript and generates the necessary files for running the app in the build directory.

The Lights Out app will be launched when you access the build directory from a web browser. However, you need to start a web server before running the app and then access it through the web server[3] (Figs. 2 and 3).

---

[2] For details on the arguments of the ccall function, see [10].
[3] This is due to the security restrictions imposed by web browsers on loading local files.

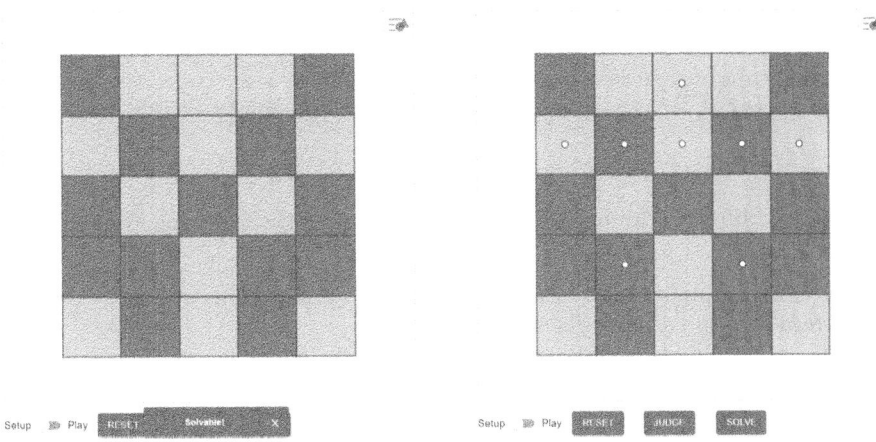

Fig. 2. Solvable pattern **Fig. 3.** A solution of Lights Out

The Lights Out app can be accessed at [11]. The source code for the Lights Out app is available in the repository [12].

6 Summary

In this article, I explain how to implement a web app using GeoGebra and React for the UI, and C language and WebAssembly for the computational engine. By using React, we can create dynamic and complex UIs with concise code. Moreover, WebAssembly allows us to create computational engines in various programming languages, including C/C++. The source code of the Light Out app consists of 200 lines of React code and 100 lines of C code.

There are two key points to consider when using this approach:

- **Preparing to Use WebAssembly from React**: Set the options for the emcc command and the configuration for ESLint appropriately[4].
- **Passing Arrays Between JavaScript and C Functions**: Arrays must be converted to a compatible format before transfer and appropriately reconverted post-transfer.

It is important to compare our implementation using WebAssembly with an implementation using JavaScript only. Table 1 shows the processing time for the event handling of the Judge button implemented with JavaScript only and using WebAssembly. The timing values in Table 1 were measured using the JavaScript `console.time()` method. The results of Using Wasm are 'the preparation time of the array for C' + 'the execution time of the C function'.

---

[4] These settings are based on [13].

Table 1. Timing Data for the Judge Button in Milliseconds (ms)

Implementation	1st Run Avg. (10 Trials)	Subsequent Avg. (100 Trials)
JavaScript only	0.4451	0.0772
Using Wasm	0.0333 + 0.5747	0.0131 + 0.0622

The experimental timing data was obtained under the following system specifications:

CPU: Intel Core i7-7600U 3.9 GHz (Base clock 2.9 GHz)
Memory: 40 GB (DDR4-2666)
OS: Windows 10 Pro 64-bit (Version 22H2)
Web Browser: Google Chrome (Version 122.0.6261.95)

The first run is faster with JavaScript only, but Using Wasm is slightly faster from the second run. Both are over 80% faster on the second run due to caching. There was no significant difference in execution time because the main part of the event handling was Gaussian elimination for a 25×26 sparse matrix. However, there are known cases where WebAssembly can be several times faster for heavy processing such as image conversion (e.g., [14]).

Additionally, WebAssembly has the advantage of leveraging existing software assets. There are many computer algebra systems implemented in C/C++ such as Macaulay2, Singular, Magma, Gap, CoCoA, Risa/Asir, and PARI/GP. The development of WebAssembly versions of these systems is expected to contribute to the advancement of mathematical software[5]. By applying this implementation method, one can develop mathematical software that integrates a more powerful CAS with GeoGebra.

Acknowledgements. This work was supported by JSPS KAKENHI Grant Number JP21K12157.

References

1. GeoGebra. https://www.geogebra.org/
2. GeoGebra API. https://wiki.geogebra.org/en/Reference:GeoGebra_Apps_API
3. React. https://react.dev/
4. WebAssembly. https://webassembly.org/
5. react-geogebra. https://www.npmjs.com/package/react-geogebra
6. Emscripten. https://emscripten.org/
7. Lights Out (game). https://en.wikipedia.org/wiki/Lights_Out_(game)
8. Material UI. https://mui.com/
9. ESLint. https://eslint.org/
10. preamble.js. https://emscripten.org/docs/api_reference/preamble.js.html
11. Fujimoto, M.: Lights out puzzle app with GeoGebra and WebAssembly. https://mitsushi-fujimoto.github.io/geo-lights/

---

[5] A WebAssembly version of PARI/GP is already available [15].

12. Fujimoto, M.: geo-lights. https://github.com/mitsushi-fujimoto/geo-lights/
13. Chen, B.: React C/C++ WASM demo. https://github.com/bobbiec/react-wasm-demo/blob/main/README-inlined-version.md
14. Oishi, S., Ishikawa, K., Nogami, H., Fukushima, N.: Performance evaluation of image convolution with WebAssembly. In: Proceedings of the SPIE, vol. 12592, International Workshop on Advanced Imaging Technology (IWAIT) 2023, pp. 125922G (2023). https://doi.org/10.1117/12.2667004
15. Run PARI/GP in your browser. https://pari.math.u-bordeaux.fr/gpwasm.html

DetGB: A Software Package for Computing Gröbner Bases of Determinantal Ideals

Chenqi Mou, Qiuye Song[✉], and Yutong Zhou

LMIB-School of Mathematical Sciences, Beihang University,
Beijing 100191, China
{chenqi.mou,qiuye.song,zpengyt}@buaa.edu.cn

1 Overview

Determinantal ideals are a fundamental concept in commutative algebra and algebraic geometry with deep connections to combinatorics and symbolic computation [3]. In particular, the Gröbner bases of various determinantal ideals have been identified and applied to the MinRank problem from cryptography [4], Schubert enumerative calculus [5,8], and low-rank matrix completion [13], etc. The software package DetGB[1] is a collection of functions we develop in the computer algebra system MAPLE for computing Gröbner bases of determinantal ideals and thus facilitates the study, analysis, and visualization of determinantal ideals in a computational way.

The developed package consists of three modules which compute the Gröbner bases of three main kinds of determinantal ideals—the normal, (mixed) ladder, and Schubert ones—respectively and one module with supporting functions like those for the RSK correspondence, the straightening law, and pipe dreams from combinatorics. The corresponding modules and the functions they contain in this software package are summarized in Fig. 1 below.

Two underlying philosophies for our design of the package are: (1) In addition to traditional results on Gröbner bases of determinantal ideals, we also want to include the latest development in this direction, like the minimal Gröbner bases of Schubert determinantal ideals via elusive minors [6], our recent algorithm for computing their reduced Gröbner bases [10], and the study on the initial ideals of products of determinantal ideals via the RSK correspondence and increasing decomposition. (2) We try to provide as many plotting functions as possible, like those for the ladders of (mixed) ladder determinantal ideals, for the essential sets to construct Fulton generators of Schubert determinantal ideals, and for reduced pipe dreams which are inherently related to the initial ideals of Schubert determinantal ideals. All the illustrative figures in this paper are the output of

---

[1] Available at www.cmou.net/DetGB.html

Supported by the National Natural Science Foundation of China (NSFC 11971050).

corresponding plotting functions in the package, and many of these functions seem not to be available in other existing software packages.

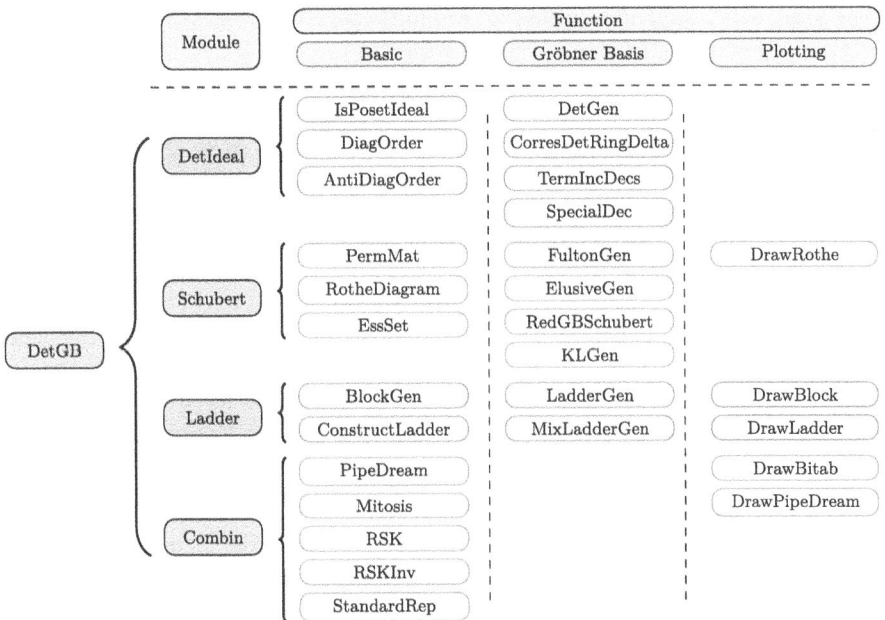

Fig. 1. Infrastructure of DetGB software package

2 Modules and Functions

For any positive integer n, we use $[n]$ to denote the set $\{1,\ldots,n\}$. Let $\mathbb{K}$ be a field and X be a generic matrix of size $m \times n$ with entries x_{ij} for $i \in [m]$ and $j \in [n]$. Without loss of generality, we assume $m \leq n$. Consider the entries $x_{11},\ldots,x_{1n},\ldots,x_{m1},\ldots,x_{mn}$ of X as variables and write them as $\boldsymbol{x}$ for short. Then determinantal ideals are those generated by specific minors of X in the ring $\mathbb{K}[\boldsymbol{x}]$. *Terms* in $\mathbb{K}[\boldsymbol{x}]$ are pure products of the variables in $\boldsymbol{x}$ and a *term order* of $\mathbb{K}[\boldsymbol{x}]$ is a total and well order of all the terms in $\mathbb{K}[\boldsymbol{x}]$ which preserves multiplication. For determinantal ideals, diagonal and anti-diagonal term orders are the most commonly used ones. A term order of $\mathbb{K}[\boldsymbol{x}]$ is said to be *(anti-)diagonal* if for each minor of X, its greatest term with respect to (w.r.t. hereafter) this order is the product of entries on the (anti-)diagonal of the corresponding submatrix of X.

Next, we briefly introduce the modules and functions of this package.

2.1 Module DetIdeal for (Normal) Determinantal Ideals

For the generic matrix X, all its minors are polynomials in $\mathbb{K}[\boldsymbol{x}]$ after expansion. For each integer $t = 2, \ldots, n$, denote the ideal generated by all the t-minors of X in $\mathbb{K}[\boldsymbol{x}]$ by I_t, and it is called a (normal) *determinantal ideal*. It is proved in [11,12] that the generating t-minors form a Gröbner basis of I_t w.r.t. any (anti-)diagonal term order. In particular, in the cases when $t = 2$ and m, the generating set of I_t also forms its universal Gröbner basis [1]. Furthermore, for any t the Gröbner bases of powers and symbolic powers of I_t are identified by using the RSK correspondence, but the Gröbner basis of the product $I_p I_q$ when $p \neq q$ is still unknown [2].

The module DetIdeal of the package includes functions for (1) constructing specific (anti-)diagonal term orders and the generating minors of I_t (which are its Gröbner basis), (2) studying generalized determinantal ideals as poset ideals, and (3) studying the initial ideals of $I_p I_q$ via the increasing decomposition and RSK correspondence. Next we introduce these functions and demonstrate some of them via examples. To utilize this module, one can either use with(DetIdeal) or DetIdeal[command] to access a command provided by this module.

(Anti-)diagonal Term Order and Gröbner Basis. To study determinantal ideals in X, we need to specify the (anti-)diagonal term orders. One kind of these diagonal term orders are the lexicographic one induced by a scanning variable order of $\boldsymbol{x}$. Take the diagonal term order for example, there are four scanning variable orders. (1 and 2) Assign the North-West corner variable x_{11} of X as the greatest and assign the next greatest variable by scanning row by row to the East (or column by column to the South): we call these two variable orders the NWE and NWS ones respectively; (3 and 4) similarly we have the SEW and SEN variable orders. One can show that the lexicographic term orders induced by these four scanning variable orders are all diagonal.

In the module DetIdeal, the commands DiagOrder and AntiDiagOrder construct lexicographic term orders induced by scanning variables which are diagonal and anti-diagonal respectively, and the command DetGen constructs the generating minors of I_t in X.

```
maple> m:=3: n:=4: t:=3:
maple> G:=DetIdeal[DetGen](m,n,t);
Output: G= [x₁₁x₂₂x₃₃ −x₁₁x₂₃x₃₂ −x₁₂x₂₁x₃₃ +x₁₂x₂₃x₃₁ +x₁₃x₂₁x₃₂ −x₁₃x₂₂x₃₁,
x₁₁x₂₂x₃₄ − x₁₁x₂₄x₃₂ − x₁₂x₂₁x₃₄ + x₁₂x₂₄x₃₁ + x₁₄x₂₁x₃₂ − x₁₄x₂₂x₃₁, x₁₁x₂₃x₃₄
−x₁₁x₂₄x₃₃ − x₁₃x₂₁x₃₄ + x₁₃x₂₄x₃₁ + x₁₄x₂₁x₃₃ − x₁₄x₂₃x₃₁, x₁₂x₂₃x₃₄ −x₁₂x₂₄x₃₃
−x₁₃x₂₂x₃₄ + x₁₃x₂₄x₃₂ + x₁₄x₂₂x₃₃ − x₁₄x₂₃x₃₂]
maple> ord:=DetIdeal[DiagOrder](NWE,m,n);
Output: ord=[x₁₁,x₁₂,x₁₃,x₁₄,x₂₁,x₂₂,x₂₃,x₂₄,x₃₁,x₃₂,x₃₃,x₃₄]
maple> Groebner[IsBasis](G,plex(op(ord)));
Output: true
```

Poset Ideal and Generalized Determinantal Ideal. Each minor of X can be identified by its row and column indices, and a partial order $\leq$ can be established among minors of X as follows: for two minors $m_1 = [a_1, \ldots, a_u | b_1, \ldots, b_u]$ and $m_2 = [c_1, \ldots, c_v | d_1, \ldots, d_v]$ of X, we say that $m_1 \leq m_2$ if $u \geq v$ and $a_1 \leq c_1, \ldots, a_v \leq c_v$, $b_1 \leq d_1, \ldots, b_v \leq d_v$. This partial order is critical in the study of determinantal ideals by viewing them as algebras with straightening law. For a poset $(P, \leq)$, a non-empty subset I is called an ideal if for every $x \in I$ and $y \in P$, $y \leq x$ implies $y \in I$. Let $\Delta(X)$ be the set of all minors of X. Then we can study poset ideals of $\Delta(X)$ with the partial order above. The command IsPosetIdeal is for identifying whether a subset of $\Delta(X)$ is a poset ideal.

In [3, Page 5] the following generalized determinantal ideal I_g is studied: Given integers $1 \leq u_1 \leq \cdots \leq u_p \leq m$, $0 \leq r_1 < \cdots < r_p < m$, $1 \leq v_1 \leq \cdots \leq v_q \leq n$, and $0 \leq s_1 < \cdots < s_q < n$, the ideal I_g is generated by the (r_i+1)-minors of the first u_i rows and the (s_j+1)-minors of the first v_j columns for $i \in [p]$ and $j \in [q]$. It is shown that this I_g can be represented by a poset ideal $I(X; \delta) = \{\pi \in \Delta(X) : \pi \not\geq \delta\}$ determined by a minor δ. The command CorresDetRingDelta constructs this minor.

```
maple> U:=[1,3]: R:=[0,2]: V:=[1,3,4]: S:=[0,1,2]:
maple> L:=[[U,R],[V,S]]: m:=3: n:=4:
maple> delta:=DetIdeal[CorresDetRingDelta](L,m,n);
Output: delta=[[2,3],[2,4]]
```

Initial Ideal of Product of Determinantal Ideals. The initial ideal $\mathrm{in}(I_p I_q)$ of products of determinantal ideals are studied in [2] by using increasing decomposition of terms and the RSK correspondence. The *increasing decomposition* of a term in $\mathbb{K}[x]$ refers to decomposing the sequence of its column indices into increasing subsequences. For example, the term $\mu = x_{11}x_{12}x_{23}x_{24}x_{32}x_{33}x_{45}x_{45}x_{57}$ has an increasing decomposition $(1, 3, 4, 5, 7)$, $(2, 4, 5)$, and (3). In the study of $\mathrm{in}(I_p I_q)$, we are particularly interested in increasing decompositions of the maximal shape, which are said to be *special*, and in the DetIdeal module the command TermIncDecs computes such special increasing decompositions of terms represented by the sequence of its column indices.

```
maple> DetIdeal[TermIncDecs]([4,1,5,2,4,6,3,7,4]);
Output: [[[1,2,4,6,7], [4,5], [3,4]], [[4,5,6,7], [1,2,3,4], [4]]]
```

The following problem is of our concern: given two determinantal ideals I_p and I_q, do the minimal generators of $\mathrm{in}(I_p I_q)$ contain a term of degree d. It turns out that this problem can be studied by using special increasing decompositions of terms of degree d corresponding to different shapes of standard bitableaux, removing the decompositions that are clearly divisible. The command SpecialDec is for finding these special increasing decompositions: if an empty table is returned, it indicates that the minimal generators of $\mathrm{in}(I_p I_q)$ do not contain a term of degree d; otherwise the returned table can be further utilized to study this problem.

```
maple> p:=4: q:=3: d:=8: print(DetIdeal[SpecialDec]([p, q], d));
Output: table([])
maple> p:=5: q:=3: d:=9: print(DetIdeal[SpecialDec]([p, q], d));
Output: table(["531" = [[[5,2,1,1],[4,4,1]], [[5,2,2],[4,4,1]]],
"621" = [[[6,1,1,1],[4,4,1]]]])
```

2.2 Module Schubert for Schubert Determinantal Ideals

Schubert determinantal ideals, the defining ideals of matrix Schubert varieties, are a fundamental object in combinatorial algebraic geometry [8]. For an integer n, let w be a permutation in the symmetric group S_n, and for Schubert determinantal ideals we work on a square generic matrix X of size $n \times n$. The Schubert determinantal ideal I_w is generated by all the minors of sizes specified by w in certain parts of X. The module Schubert in the package are developed to facilitate the study on the Gröbner basis of I_w.

Gröbner Bases of Schubert Determinantal Ideals. For a pair (p, q) of integers with $p, q \in [n]$, let X_{pq} represent the submatrix of X consisting its northwest $p \times q$ entries. For an arbitrary permutation $w \in S_n$, the *Rothe diagram* $D(w)$ is constructed by including all boxes in the $n \times n$ grid that are neither to the south nor the east of a nonzero entry in w^T, the permutation matrix of w. The *essential set* of w, denoted by Ess(w), is defined as $\{(p, q) \in D(w) :$ neither $(p, q+1)$ nor $(p+1, q)$ is in $D(w)\}$. Then the *Fulton generators* are the minors of X_{pq} of size rank(w_{pq}^T) + 1 for all $(p, q) \in$ Ess(w). It is proved in [8] that Fulton generators form a Gröbner basis of I_w w.r.t. any anti-diagonal term order. Furthermore, they form a Gröbner basis of I_w w.r.t. any diagonal term order if and only if w is vexillary (namely 2143-avoiding) [9].

Given an arbitrary permutation w, the commands EssSet and DrawRothe compute the essential set and plot the Rothe diagram with the essential set (shown in Fig. 2(left) below) respectively, and the command FultonGen constructs the Fulton generators.

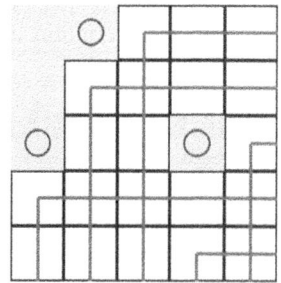
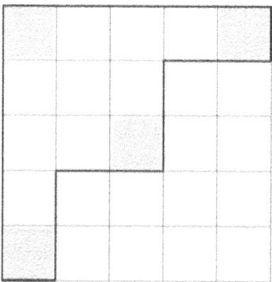

Fig. 2. Rothe diagram of $w = 32514$ (left); A one-sided ladder (right)

```
maple> w:=[3,2,5,1,4]:
maple> Ess:=Schubert[EssSet](w);
Output: Ess={(1,2),(3,1),(3,4)}
maple> Schubert[DrawRothe](w);
maple> Schubert[FultonGen](w);
```
Output: $[x_{11}, x_{12}, x_{21}, x_{31}, x_{11}x_{22}x_{33} - x_{11}x_{23}x_{32} - x_{12}x_{21}x_{33} + x_{12}x_{23}x_{31} +$
$x_{13}x_{21}x_{32} - x_{13}x_{22}x_{31}, x_{11}x_{22}x_{34} - x_{11}x_{24}x_{32} - x_{12}x_{21}x_{34} + x_{12}x_{24}x_{31} + x_{14}x_{21}x_{32}$
$-x_{14}x_{22}x_{31}, x_{11}x_{23}x_{34} - x_{11}x_{24}x_{33} - x_{13}x_{21}x_{34} + x_{13}x_{24}x_{31} + x_{14}x_{21}x_{33} - x_{14}x_{23}x_{31},$
$x_{12}x_{23}x_{34} - x_{12}x_{24}x_{33} - x_{13}x_{22}x_{34} + x_{13}x_{24}x_{32} + x_{14}x_{22}x_{33} - x_{14}x_{23}x_{32}]$

Minimal and Reduced Gröbner Bases. Fulton generators form Gröbner bases of Schubert determinantal ideals, but in general they are not minimal. In [6], the concept of elusive minors is introduced to study the minimality of Fulton generators.

Let m be a minor of size $\text{rank}(w_{pq}^T) + 1$ for some $(p, q) \in \text{Ess}(w)$ and $(\tilde{p}, \tilde{q})$ be another pair in $\text{Ess}(w)$. Then m is said to *attend* the submatrix $X_{\tilde{p}\tilde{q}}$ if the intersection of m and the submatrix $X_{\tilde{p}\tilde{q}}$ contains at least $\text{rank}(w_{\tilde{p}\tilde{q}}^T) + 1$ full rows or $\text{rank}(w_{\tilde{p}\tilde{q}}^T) + 1$ full columns of m. A Fulton generator of I_w for some $(p, q) \in \text{Ess}(w)$ is said to be *elusive* if it does not attend any $X_{\tilde{p}\tilde{q}}$ for each $(\tilde{p}, \tilde{q}) \in \text{Ess}(w) \setminus \{(p, q)\}$ such that $\text{rank}(w_{\tilde{p}\tilde{q}}^T) < \text{rank}(w_{pq}^T)$. Then it is proved in [6, Corollary 1.8] that for any anti-diagonal term order, all the elusive minors form a minimal Gröbner basis of I_w. In particular, if w is vexillary, the same also holds for any diagonal term order.

Then in [10] it is shown that for any vexillary permutation, elusive minors form the reduced Gröbner basis of I_w w.r.t. any anti-diagonal or diagonal term order, and if elusive minors form the reduced Gröbner basis of I_w w.r.t. an anti-diagonal term order, then w is vexillary. As one can see, the reduced Gröbner bases of I_w for a non-vexillary permutation w w.r.t. anti-diagonal term orders are still unknown. Thus, the first two authors of this paper developed a method in [10] for computing their reduced Gröbner bases which avoids all algebraic polynomial operations. In the case when n is large, the process of expanding all the elusive minors of $w \in S_n$ and performing reduction among them can be very costly, and thus this new method is of computational significance. Continue with the permutation $w = 32514$, the command `RedGBSchubert` computes the reduced Gröbner basis of I_w.

```
maple> Schubert[ElusiveGen](w);
```
Output: $[x_{11}, x_{12}, x_{21}, x_{31}, x_{12}x_{23}x_{34} - x_{12}x_{24}x_{33} - x_{13}x_{22}x_{34} + x_{13}x_{24}x_{32} +$
$x_{14}x_{22}x_{33} - x_{14}x_{23}x_{32}]$
```
maple> ord:=DetIdeal[AntiDiagOrder](NEW,nops(w),nops(w)):
maple> Schubert[RedGBSchubert](w,plex(op(ord)));
```
Output: $[x_{11}, x_{12}, x_{21}, x_{31}, x_{13}x_{22}x_{34} - x_{13}x_{24}x_{32} - x_{14}x_{22}x_{33} + x_{14}x_{23}x_{32}]$

2.3 Module Ladder for Ladder Determinantal Ideals

Ladder determinantal ideals are generated by minors inside specific subsets of X called ladders due to their shapes. In the Ladder module, we develop functions for constructing ladder determinantal ideals and further blockwise determinantal ideals which are generalized ladder ones introduced in [10].

A (two-sided) *ladder* in X is a subset L of X such that whenever $x_{ij}, x_{kl} \in L$ with $i \leq k$ and $j \leq l$, we have $x_{pq} \in L$ for all $i \leq p \leq k$ and $j \leq q \leq l$. A ladder L in X can be identified with all its *lower corners* $(x_{a_1}, x_{b_1}), \ldots, (x_{a_l}, x_{b_l})$ and *upper corners* $(x_{c_1}, x_{d_1}), \ldots, (x_{c_u}, x_{d_u})$ such that $1 \leq a_1 < \cdots < a_l \leq m$, $n \geq b_1 \geq \cdots \geq b_l \geq 1$, $1 \leq c_1 < \cdots < c_u \leq m$, and $n \geq d_1 \geq \cdots \geq d_u \geq 1$. Furthermore, a ladder L in X is said to be *one-sided* if all its upper corners lie in the first row (or column) of L.

With the module Ladder, one can construct a ladder by specifying its corners with the command ConstructLadder (returning False if they do not form one) and plot ladders with DrawLadder. For the following example, the plotted ladder is shown in Fig. 2(right) above.

```
maple> U:=[[1,1],[4,4]]: L:=[[5,1],[3,3],[1,5]]:
maple> Ladder[ConstructLadder](U,L);
Output: False;
maple> U:=[[1,1]]: L:=[[5,1],[3,3],[1,5]]:
maple> A:=Ladder[ConstructLadder](U,L):
maple> DrawLadder(A);
```

Consider a one-sided ladder L with one upper corner (x_{c_u}, x_{d_u}) and lower corners $(x_{a_1}, x_{b_1}), \ldots, (x_{a_l}, x_{a_l})$. Then for each lower corner (x_{c_i}, x_{d_i}), it determines a submatrix R_i of X with (x_{c_1}, x_{d_1}), and thus $L = \cup_{i=1}^{l} R_i$. Given a list of non-negative integers $\boldsymbol{r} = (r_1, \ldots, r_l)$, then the ideal in $\mathbb{K}[\boldsymbol{x}]$ generated by all the r_i-minors contained in R_i for $i = 1, \ldots, l$ is called a *ladder determinantal ideal* and denoted by $I(L, \boldsymbol{r})$. It is proved in [5] that for any ladder determinantal ideal $I(L, \boldsymbol{r})$, there exists a Schubert determinantal ideal I_w such that $I(L, \boldsymbol{r}) = I_w$ and w is vexillary and unique under certain conditions. Then the Gröbner bases, including the minimal and reduced ones, of ladder determinantal ideals can be constructed as Schubert ones of vexillary permutations via this transformation. In the Ladder module, there is a command LadderGen for constructing the generating minors of a ladder determinantal ideal $I(L, \boldsymbol{r})$ and a command OneLadder2Perm for computing such a corresponding permutation from $I(L, \boldsymbol{r})$.

```
maple> L:=[[5,1],[3,3],[1,5]]: r:=[1,2,1]:
maple> w:=Ladder[OneLadder2Perm](L,r);
Output: w=[6,2,4,3,5,1]
maple> Schubert[FultonGen](w);
```
Output: $[x_{11}, x_{12}, x_{13}, x_{14}, x_{15}, x_{21}, x_{31}, x_{41}, x_{51}, x_{11}x_{22} - x_{12}x_{21}, x_{11}x_{23} - x_{13}x_{21},$
$x_{12}x_{23} - x_{13}x_{22}, x_{11}x_{32} - x_{12}x_{31}, x_{11}x_{33} - x_{13}x_{31}, x_{12}x_{33} - x_{13}x_{32}, x_{21}x_{32} - x_{22}x_{31},$

$x_{21}x_{33} - x_{23}x_{31}, x_{22}x_{33} - x_{23}x_{32}]$

The determinantal ideals corresponding to general ladders are called *mixed ladder determinantal ideals*. It is proved in [7] that the generating minors of any mixed ladder determinantal ideal form its Gröbner basis. In the `Ladder` module, there is a command `MixLadderGen` for constructing the generating minors of such an ideal.

2.4 Module `Combin` for Underlying Combinatorial Tools

Two fundamental tools to study (normal) determinantal ideals are the straightening law and Robinson-Schensted-Knuth (RSK) correspondence [3]. For Schubert determinantal ideals, pipe dreams of permutations are closely related to their initial ideals. In the module `Combin` we develop functions for these combinatorial tools for the study of determinantal ideals.

RSK Correspondence. The RSK correspondence sets up a bijection between standard bitableaux and terms in $\mathbb{K}[x]$ which can be represented by two-row arrays. The module `Combin` contains the commands `RSK` for sending a bitableaux to a term, `RSKInv` for the inverse mapping, and `DrawBitab` for plotting bitableaux.

```
maple> T:=[[1,2,3,4,5,5,5,6,6,6,8,9,9,9,10,10],
[2,5,6,4,6,3,3,8,8,1,5,7,4,2,5,2]]:
maple> B:=Combin[RSKInv](T);
Output: B=[[[1,2,3,6],[4,5,6],[5,8,9],[5,9,10],[6,9],
[10]],[[1,2,4,5],[2,3,7],[2,5,8],[3,6,8],[4,6],[5]]]
maple> Combin[DrawBitab](B);
```

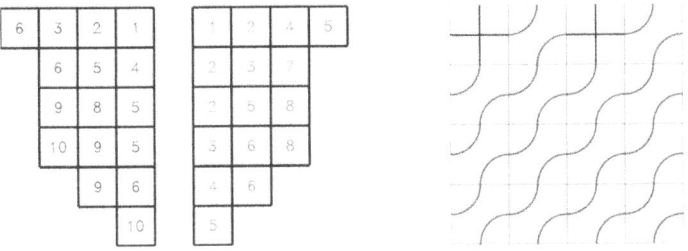

Fig. 3. A bitableaux (left); A pipe dream of $w = 2143$ (right)

Pipe Dreams. A *(reduced) pipe dream* of $w \in S_n$ is a tiling of an $n \times n$ grid with crosses and elbow joints such that any two pipes crosses at most once.

In the `Combin` module, the mitosis algorithm in [8] is implemented as the command `Mitosis` to compute all the pipe dreams of a permutation. The underlying structures of the pipe dreams are matrices with 0–1 entries, representing elbow joints and crosses respectively. There is also a command `DrawPipeDream` available for visualizing pipe dreams. For the following example one of the plotted pipe dreams is shown in Fig. 3(right).

```
maple> w:=[2,1,4,3]: Combin[Mitosis](w);
```

$$\text{Output:} \quad \left[\begin{bmatrix} 1 & 0 & 1 & 0 \\ 0 & 0 & 0 & 0 \\ 0 & 0 & 0 & 0 \\ 0 & 0 & 0 & 0 \end{bmatrix}, \begin{bmatrix} 1 & 0 & 0 & 0 \\ 0 & 1 & 0 & 0 \\ 0 & 0 & 0 & 0 \\ 0 & 0 & 0 & 0 \end{bmatrix}, \begin{bmatrix} 1 & 0 & 0 & 0 \\ 0 & 0 & 0 & 0 \\ 1 & 0 & 0 & 0 \\ 0 & 0 & 0 & 0 \end{bmatrix} \right]$$

```
maple> Combin[DrawPipeDream](w);
```

Straightening Law. With the partial order defined in Sect. 2.1, the product of incomparable minors of X can be expressed as a linear combination of products of comparable minors, and this representation is called its *standard representation*. In the module `Combin` there is a command `StandardRep` which computes the standard representation of incomparable minors of X based on the Plücker relations. In the simple example of the straightening law below, a and b are two maximal minors of the $m \times n$ matrix X.

```
maple> a:=[1,4,6]:b:=[2,3,5]:m:=3:n:=6:
maple> Combin[StandardRep](a,b,m,n);
Output:
[2,3,5][1,4,6]=-[1,2,5][3,4,6]-[1,2,3][4,5,6]+[1,3,5][2,4,6]
```

3 Experimental Results for Computing Reduced Gröbner Bases of Schubert Determinantal Ideals

One of the new algorithms implemented in the DetGB package is that for computing reduced Gröbner bases of Schubert determinantal ideals for non-vexillary permutations w.r.t. anti-diagonal term orders based on our constructive method presented in [10]. Essentially this algorithm avoids all the algebraic operations in the transformation from Gröbner bases to the reduced ones by means of reduction among them. We compared the computational efficiency of the command `RedGBSchubert` in the DetGB package for computing reduced Gröbner bases of Schubert determinantal ideals with the built-in function `InterReduce` in MAPLE which turns input Gröbner bases, in this case the Fulton generators constructed by `FultonGen` in the package, into the reduced ones. The benchmark permutations in the experiments are randomly generated and then verified not to be vexillary. The experimental results are summarized in Table 1 below. All the experiments were conducted in MAPLE 2019 on a laptop computer with AMD Ryzen 5 3500U CPU at 2.10 GHz×4 with 8 GB RAM under Windows 10.

In Table 1, the column "Permutation" records the specific non-vexillary permutation, "n" its length, and "N" and "d" the number of polynomials and their maximum degree in the Fulton generators respectively. The timings (in second) by the methods in MAPLE and of ours are presented in the two columns "Maple" and "Ours" respectively. The symbol "—" means that the computation does not terminate within 3600 seconds.

Table 1. Timings (in second) for computing reduced Gröbner bases of Schubert determinantal ideals

Permutation	n	N	d	Maple	Ours	Permutation	n	N	d	Maple	Ours
[6, 2, 1, 5, 3, 4]	6	22	3	0.02	0.08	[3, 2, 1, 6, 5, 4, 9, 8, 7]	9	27	7	318.53	26.53
[5, 3, 1, 6, 2, 4]	6	28	3	0.03	0.11	[7, 10, 8, 2, 4, 6, 1, 9, 3, 5]	10	685	4	1.11	2.69
[3, 2, 4, 6, 5, 1]	6	11	4	0.02	0.06	[2, 4, 10, 9, 7, 3, 6, 8, 5, 1]	10	761	4	2.19	11.00
[3, 4, 5, 2, 7, 6, 1]	7	15	1	0.06	0.27	[5, 4, 10, 2, 1, 6, 9, 3, 7, 8]	10	325	6	100.22	5.94
[4, 5, 2, 7, 3, 1, 6]	7	42	4	0.05	0.16	[8, 10, 3, 1, 6, 5, 2, 9, 4, 7]	10	472	6	108.20	3.57
[1, 3, 2, 5, 6, 7, 4]	7	16	5	0.02	0.09	[1, 9, 4, 2, 7, 6, 3, 5, 10, 8]	10	188	8	—	721.20
[4, 1, 7, 3, 5, 6, 2, 8]	8	44	3	0.02	0.27	[11, 8, 4, 7, 2, 10, 3, 6, 5, 9, 1]	11	1652	5	25.28	7.36
[2, 4, 3, 7, 6, 5, 1, 8]	8	44	4	0.08	0.55	[11, 4, 7, 2, 9, 8, 1, 3, 10, 6, 5]	11	1732	5	47.42	13.84
[1, 2, 5, 4, 6, 3, 8, 7]	8	14	7	17.38	3.99	[2, 7, 4, 5, 1, 10, 3, 9, 11, 6, 8]	11	545	7	—	438.47
[3, 1, 5, 8, 7, 6, 2, 9, 4]	9	257	4	0.88	3.28	[11, 12, 3, 10, 6, 8, 9, 1, 4, 7, 5, 2]	12	2705	4	7.39	12.22
[8, 3, 7, 9, 4, 2, 1, 6, 5]	9	162	5	0.55	0.89	[9, 12, 1, 6, 4, 10, 2, 11, 7, 3, 5, 8]	12	5153	5	525.77	23.70
[9, 3, 6, 2, 1, 8, 5, 7, 4]	9	299	5	3.13	1.77	[8, 12, 9, 5, 7, 3, 11, 2, 4, 1, 10, 6]	12	2916	6	1196.48	20.41
[3, 2, 9, 4, 5, 6, 8, 7, 1]	9	114	6	20.91	1.17	[10, 1, 13, 6, 3, 11, 12, 9, 4, 7, 8, 2, 5]	13	9126	4	794.47	100.69
[1, 6, 3, 2, 4, 9, 5, 8, 7]	9	51	7	137.23	24.75	[1, 8, 13, 9, 6, 3, 4, 5, 12, 2, 7, 10, 11]	13	2485	8	—	—
[2, 5, 1, 4, 6, 3, 9, 8, 7]	9	33	7	339.67	42.97						

From Table 1 one can easily find that when the Schubert determinantal ideals are complicated ("N" and "d" are large), the implementation of our new method in [10] outperforms the built-in function in MAPLE w.r.t. the computational efficiency. See the timings for the permutation $w = [8, 12, 9, 5, 7, 3, 11, 2, 4, 1, 10, 6] \in S_{12}$ for example. When the ideals are simply structured ("N" or "d" is small), the built-in function in MAPLE, which has been polished for a couple of decades, performs better. In conclusion, when the lengths of permutations become large, the Schubert determinantal ideals can grow very complicated due to their combinatorial inherence. In this case, the reduction to get the reduced Gröbner bases, which is conventionally considered to be much simpler than computation of Gröbner bases, becomes computationally hard, and this is when our implementation excels.

4 Concluding Remarks

There is one additional kind of determinantal ideals called the Kazhdan-Lusztig ideals. They are similar to Schubert determinantal ideals but are defined w.r.t. two permutations, and their Gröbner bases have also been identified. We plan to further investigate this kind of determinantal ideals from the viewpoint of Gröbner bases and include functions related to them in the package.

Currently the package has weak dependence on the proprietary computer algebra system MAPLE in the sense that some underlying built-in functions in MAPLE are needed in the package and in our own study on determinantal ideals. We will further evaluate the possibility to port the package to other free platforms so that it becomes more accessible to interested users.

The authors would like to thank the reviewers for their helpful comments which led to enrichment and improvement of the paper.

References

1. Bernstein, D., Zelevinsky, A.: Combinatorics of maximal minors. J. Algebr. Combin. **2**, 111–121 (1993)
2. Bruns, W., Conca, A., Raicu, C., Varbaro, M.: Determinants, Gröbner Bases and Cohomology. Springer, Cham (2022). https://doi.org/10.1007/978-3-031-05480-8
3. Bruns, W., Vetter, U.: Determinantal Rings. Springer, Heidelberg (2006)
4. Faugère, J.C., Safey El Din, M., Spaenlehauer, P.J.: On the complexity of the generalized MinRank problem. J. Symb. Comput. **55**, 30–58 (2013)
5. Fulton, W.: Flags, Schubert polynomials, degeneracy loci, and determinantal formulas. Duke Math. J. **65**(3), 381–420 (1992)
6. Gao, S., Yong, A.: Minimal equations for matrix Schubert varieties. arXiv preprint arXiv:2201.06522 (2022)
7. Gorla, E.: Mixed ladder determinantal varieties from two-sided ladders. J. Pure Appl. Algebra **211**(2), 433–444 (2007)
8. Knutson, A., Miller, E.: Gröbner geometry of Schubert polynomials. Ann. Math. 1245–1318 (2005)
9. Knutson, A., Miller, E., Yong, A.: Gröbner geometry of vertex decompositions and of flagged tableaux. J. für die Reine und Angewandte Mathematik **2009**(630), 1–31 (2009)
10. Mou, C., Song, Q.: On the reduced Gröbner bases of blockwise determinantal ideals. arXiv preprint arXiv:2309.15035 (2023)
11. Narasimhan, H.: The irreducibility of ladder determinantal varieties. J. Algebra **102**(1), 162–185 (1986)
12. Sturmfels, B.: Gröbner bases and Stanley decompositions of determinantal rings. Math. Z. **205**(1), 137–144 (1990)
13. Tsakiris, M.: Results on the algebraic matroid of the determinantal variety. Trans. Am. Math. Soc. **377**(01), 731–751 (2024)

Extrapolating Solution Paths of Polynomial Homotopies Towards Singularities with PHCpack and Phcpy

Jan Verschelde[(✉)] and Kylash Viswanathan

Department of Mathematics, Statistics, and Computer Science,
University of Illinois at Chicago, 851 S. Morgan St. (m/c 249), Chicago, IL
60607-7045, USA
{janv,kviswa5}@uic.edu
http://www.math.uic.edu/~jan

Abstract. PHCpack is a software package for polynomial homotopy continuation, which provides a robust path tracker [Telen, Van Barel, Verschelde, SISC 2020]. This tracker computes the radius of convergence of Newton's method, estimates the distance to the nearest path, and then applies Padé approximants to predict the next point on the path. A priori step size control is less sensitive to finely tuned tolerances than a posteriori step size control, and is therefore robust. The Python interface phcpy is extended with a new step-by-step tracker and is applied to experiment with extrapolation methods to accurately locate the singular points at the end of solution paths.

Keywords: extrapolation · numerical analytic continuation · pole · polynomial homotopy · singularity

1 Introduction

A polynomial homotopy is a family of polynomial systems which depend on one parameter t. Regular solutions of a polynomial homotopy have Taylor series developments in t. The solutions paths are analytic functions of t. Application of the theorem of Fabry [5] enables the location of the nearest singular solution. The calculations in this paper can be considered in the area of numerical analytic continuation [7,20,21].

Extrapolation methods [2,18,28], in particular Aitken's algorithm and the rho algorithm [29], are effective in accelerating logarithmically converging series towards a single pole of a polynomial homotopy. We demonstrate the interplay between compiled code in a library, for the computationally intensive calculation of the Taylor series, and the interactive Python scripts for the extrapolation methods. Our calculations happen in Jupyter notebooks [12] with phcpy [17,23] in a Python kernel. The interface to the compiled code in PHCpack [22] does not

---

Supported by the National Science Foundation under grant DMS 1854513.

© The Author(s), under exclusive license to Springer Nature Switzerland AG 2024
K. Buzzard et al. (Eds.): ICMS 2024, LNCS 14749, pp. 365–374, 2024.
https://doi.org/10.1007/978-3-031-64529-7_37

require compilation as it is done through the Ctypes module of Python which allows to call functions in libraries directly, similar to ccall in Julia.

Since version 2.4.88, PHCpack became a crate of alire[1] the package manager of the gnu-ada compiler, which builds the executable phc and libPHCpack, integrating code of MixedVol [6] and DEMiCs [15], for fast mixed volume computation, and algorithms for multiple double arithmetic of QDlib [8] and CAMPARY [10]. The development of phcpy was motivated in part by SageMath [4]. All software is free and open source, released under version 3 of the GNU GPL license. Computations are done with phcpy 1.1.4 and version 2.4.90 of PHCpack.

The main contribution of this paper is the illustration of phcpy to visualize solution paths and to experiment with extrapolation methods.

2 Extrapolation Experiments

The polynomial homotopy

$$x^2 - \frac{4}{5}\left(\frac{1}{2} - I\right)(1-t)\left(\frac{1}{2} + I - t\right) = 0, \quad I = \sqrt{-1}. \tag{1}$$

has two singularities: one at $t = 1$ and another at $t = 1/2 + I$. The coefficient $4/5(1/2 - I)$ makes that at $t = 0$, the solution paths $x(t)$ start at $x(0) = \pm 1$. The phase portrait [27], made with complexplorer-0.1.2 [14] (which depends on matplotlib [9]), is shown in Fig. 1.

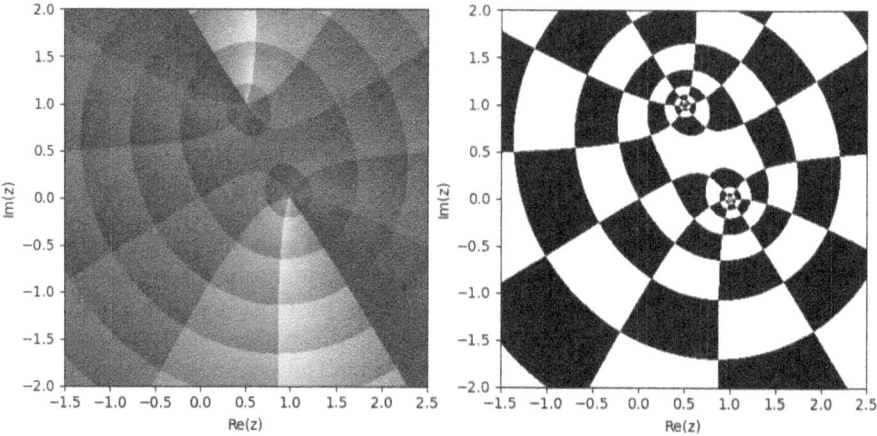

Fig. 1. A phase portrait of $f(z) = \sqrt{4/5(1/2 - I)(1 - z)(1/2 + I - z)}$, depicting at the left the color-coded phase $f/|f|$ on the domain of f. At the right is a polar chessboard

Using the algorithms in [1] to compute the Taylor series is done with phcpy in the code snippet below:

---

[1] Version 1.2.2 of alr, GNAT 12.2.1 and gprbuild 22.0.1.

```
from phcpy.solutions import make_solution
from phcpy.series import double_newton_at_point

pol = ['x^2 - (4/5)*(1/2 - I)*(1 - t)*(1/2 + I - t);']
variables = ['x', 't']
sols = [make_solution(variables, [1, 0])]
deg = 33   # degree of truncation
nit = 8    # number of iterations of Newton's method
srs = double_newton_at_point(pol, sols, idx=2, maxdeg=deg, nbr=nit)
```

The srs contains the string representation of the Taylor series and the coefficients can be extracted with the help of SymPy [11]. Denoting c_n as the coefficient of t^n in the Taylor series, the ratio c_n/c_{n+1} equals (1.0362677867627397 -0.03656143770249911j), for $n = 31$, illustrating the very slow convergence to 1. Sequences such as this are said to be converge logarithmically, as it may take about twice as many terms in the series to gain one bit of accuracy.

The rho algorithm [29] performs spectacularly well on the Taylor series of $x(t) = \sqrt{1-t}$, where there is only one pole. In the definition of the function rhoComplex below, observe the inverse divided differences. The connection with Thiele interpolation [3] is one possible justification of its good performance on $x(t) = \sqrt{1-t}$. The case of nearby poles is analyzed in [25].

```
def rhoComplex(nbr):
    """
    Runs the rho algorithm in complex double arithmetic,
    on the numbers given in the list nbr, using x(n) = n+1.
    Returns the last element of the table of extrapolated numbers.
    """
    rho1 = [1.0/(nbr[n] - nbr[n-1]) for n in range(1, len(nbr))]
    rho = [nbr, rho1]
    for k in range(2, len(nbr)):
        nextrho = []
        for n in range(k, len(nbr)):
            invrho1 = complex(k)/(rho[k-1][n-k+1] - rho[k-1][n-k])
            nextrho.append(rho[k-2][n-k+1] + invrho1)
        rho.append(nextrho)
    return nextrho[-1]
```

Because the two singularities of (1) are relatively too close to each other, the rho algorithm fails to improve the sequence of Fabry ratios. In the second homotopy:

$$x^2 - \left(\frac{1}{272}\right)\left(-4 - 16I\right)\left(1-t\right)\left(-4 + 16I - t\right) = 0, \qquad (2)$$

the singularity at $-4 + 16I$ is much farther from the other singularity at 1, and then the rho algorithm ends at (1.0000000000014202+

4.354118985724171e-14j), using 32 coefficients of the Taylor series, with an error of 1.42e-12.

The differences between the location of the singularities in the homotopies (1) and (2) is shown in the schematic of Fig. 2. The P at the right of Fig. 2 illustrates the notion of *the last pole*, introduced in [26].

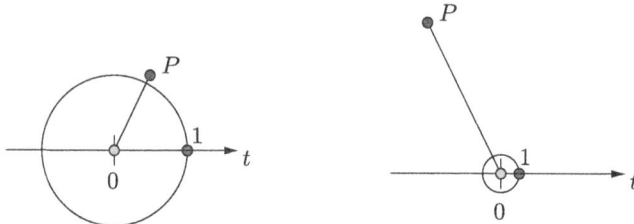

Fig. 2. At the left, the pole $P = 1/2 + I$ is relatively close to 1, while at the right, the pole $P = -4 + 16I$ is much farther from 1

The right of Fig. 2 is a representative schematic for a solution path converging to a singular solution at 1. The effect a nearby and far away poles on extrapolation method is the subject of [25].

3 Path Tracking Towards a Singular Solution

One important innovation in the modernization [23] of PHCpack with a scripting interface was the introduction of a step-by-step path tracker, which allows the user to ask for the next point on a path. Recently, another step-by-step tracker which applies the a priori step size control algorithms of [19] was added to phcpy.

The homotopy

$$\gamma(1-t)\begin{pmatrix} x^2 - 1 = 0 \\ y^2 - 1 = 0 \end{pmatrix} + t\begin{pmatrix} x^2 + y - 3 = 0 \\ x + 0.125y^2 - 1.5 = 0 \end{pmatrix} \qquad (3)$$

ends at $t = 1$ at an example of [16], which has a triple root at $(x, y) = (1, 2)$. Figure 3 shows one path defined by the homotopy (3), converging to $(1, 2)$.

The code below collects the points on the solution path in the list **path**. The constant γ in the homotopy (3) is set to the value $-0.917 - 0.398I$. Expecting a singularity at the end of the path, as long as the distance between the closest pole and 1.0 is larger than 1.0e-4, the path tracker continues. Three lists are produced: in **path** are the points on the path, in **predicted** are the predicted solutions, and **poles** contains the list of poles closest to the path.

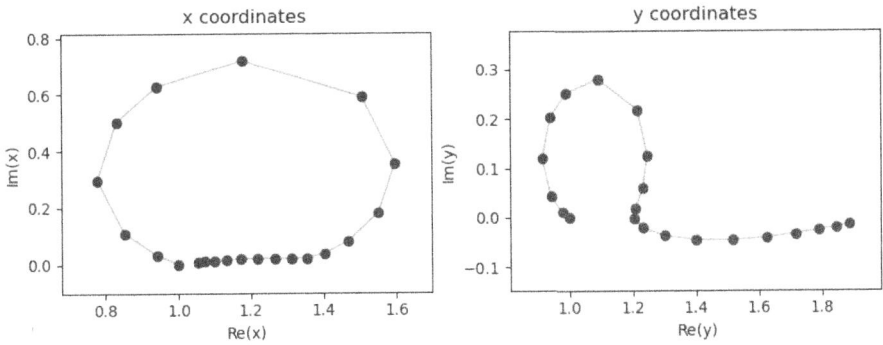

Fig. 3. Points on one solution path starting at $(1,1)$ and converging to $(1,2)$

```
from phcpy.solutions import make_solution
from phcpy.curves import set_gamma_constant
from phcpy.curves import initialize_double_artificial_homotopy
from phcpy.curves import set_double_solution, get_double_solution
from phcpy.curves import double_predict_correct, double_t_value
from phcpy.curves import double_closest_pole

pols = ['x**2+y-3;', 'x+0.125*y**2-1.5;']
start = ['x**2 - 1;', 'y**2 - 1;']
startsol = make_solution(['x', 'y'], [1, 1])
set_gamma_constant(complex(-0.917, -0.398))
initialize_double_artificial_homotopy(pols, start)
set_double_solution(2, startsol)
path = [startsol]
predicted = []
poles = []
cfp = complex(0.0)
while abs(cfp - 1.0) > 1.0e-4:
    double_predict_correct()
    (repole, impole) = double_closest_pole()
    tval = double_t_value()
    cfp = complex(tval + repole, impole)
    poles.append(cfp)
    sol = get_double_solution()
    predsol = get_double_predicted_solution()
    path.append(sol)
    predicted.append(predsol)
print(sol)
```

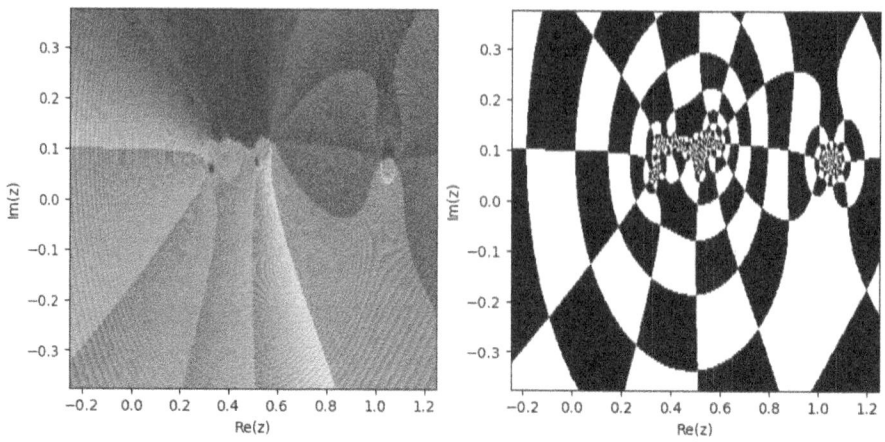

Fig. 4. A phase portrait of $f(z)$, defined in (4) by the poles of one path converging to a singular solution.

The code prints

```
t :  9.99968379127468E-01    0.00000000000000E+00
m : 1
the solution for t :
 x :  1.05353846638456E+00    6.88135807075172E-03
 y :  1.89010705713382E+00   -1.44977473137858E-02
== err :  6.369E-14 = rco :  7.184E-04 = res :  2.849E-17 =
```

The value 7.184E-04 next to rco is the estimate for the inverse condition number of the Jacobian matrix at the point, which gives an upper bound on the error of the update in Newton's method. In particular, the update of Newton's method is correct up to the last four decimal places. Even as 0.999684 is already close to 1, the solution is thus still well conditioned.

The number (1.0000808949264557-6.886127445687259e-08j) ends the list poles. Observe how small the imaginary part is. Only seven terms in the Taylor series were used to compute this approximation for 1.0.

To visualize the poles with complexplorer, consider the function

$$f(z) = \sum_{p \in L} \frac{1}{z-p}, \tag{4}$$

where L is the list of the first twelve poles. The phase portrait of f is shown in Fig. 4. The phase portrait starts to look interesting at the first closest pole, around $t \approx 0.471 + 0.101I$.

Figure 5 is constructed using the information of the points on the path and the predicted solutions. The predicted solutions are evaluated Padé approximants, evaluated in the radius of convergence for Newton's method, as estimated by the application of the theorem of Fabry. The plots in Fig. 5 show a sequence

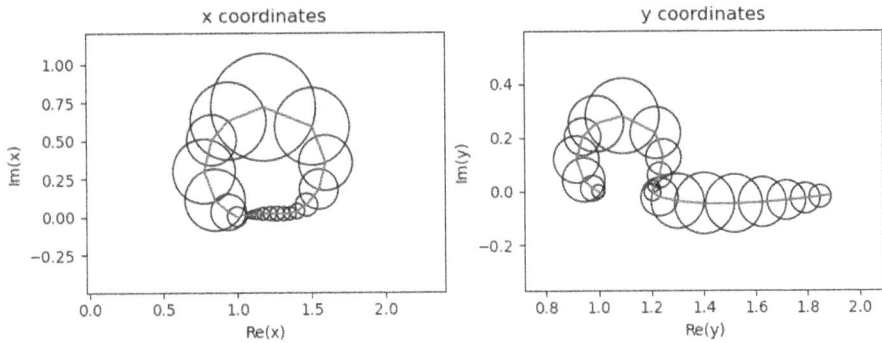

Fig. 5. At each point on the path, a circle is drawn, with center at the point and the radius is the distance to the predicted solution

of circles with overlapping interiors. Towards the end of the path, observe that the radii of the circles decrease, as we approach the singularity at the end. The smaller circles in the middle of the path indicate nearby poles.

4 Iterated Aitken Extrapolation

Looking at the list **path**, the solution at the end is not very accurate: the 1-norm of its solution components equals 2.27e-01, which means that we have only one decimal place of accuracy. If we consider the last seven points on the path, then we observe a very slowly converging sequence of points. Aitken extrapolation is effective in accelerating logarithmically converging sequences [13]. The function below contains the definition of Aitken extrapolation:

```
def Aitken(x):
    """
    Applies Aitken extrapolation to the sequence x.
    Returns the transformed sequence.
    """
    y = [0.0 for _ in range(len(x)-2)]
    for k in range(len(x)-2):
        dxk = x[k+1] - x[k]
        ddx = x[k+2] - 2*x[k+1] + x[k]
        y[k] = x[k] - dxk*dxk/ddx
    return y
```

The iterated Aitken extrapolation applies Aitken extrapolation repeatedly [28], as defined in the next function.

```
def repeatedAitken(x, exa, verbose=True):
    """
    Applies Aitken extrapolation repeatedly on the sequence x.
    If verbose, then the error with the exact value in exa is shown.
    Returns the last element.
    """
    cffs = x
    while len(cffs) > 2:
        a = Aitken(cffs)
        if verbose:
            print('on', len(cffs), ':', a[len(a)-1], end='')
            err = abs(a[len(a)-1] - exa)
            print(f' error :{err: .2e}')
        cffs = a
    return cffs[0]
```

Executing `repeatedAitken(ypt, 2.0)` produces the sequence

```
on 7 : (2.0137784911225802+0.004510752594407998j) error : 1.45e-02
on 5 : (2.0026132752890415+0.001430742263173204j) error : 2.98e-03
on 3 : (2.0008714460040284+0.000612101736300566j) error : 1.06e-03
```

with similar results as `repeatedAitken(xpt, 1.0)`. The error (sum of errors on both x and y coordinates) has decreased from 2.27e-01 to 1.69e-03. With relatively little effort, we gained two decimal places in the solution.

The rho algorithm and its iterated version produce similar results, but its application is more delicate, as the default values of the interpolation points need to take the values of the t parameter into account.

5 Building and Installing Phcpy

Thanks to the efforts of Doug Torrance, both PHCpack and phcpy can be installed with the package managers of Ubuntu.

Ad hoc makefiles, customized for Linux, Windows, and Mac OS X, were written during the development of PHCpack, but became too cumbersome to maintain. GPRbuild is the project manager of the GNAT toolchain, which made the building of the executable `phc` and the library `libPHCpack`, along with its components in C and C++ (in particular: DEMiCs) more portable, since version 2.4.85 [24]. Since version 2.4.88, PHCpack can be built via `"alr get phcpack"`.

The development of phcpy started in python2, with the writing of extension modules, which needed compilation, and later adjustments for python3. The compilation imposed restrictions for Windows, as the Python interpreters on Windows are typically built with the Microsoft compilers which are *not* interoperable with gcc. Version 1.1.3 of phcpy (the third major rewrite of phcpy) used the Ctypes module. Once the `libPHCpack` is in its proper place, the instruction `"pip install ."` in the folder where the `setup.py` is located will extend the Python interpreter with phcpy, also on native Windows systems.

6 Conclusions

Extrapolating on Taylor series towards singular solutions, this paper illustrates the application of numerical analytic continuation to solving polynomial systems. With `phcpy`, the algorithms of [19] and [26] have become better accessible.

References

1. Bliss, N., Verschelde, J.: The method of Gauss-Newton to compute power series solutions of polynomial homotopies. Linear Algebra Appl. **542**, 569–588 (2018)
2. Brezinski, C., Redivo Zaglia, M.: Extrapolation Methods. Studies in Computational Mathematics, vol. 2. North-Holland, Amsterdam (1991)
3. Cuyt, A., Wuytack, L.: Nonlinear Methods in Numerical Analysis. North-Holland, Elsevier Science Publishers, Amsterdam (1987)
4. Eröcal, B., Stein, W.: The Sage project: unifying free mathematical software to create a viable alternative to magma, maple, mathematica and MATLAB. In: Fukuda, K., Hoeven, J., Joswig, M., Takayama, N. (eds.) ICMS 2010. LNCS, vol. 6327, pp. 12–27. Springer, Heidelberg (2010). https://doi.org/10.1007/978-3-642-15582-6_4
5. Fabry, E.: Sur les points singuliers d'une fonction donnée par son développement en série et l'impossibilité du prolongement analytique dans des cas très généraux. Annales scientifiques de l'École Normale Supérieure **13**, 367–399 (1896)
6. Gao, T., Li, T.Y., Wu, M.: Algorithm 846: MixedVol: a software package for mixed-volume computation. ACM Trans. Math. Softw. **31**(4), 555–560 (2005)
7. Henrici, P.: An algorithm for analytic continuation. SIAM J. Numer. Anal. **3**(1), 67–78 (1966)
8. Hida, Y., Li, X.S., Bailey, D.H.: Algorithms for quad-double precision floating point arithmetic. In: 15th IEEE Symposium on Computer Arithmetic (Arith-15 2001), pp. 155–162. IEEE Computer Society (2001)
9. Hunter, J.D.: Matplotlib: a 2D graphics environment. Comput. Sci. Eng. **9**(3), 90–95 (2007)
10. Joldes, M., Muller, J.-M., Popescu, V., Tucker, W.: CAMPARY: Cuda multiple precision arithmetic library and applications. In: Greuel, G.-M., Koch, T., Paule, P., Sommese, A. (eds.) ICMS 2016. LNCS, vol. 9725, pp. 232–240. Springer, Cham (2016). https://doi.org/10.1007/978-3-319-42432-3_29
11. Joyner, D., Čertík, O., Meurer, A., Granger, B.E.: Open source computer algebra systems: SymPy. ACM Commun. Comput. Algebra **45**(4), 225–234 (2011)
12. Kluyver, T., et al.: Jupyter notebooks–a publishing format for reproducible computational workflows. In: Loizides, F., Schmidt, B. (eds.) Positioning and Power in Academic Publishing: Players, Agents, and Agendas, pp. 87–90. IOS Press, Amsterdam (2016)
13. Kowalewski, C.: Acceleration de la convergence pour certaines suites a convergence logarithmique. In: de Bruin, M.G., van Rossum, H. (eds.) Padé Approximation and its Applications Amsterdam 1980. LNM, vol. 888, pp. 263–272. Springer, Heidelberg (1981). https://doi.org/10.1007/BFb0095592
14. Kuvychko, I.: Complexplorer. https://github.com/kuvychko/complexplorer

15. Mizutani, T., Takeda, A.: DEMiCs: a software package for computing the mixed volume via dynamic enumeration of all mixed cells. In: Stillman, M., Takayama, N., Verschelde, J. (eds.) Software for Algebraic Geometry. The IMA Volumes in Mathematics and its Applications, vol. 148, pp. 59–79. Springer, New York (2008). https://doi.org/10.1007/978-0-387-78133-4_5
16. Ojika, T.: Modified deflation algorithm for the solution of singular problems. I. A system of nonlinear algebraic equations. J. Math. Anal. Appl. **123**, 199–221 (1987)
17. Otto, J., Forbes, A., Verschelde, J.: Solving polynomial systems with phcpy. In: Proceedings of the 18th Python in Science Conference, pp. 563–582 (2019)
18. Sidi, A.: Practical Extrapolation Methods. Theory and Applications, Cambridge Monographs on Applied and Computational Mathematics, vol. 10. Cambridge University Press, Cambridge (2003)
19. Telen, S., Van Barel, M., Verschelde, J.: A robust numerical path tracking algorithm for polynomial homotopy continuation. SIAM J. Sci. Comput. **42**(6), A3610–A3637 (2020)
20. Trefethen, L.N.: Approximation Theory and Approximation Practice. Extented edn. SIAM (2020)
21. Trefethen, L.N.: Numerical analytic continuation. Jpn. J. Ind. Appl. Math. **40**(3), 1587–1636 (2023)
22. Verschelde, J.: Algorithm 795: PHCpack: a general-purpose solver for polynomial systems by homotopy continuation. ACM Trans. Math. Softw. **25**(2), 251–276 (1999). https://github.com/janverschelde/PHCpack
23. Verschelde, J.: Modernizing PHCpack through phcpy. In: de Buyl, P., Varoquaux, N. (eds.) Proceedings of the 6th European Conference on Python in Science (EuroSciPy 2013), pp. 71–76 (2014)
24. Verschelde, J.: Exporting Ada software Julia and Python. Ada User J. **43**(1), 75–77 (2022)
25. Verschelde, J., Viswanathan, K.: Extrapolating on Taylor series solutions of homotopies with nearby poles. arXiv:2404.17681v1 [math.NA]. Accessed 26 Apr 2024
26. Verschelde, J., Viswanathan, K.: Locating the closest singularity in a polynomial homotopy. In: Boulier, F., England, M., Sadykov, T.M., Vorozhtsov, E.V. (eds.) CASC 2022. LNCS, vol. 13366, pp. 333–352. Springer, Cham (2022). https://doi.org/10.1007/978-3-031-14788-3_19
27. Wegert, E.: Visual Complex Functions: An Introduction with Phase Portraits. Birkhäuser (2012)
28. Weniger, E.: Nonlinear sequence transformations for the acceleration of convergence and the summation of series. Comput. Phys. Rep. **10**, 189–371 (1989)
29. Wynn, P.: On a procrustean technique for the numerical transformation of slowly convergent sequences and series. Proc. Camb. Phil. Soc. **52**, 663–672 (1956)

Author Index

A
Abbott, John 29
Aichmayr, Marcus S. 155
Akian, Marianne 299
Alazemi, Abdullah 89
Azzouz-Thuderoz, Maxence 225

B
Backeljauw, Franky 215
Barket, Rashid 167
Becuwe, Stefan 215
Beentjes, Casper 207
Béreau, Antoine 299
Betten, Anton 89
Brooks, Sean 215
Brysiewicz, Taylor 265
Buckmire, Ron 215

C
Cuyt, Annie 215

D
Daisey, Oliver 331
Davenport, James 176
Deb, Madhurima 225
Della Vecchia, Antony 234

E
E. Vincent-Finley, Rachel 215
Enge, Andreas 36
England, Matthew 167, 186

F
Farabella, Ivan 57
Feng, Yichen 78
Fieker, Claus 29
Florescu, Dorian 186
Fujimoto, Mitsushi 343

G
Gaubert, Stéphane 299
Gerhard, Jürgen 167
Giles, Michael B. 207
Glasheen, Jou 63
Gnawali, Santosh 127

J
Jensen, Anders Nedergaard 313
John, Rohit 176
Joswig, Michael 234

K
Kastner, Lars 322
Kaushik, Aaruni 46, 245
Köppe, Matthias 3
Kurz, Sascha 97

L
Lee, Kisun 275
Lipinski, Dawid 72
Lorenz, Benjamin 234

M
Macbeth, Heather 12
McClain, Marjorie 215
Miller, Bruce 215
Mirgain, Benjamin 135
Mohammadi, Fatemeh 285
Mou, Chenqi 354
Müller, Stefan 155
Muratore, Giosuè 115

O
Oliver, Thomas 196

P
Petrera, Matteo 225

© The Editor(s) (if applicable) and The Author(s), under exclusive license to Springer Nature Switzerland AG 2024
K. Buzzard et al. (Eds.): ICMS 2024, LNCS 14749, pp. 375–376, 2024.
https://doi.org/10.1007/978-3-031-64529-7

R
Regensburger, Georg 155
Reidelbach, Marco 254
Ren, Yue 331

S
Saunders, Bonita V. 215
Schembera, Björn 254
Schubotz, Moritz 225
Soicher, Leonard H. 106
Song, Qiuye 354

T
Teschke, Olaf 225
Traore, Ali 145

V
Verschelde, Jan 365
Viswanathan, Kylash 365

W
Weber, Marcus 254

X
Xie, Yunzhou 78

Z
Zhang, Jujian 78
Zhou, Yanqiao 78
Zhou, Yutong 354

SPRINGER NATURE

GPSR Compliance

The European Union's (EU) General Product Safety Regulation (GPSR) is a set of rules that requires consumer products to be safe and our obligations to ensure this.

If you have any concerns about our products, you can contact us on ProductSafety@springernature.com

In case Publisher is established outside the EU, the EU authorized representative is:

Springer Nature Customer Service Center GmbH
Europaplatz 3
69115 Heidelberg, Germany

The manufacturer's authorised representative in the EU is Springer Nature Customer Service Centre GmbH, Europaplatz 3, 69115 Heidelberg, Germany. If you have any concerns regarding our products, please contact ProductSafety@springernature.com

Printed and bound by CPI Group (UK) Ltd, Croydon, CR0 4YY
25/03/2026
02078185-0015